Deep Energy Retrofit—Case Studies

Ruediger Lohse · Ove C. Mørck ·
Alexander Zhivov
Editors

Deep Energy Retrofit—Case Studies

Business and Technical Concepts for Deep Energy Retrofit of Public Buildings; Energy in Buildings and Communities Programme; Annex 61, Subtask A

Springer

Editors
Ruediger Lohse
DENEFF EDL_HUB gGmbH
Berlin, Germany

Ove C. Mørck
Vanløse, Denmark

Alexander Zhivov
Engineer Research and Development Center
United States Army Corps of Engineers
Champaign, IL, USA

ISBN 978-3-031-17519-0 ISBN 978-3-031-17517-6 (eBook)
https://doi.org/10.1007/978-3-031-17517-6

This Springer imprint is published by the registered company Springer Nature Switzerland AG
The registered company address is: Gewerbestrasse 11, 6330 Cham, Switzerland

Preface

The International Energy Agency

The International Energy Agency (IEA) was established in 1974 within the framework of the Organization for Economic Cooperation and Development (OECD) to implement an international energy programme. A basic aim of the IEA is to foster international cooperation among the 29 IEA participating countries and to increase energy security through energy research, development, and demonstration in the fields of technologies for energy efficiency and renewable energy sources.

The IEA Energy in Buildings and Communities Programme

The IEA coordinates international energy research and development (R&D) activities through a comprehensive portfolio of Technology Collaboration Programmes. The mission of the IEA Energy in Buildings and Communities (IEA EBC) Programme is to develop and facilitate the integration of technologies and processes for energy efficiency and conservation into healthy, low emission, and sustainable buildings and communities through innovation and research. (Until March 2013, the IEA EBC Programme was known as the IEA Energy in Buildings and Community Systems Programme, ECBCS.)

The R&D strategies of the IEA EBC Programme are derived from research drivers, national programmes within IEA countries, and the IEA Future Buildings Forum Think Tank Workshops. These R&D strategies aim to exploit technological opportunities to save energy in the buildings sector, and to remove technical obstacles to market penetration of new energy-efficient technologies. The R&D strategies apply to residential, commercial, office buildings and community systems and will impact the building industry in five areas of focus for R&D activities:

- Integrated planning and building design
- Building energy systems
- Building envelope
- Community scale methods
- Real building energy use.

The Executive Committee

Overall control of the IEA EBC Programme is maintained by an Executive Committee, which not only monitors existing projects, but also identifies new strategic areas in which collaborative efforts may be beneficial. As the Programme is based on a contract with the IEA, the projects are legally established as Annexes to the IEA EBC Implementing Agreement. At the present time, the following projects have been initiated by the IEA EBC Executive Committee, with completed projects identified by (*):

Annex 1: Load Energy Determination of Buildings (*)
Annex 2: Ekistics and Advanced Community Energy Systems (*)
Annex 3: Energy Conservation in Residential Buildings (*)
Annex 4: Glasgow Commercial Building Monitoring (*)
Annex 5: Air Infiltration and Ventilation Centre
Annex 6: Energy Systems and Design of Communities (*)
Annex 7: Local Government Energy Planning (*)
Annex 8: Inhabitants Behavior with Regard to Ventilation (*)
Annex 9: Minimum Ventilation Rates (*)
Annex 10: Building HVAC System Simulation (*)
Annex 11: Energy Auditing (*)
Annex 12: Windows and Fenestration (*)
Annex 13: Energy Management in Hospitals (*)
Annex 14: Condensation and Energy (*)
Annex 15: Energy Efficiency in Schools (*)
Annex 16: BEMS 1—User Interfaces and System Integration (*)
Annex 17: BEMS 2—Evaluation and Emulation Techniques (*)
Annex 18: Demand Controlled Ventilation Systems (*)
Annex 19: Low-Slope Roof Systems (*)
Annex 20: Air Flow Patterns within Buildings (*)
Annex 21: Thermal Modeling (*)
Annex 22: Energy-Efficient Communities (*)
Annex 23: Multi-zone Air Flow Modeling (COMIS) (*)
Annex 24: Heat, Air, and Moisture Transfer in Envelopes (*)
Annex 25: Real-time HVAC Simulation (*)
Annex 26: Energy-Efficient Ventilation of Large Enclosures (*)
Annex 27: Evaluation and Demonstration of Domestic Ventilation Systems (*)
Annex 28: Low Energy Cooling Systems (*)

Annex 29: Daylight in Buildings (*)
Annex 30: Bringing Simulation to Application (*)
Annex 31: Energy-Related Environmental Impact of Buildings (*)
Annex 32: Integral Building Envelope Performance Assessment (*)
Annex 33: Advanced Local Energy Planning (*)
Annex 34: Computer-Aided Evaluation of HVAC System Performance (*)
Annex 35: Design of Energy-Efficient Hybrid Ventilation (HYBVENT) (*)
Annex 36: Retrofitting of Educational Buildings (*)
Annex 37: Low Exergy Systems for Heating and Cooling of Buildings (LowEx) (*)
Annex 38: Solar Sustainable Housing (*)
Annex 39: High-Performance Insulation Systems (*)
Annex 40: Building Commissioning to Improve Energy Performance (*)
Annex 41: Whole Building Heat, Air, and Moisture Response (MOIST-ENG) (*)
Annex 42: The Simulation of Building-Integrated Fuel Cell and Other Cogeneration Systems (FC+COGEN-SIM) (*)
Annex 43: Testing and Validation of Building Energy Simulation Tools (*)
Annex 44: Integrating Environmentally Responsive Elements in Buildings (*)
Annex 45: Energy-Efficient Electric Lighting for Buildings (*)
Annex 46: Holistic Assessment Tool-kit on Energy-Efficient Retrofit Measures for Government Buildings (EnERGo) (*)
Annex 47: Cost-Effective Commissioning for Existing and Low-Energy Buildings (*)
Annex 48: Heat Pumping and Reversible Air-conditioning (*)
Annex 49: Low Exergy Systems for High-Performance Buildings and Communities (*)
Annex 50: Prefabricated Systems for Low-Energy Renovation of Residential Buildings (*)
Annex 51: Energy-Efficient Communities (*)
Annex 52: Toward Net Zero Energy Solar Buildings (*)
Annex 53: Total Energy Use in Buildings: Analysis and Evaluation Methods (*)
Annex 54: Integration of Micro-Generation and Related Energy Technologies in Buildings (*)
Annex 55: Reliability of Energy-Efficient Building Retrofitting—Probability Assessment of Performance and Cost (RAP-RETRO) (*)
Annex 56: Cost-Effective Energy and CO_2 Emissions Optimization in Building Renovation
Annex 57: Evaluation of Embodied Energy and CO_2 Equivalent Emissions for Building Construction (*)
Annex 58: Reliable Building Energy Performance Characterization Based on Full-Scale Dynamic Measurements (*)
Annex 59: High-Temperature Cooling and Low-Temperature Heating in Buildings (*)

Annex 60:	New-Generation Computational Tools for Building and Community Energy Systems
Annex 61:	Business and Technical Concepts for Deep Energy Retrofit of Public Buildings
Annex 62:	Ventilative Cooling
Annex 63:	Implementation of Energy Strategies in Communities
Annex 64:	LowEx Communities—Optimized Performance of Energy Supply Systems with Exergy Principles
Annex 65:	Long-Term Performance of Super-Insulating Materials in Building Components and Systems
Annex 66:	Definition and Simulation of Occupant Behavior in Buildings
Annex 67:	Energy Flexible Buildings
Annex 68:	Indoor Air Quality Design and Control in Low-Energy Residential Buildings
Annex 69:	Strategy and Practice of Adaptive Thermal Comfort in Low-Energy Buildings
Annex 70:	Energy Epidemiology: Analysis of Real Building Energy Use at Scale
Annex 71:	Building Energy Performance Assessment Based on In situ Measurements
Annex 72:	Assessing Life Cycle-Related Environmental Impacts Caused by Buildings
Annex 73:	Toward Net Zero Energy Public Communities
Annex 74:	Energy Endeavor
Annex 75:	Cost-effective Building Renovation at District Level Combining Energy Efficiency and Renewables
Working Group	Energy Efficiency in Educational Buildings (*)
Working Group	Indicators of Energy Efficiency in Cold Climate Buildings (*)
Working Group	Annex 36 Extension: The Energy Concept Adviser (*)
Working Group	HVAC Energy Calculation Methodologies for Non-residential Buildings

Berlin, Germany — Ruediger Lohse
Vanløse, Denmark — Ove C. Mørck
Champaign, IL, USA — Alexander Zhivov

Acknowledgements

The "Deep Energy Retrofit—Case Studies" report was developed within the IEA EBC Programme Annex 61, "Business and technical Concepts for Deep Energy Retrofit of Public Buildings" as the result of a joint effort by researchers and practitioners from Austria, Canada, China, Denmark, Estonia, Germany, Latvia, UK, and the USA. The authors express their appreciation to the many international contributors and organizations whose volunteer efforts contributed to development of a series of Guides developed under this project. Special gratitude to the ASHRAE and its Technical Committees TC 7.6 "Building Energy Performance," TC 5.2 "Duct Design," and a Standing Standard Project Committee 90.1 for providing a platform for public discussion of the project progress and to Chartered Institution of Building Services Engineers (CIBSE) for sharing its technical information and providing valuable input into these documents. The authors would like to personally thank the members of the EBC Programme Executive Committee for their directions, guidance, and support to the project. Special appreciation to the following reviewers, who provided their valuable comments and suggested improvements to this Guide: Michele Zinzi (ENEA, Italian National Agency for New Technologies, Energy and Sustainable Economic Development), Prof. Michael Donn (Victoria University of Wellington, New Zealand), and Daniël van Rijn (Netherlands Enterprise Agency). The authors gratefully acknowledge William J. Wolfe, Writer-Editor, ERDC-CERL for his help in coordinating the preparation of this document. The authors would like to acknowledge the financial and other support from the following: the office of the Assistant Secretary of the U.S. Army for Installations, Energy, and Environment; U.S. Army Corps of Engineers; U.S. Department of Energy (Federal Energy Management Program), The Austrian Ministry for Transport, Innovation and Technology, the EUDP-programme of the Danish Energy Agency, and Bundesministerium für Wirtschaft und Energie, Germany.

Case Study Overview

Number of case	Country	Site	Building type	Pictures
1	Austria	Kapfenberg	Social Multi-family building	
2	Denmark	Egedal, Copenhagen	School	
3	Denmark	Vester Voldgade	Multi-family building	
4	Estonia	Kindergarten in Valga	Kindergarten	

(continued)

(continued)

Number of case	Country	Site	Building type	Pictures
5	Germany	Ludwigshafen-Mundenheim	Multi-family building	
6	Germany	Nürnberg, Bavaria	Multi-family building	
7	Germany	Ostfildern	Gymnasium	
8	Germany	Baden-Württemberg	School	
9	Germany	Osnabrueck	School	
10	Germany	Olbersdorf	School	

(continued)

(continued)

Number of case	Country	Site	Building type	Pictures
11	Germany	Darmstadt	Office building	
12	Germany	Town Hall-Baviera	Office building	
13	Germany	Nordrhein-Westfalen	High school	
14	Ireland	Dun Laoghaire Rathdown	Social Housing	
15	Latvia	Riga	Multi-family building	
16	Montenegro	Plevlja, Montenegro	Primary School	
17	Montenegro	Kotor, Montenegro	Student Dormitory	

(continued)

(continued)

Number of case	Country	Site	Building type	Pictures
18	The Netherlands	Leeuwarden	Shelter home	
19	UK	London	Mildmay Center	
20	USA	Grand Junction, CO	Office Building/Courthouse	
21	USA	Silver Spring and Lanham, MD	Office/Federal Building	
22	USA	Bethesda-MD, DIA ICC	Intelligence Community Campus	
23	USA	Seattle WA. JBDG	Office	

(continued)

(continued)

Number of case	Country	Site	Building type	Pictures
24	USA	Priest River, ID	Beardmore Building Office	
25	USA	435 Indio, Sunnyvale, CA	Office/Warehouse	
26	USA	Byron Rogers Federal Office Building, Denver, CO	Office	

Contents

Contributors

Ruediger Lohse DENEFF EDL_HUB gGmbH, Berlin, Germany

Ove C. Mørck Vanløse, Denmark

Alexander Zhivov Engineer Research and Development Center, United States Army Corps of Engineers, Champaign, IL, USA

Abbreviations

AUS	Austria
DH	District heating
DHW	Domestic Hot Water
DK	Denmark
EN	European Norm
EPBD	Energy Performance of Buildings Directive
EST	Estonia
GE	Germany
HP	Heat pump
HVAC	Heating, Ventilation, and Air-conditioning
IEA	Energy in Buildings and Communities Programme of the International
EBC	Energy Agency
IRL	Ireland
kWh	Kilowatt hours: 1 kWh = 3.6 MJ
LAT	Latvia
MNE	Montenegro
MVHR	Mechanical ventilation with heat recovery
NED	The Netherlands
NZEB	Nearly zero energy building or nearly zero emissions building
PV	Photovoltaic
Ref	Reference
RES	Renewable energy sources
SOW	Scope of work
UK	United Kingdom
USA	The United States of America
Λ	Lambda value (value for the insulating capacity of a material)

Units of Measurement and Currency

1 kBTU	0.293 KWh
1 Therm	100,066.96 BTU
1 Therm	29.32673126 kWh
1 $KBTU/ft^2$	3.15458 kWh/m^2
1 sqf (ft^2)	0.09290 m^2
1 m^3	10.557 kWh
1 ft^3	0.29 kWh
1 British pound (£)	1.42 Euros (€)
1 USA dollar ($)	0.94 Euros (€)

List of Figures

List of Tables

Chapter 1
Project Summary and Analysis

Ruediger Lohse, Ove C. Mørck, and Alexander Zhivov

Abstract The book "DER Energy Retrofit—Case Studies" contains 26 well-documented case studies from Europe (Austria, Denmark, Estonia, Germany, Ireland, Latvia, Montenegro, The Netherlands, United Kingdom) and the United States. After data were collected, the case studies were analyzed with respect to energy use (before and after renovation), reasons for undertaking the renovation, co-benefits achieved, project cost-effectiveness, and the business models used in project implementation. Finally, "lessons learned" were compiled and compared. Based on an extensive literature review and lessons learned from these case studies, the IEA-EBC Annex 61 team has proposed the following definition of the DER:

> Deep Energy Retrofit (DER) is a major building renovation project in which site energy use intensity (including plug loads) has been reduced by at least 50% from the pre-renovation baseline with a corresponding improvement in indoor environmental quality and comfort.

Lessons learned from the case studies and experiences of the project team clearly indicate that DER can be achieved with the application of "bundles" of a limited number of core technologies readily available on the market. Specific characteristics of some of these core technology bundles generally depend on the technologies available in an individual nation's market, on the minimum requirements of national standards, and on economics (as determined by a life cycle cost [LCC] analysis). Also, requirements for building envelope-related technologies (e.g., insulation levels, windows, vapor and water barriers, and requirements for building airtightness) will depend on specific climate conditions. Case studies documented in

---

R. Lohse
DENEFF EDL_HUB gGmbH, Berlin, Germany
e-mail: ruediger.lohse@edlhub.org

O. C. Mørck
Vanløse, Denmark
e-mail: ovemorck@gmail.com

A. Zhivov (✉)
Engineer Research and Development Center, United States Army Corps of Engineers, Champaign, IL, USA
e-mail: Alexander.M.Zhivov@usace.army.mil

R. Lohse et al. (eds.), *Deep Energy Retrofit—Case Studies*,
https://doi.org/10.1007/978-3-031-17517-6_1

this book were combined with original research conducted under Annex 61 and information collected from literature research and expert discussions conducted at Annex 61 Industry Forums to develop several guidelines that were published by Springer: "Deep Energy Retrofit—A Guide to Achieving Significant Energy Use Reduction with Major Renovation Projects", Deep Energy Retrofit Guide for Public Buildings—Business and Financial Models", which describes business models for DER, project structuring and project financing options, direct cost savings benefits and monetizing of indirect benefits, and macro- and microeconomic barriers for implementation of DER projects in the public building sector and "Deep Energy Retrofit—a Guide for Decision Makers".

1.1 Introduction

IEA EBC Program Annex 61 has conducted research with the goal of providing a framework, selected tools, and guidelines to significantly reduce energy use (by more than 50%) in government and public buildings. The scope of the project was limited to public buildings that were constructed before the 1980s with low internal loads (e.g., office buildings, dormitories, barracks, public housing and educational buildings), and that were undergoing major renovation. One of the Annex 61 deliverables is the book of "DER Energy Retrofit—Case Studies," which contains 26 well documented case studies from Europe (Austria, Denmark, Estonia, Germany, Ireland, Latvia, Montenegro, The Netherlands, United Kingdom) and the United States. After these data were collected, the case studies were analyzed with respect to energy use (before and after renovation), reasons for undertaking the renovation, co-benefits achieved, resulting cost effectiveness, and the business models followed. Finally, "lessons learned" were compiled and compared.

Based on an extensive literature review and lessons learned from these case studies, the IEA-EBC Annex 61 team has proposed the following definition of the DER:

> Deep Energy Retrofit (DER) is a major building renovation project in which site energy use intensity (including plug loads) has been reduced by at least 50% from the pre-renovation baseline with a corresponding improvement in indoor environmental quality and comfort.

Lessons learned from the case studies and experiences of the team clearly indicate that DER can be achieved with the application of "bundles" of a limited number of core technologies readily available on the market. Specific characteristics of some of these core technology bundles generally depend on the technologies available on an individual nation's market, on the minimum requirements of national standards, and on economics (as determined by a life cycle cost [LCC] analysis). Also, requirements for building envelope-related technologies (e.g., insulation levels, windows, vapor and water barriers, and requirements for building airtightness) will depend on specific climate conditions.

Another Annex 61 deliverable, *Deep Energy Retrofit—A Guide to Achieving Significant Energy Use Reduction with Major Renovation Projects*, describes the characteristics of these technologies and documents examples of best practices in their application in different construction situations.

Case studies documented in this book were combined with original research conducted under Annex 61 and information collected from literature research and expert discussions conducted at Annex 61 Industry Forums to develop the Annex 61 document, *Deep Energy Retrofit—Business Guide*, which describes business models for DER, project structuring and project financing options, direct cost savings benefits and monetizing of indirect benefits, and macro- and microeconomic barriers for implementation of DER projects in the public building sector. Both Guides are included in the series of Annex 61 Deep Energy Retrofit books, published by New Buildings Institute.

Twenty-six deep energy retrofit (DER) case studies were collected under the IEA EBC Annex 61 project using a common template. The template had to be filled in by the building owners who had recently carried out a DER project. The data pieces considered the ex-ante and ex-post data including descriptive information on the buildings' age and usage, construction specifics, the energy related constructive details such as U-values of the major building parts, technical key performance indicators of HVAC equipment, energy, and other life cycle costs. In total, 26 buildings were collected, which provided a sufficient data base.

The objectives of this work were:

- To show successful renovation projects as inspirations to motivate decision makers and stimulate the market.
- To support decision makers and experts with relevant information to support their future decisions.
- To learn from the experience and to create lessons learned from these early cutting-edge projects.

To achieve these objectives, the case studies were analyzed to extract all relevant information. The analyses focused on:

- Climate zone.
- Energy saving strategies.
- Energy savings/reduction levels.
- Plotted comparison of energy use before and after renovation.
- Energy use intensity pre-DER.
- Energy measured.
- Reasons for renovation/anyway measures.
- Co-benefits.
- Business models and funding sources.
- Cost effectiveness.
- Experiences/lessons learned.
- Renovation cost.

Analyses were carried out by reviewing the collected descriptions and extracting the relevant information. The following sections summarize the findings.

1.2 Climate Zones

Climate conditions (combined with building design, type, and occupancy patterns) define peak heating and cooling loads, and also determine the window energy performance variables that should be prioritized.

ASHRAE has developed a standard for climate zone classification that is derived from accumulated weather data from all over the world. ASHRAE Standard 169 entitled *Weather Data for Building Design Standards* includes dry-bulb, dew-point, and wet-bulb temperatures; enthalpy; humidity ratio; wind conditions; solar irradiation; latitude; longitude; and elevation for locations worldwide. These data have been compiled into broad climate zones to generally characterize climates for building codes and energy analysis. This standard also identifies representative cities for each broad climate zone to aid in building modeling and energy analyses.

Figure 1.1 shows these broad climate zones for the world. More detail for each country and representative city can be found in Standard 169-2013.

Table 1.1 lists case studies, sorted by country and climate zone according to ASHRAE Standard 169.

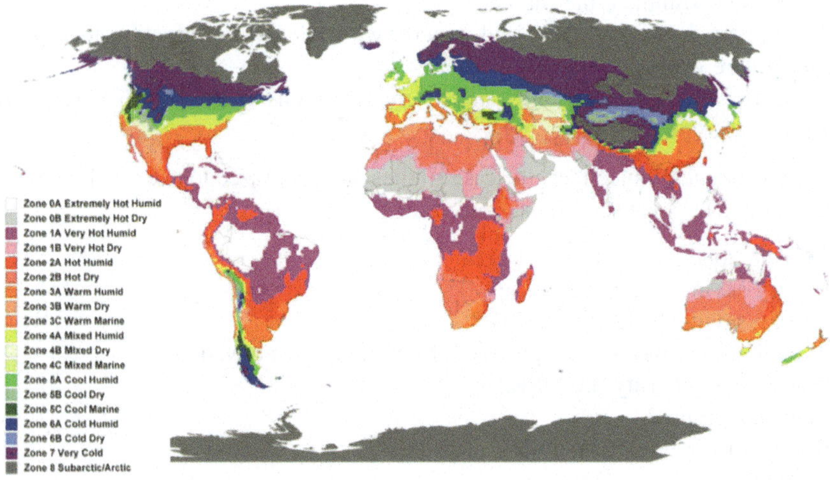

Fig. 1.1 Climate Zones Worldwide in Standard 169-2013

Table 1.1 International location climate zones (based on ASHRAE classification) used in the Annex 61 DER building envelope analysis

Case study	Country	Climate zone(s)	Representative City
25. Office/Warehouse Indio. USA	USA	**3c** (Warm-Marine), (3600 ≥ HDD65 °F)	California
14. Social housing Dún Laoghaire. IE 18. Shelter home. Leeuwarden. NL 19. Mildmay Center London. UK 21. Office/Federal building Maryland. USA 22. Intelligence Community Maryland. USA	Ireland Netherland UK USA	**4a** (Mixed-Humid), (3600 < HDD65 °F[a] ≤ 5400)	Dublin, Galway, Cork, Donegal. London
23. Office. Seattle WA. USA	USA	**4c** (Mixed-Marine) (3600 < HDD65 °F ≤ 5400)	Washington, Oregon
1. Social house Kapfenberg. AT 2. School Egedal. DK 3. OfficeVester Voldgade. DK 5. Passivehaus LudMun. GE 6. Apartments Nûrnberg. GE 7. Gym Ostildern. GE 8. School BaWû. GE 9. School Osnabrueck. GE 10. School Olbersdorf. GE 11. Passivehaus Office Darmstadt. GE 12. Town Hall – Baviera. GE 13. Passivehaus High school NordWest. GE 16. Primary school Plevlja. MON 17. Student Dormitory Kontor. MON	Denmark Germany USA	**5a** (Cool-Humid), (5400 < HDD65 °F ≤ 7200)	Copenhagen. Wuerzburg. Braganza, Innsbruck, Klagenfurt, Linz, Wien, Eisenstaedt, Graz. Pennsylvania, Nebraska, Massachusetts, Indiana, New York
20. Federal building Grand Junction. USA 24. Beardmore Priest River. USA 26. Federal building Denver, CO. USA	USA	**5b** (Dry) (5400 < HDD65 °F ≤ 7200)	Colorado, Idaho
4. Kindergarten Valga. EE 15. Apartments. Riga. LV	Estonia Latvia	**6a** (Cold-Humid), (7200 < HDD65 °F ≤ 9000)	Tartu Riga

[a] 183 °C

1.3 DER Measure Bundles

The implemented DER measure bundles were grouped into 16 categories. Table 1.2 lists the core bundles of technologies used to achieve DER with regard to the building usage and climate zone. The technologies can be classified into four groups: (1) building envelope, (2) lighting/electrical systems, (3) HVAC, building automation, and (4) renewable energy systems.

Those renovation technologies that have been applied in more than half of the case studies are outlined in red. These data show that the majority of DER cases were implemented with energy saving bundles of technologies that improve the

Table 1.2 Core bundles of technologies implemented in DER related to the climate zones (light green = cz 5a; dark green= cz 6a; light pink = cz 4a; dark blue= cz 4c; indian yellow= cz 3a)

CORE BUNDLES OF TECHNOLOGIES IMPLEMENTED IN DER	Building Envelope					Lighting and Electrical systems			HVAC					Renewable energy systems		
Case study	Wall insulation	Roof insulation	Floor insulation	New window/ door	Roof lights	Daylight Strategy/external shading	Efficiency lighting/control	BEMS	MVHR	New ventilation system	New heat-cooling supplier/distribution system	New heat supply: radiators, floor heating	Air source heat pump	Ground coupled heat pump	Solar thermal system	Photovoltaic panels
1. Social house Kapfenberg. AT	√	√	√	√					√		√				√	√
2. School Egedal. DK	√			√			√	√	√	√	√					√
3. OfficeVester Voldgade. DK	√			√		√	√		√					√	√	
4. Kindergarten Valga. EE	√	√	√			√	√		√						√	
5. Dwelling passive house LudMun. GE	√	√	√	√						√		√				√
6. Apartments Nuernberg. GE	√	√	√	√					√		√				√	
7. High School Ostildern. GE	√	√		√	√		√		√							
8. School BaWue. GE	√	√		√			√			√	√					√
9. School Osnabrueck. GE	√	√	√	√			√		√	√	√			√		
10. School Olbersdorf. GE	√	√	√	√	√	√	√			√				√		
11. Passive house Office Darmstadt. GE	√	√	√	√			√		√	√						
12. Town Hall – Baviera. GE	√	√	√	√	√		√		√							
13. Passive house High school NordWest. GE	√	√	√	√		√	√		√		√			√		√
14. Social housing Dún Laoghaire. IE	√	√	√	√					√		√					
15. Apartments Riga. LV	√	√	√	√			√		√			√				
16. Primary school Plevlja. MON	√			√			√	√			√					
17. Student Dormitory Kontor. MON	√	√		√			√				√	√			√	
18. Shelter home. Leeuwarden. NL	√	√	√	√		√	√		√	√	√				√	
19. Mildmay Center London. UK		√	√	√		√	√		√	√				√	√	√
20. Federal building Grand Junction. USA	√	√			√		√			√	√	√		√		√
21. Office/Federal building Maryland. USA		√		√			√	√	√	√	√	√	√	√	√	√
22. Intelligence Community Maryland. USA	√	√	√	√		√	√		√	√	√					√
23. Office. Seattle WA. USA				√		√	√		√				√			√
24. Beardmore Priest River. USA	√	√		√	√		√			√			√			
25. Office/Warehouse Indio. USA	√	√		√	√	√	√		√	√			√			√
26. Federal building Denver-Colorado. USA	√	√	√	√			√				√		√		√	

envelope (wall, roof, floor, and windows), lighting, ventilation system, and supply and/or distribution system. One case study (the office in Maryland, U.S.) implemented 12 of the 16 technologies.

The savings are defined as reduction of consumption by the application of energy efficiency measures or rational use of energy (RUE), fuel switching, and the implementation of renewable energy targeted to reduce the portion of energy demand provided by the utility grid.

1.4 Investment Costs

Figures 1.2 and 1.3 show total, non-energy- and energy-related renovation costs for some of the case studies. Note that cases that indicate only "energy-related investment costs" provided no other information (Nos. 16, 17, 19, 21a, 21b, 23, 24, 25, 26).

For those case studies in which both the non-energy and energy-related costs were available, it can be seen that the non-energy investment costs, which include measures from the general refurbishment, are in most of the cases two or three times higher than the energy-related investment costs. Some case studies show that this factor is higher when major construction measures are carried out in the general refurbishment.

A review of the investment costs for the energy-related costs shows that these costs vary widely for comparable building usages in comparable climate zones and (mostly) comparable concepts:

- In cz 5a (Case Nos. 1, 2, 3, 5-14, 16-19), the investment costs in schools show a larger cluster with investment costs between 150 and 250 €/m^2 and one school with investment costs of 540 €/m^2 (No. 9). In most of the cases, the scope of measures included roof insulation, wall insulation, new fenestration, and ventilation system.
- In cz 5a, the investment costs in a passive house refurbishment of an office building (No. 11 with 490 €/m^2, No. 12 with 160 €/m^2, and No. 3 with 240 €/m^2). Case No. 11 included additional measures (flat roof insulation and ground floor insulation) that do not fully explain the cost difference of more than 100%.
- In cz 5a, the energy-related DER investment costs in multifamily dwelling houses show two cases with investment costs around 400 €/m^2 (Case Nos. 1 and 5), 320 €/m^2 (No. 18), and 200 €/m^2 (No. 6). The difference may be explained by higher costs for a passive house-related concepts at that point in time (Nos. 1 and 5).

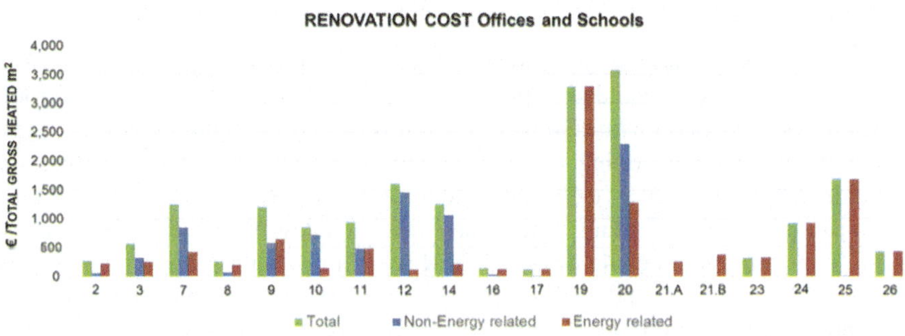

Fig. 1.2 Renovation cost for public buildings

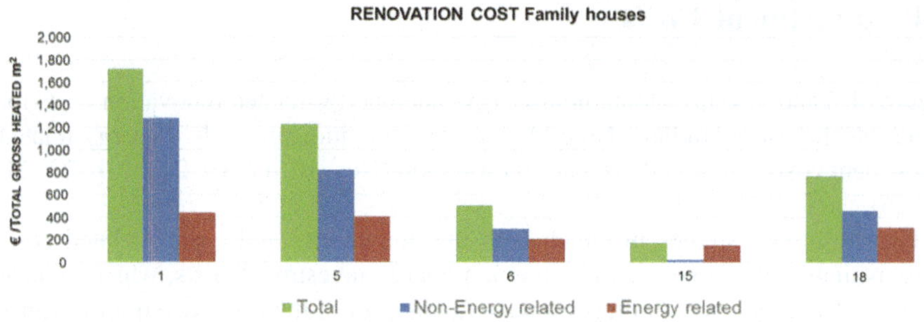

Fig. 1.3 Renovation cost for residential buildings

These differences in investment costs can be explained by the different allocation of energy-related and non-energy-related investments (obviously in Case No. 11), different costs for labor and components, and different execution of the DER concept in detail. In some DER cases, the implementation of passive house windows, thicker or more expensive insulation materials, and higher costs for thermal bridge mitigation can explain the higher investment costs. In the past 5-8 years since these case studies have been started, some of the investment costs e.g., the specific costs for passive house windows, may have decreased so that the significant difference between more ambitious passive house concepts and a "normal" DER may be smaller today than it is seen here.

1.5 Impact of DER Measure Bundles

The impact of DER measure bundles, i.e., the incorporation of the technologies described above, was analyzed with regard to: (1) the energy use before and after DER and (2) the net energy consumption after the incorporation of solar energy systems (thermal and/or PV systems). By definition, the baseline includes and distinguishes heating fuels, district heating, and electricity including the plug loads.

In most of the climate zones, the rational use of energy (RUE) was achieved by refurbishing major parts of the thermal envelope:

- The majority of buildings were located in cz 5a. In most of these cases, the DER technology bundle included improvements to the thermal envelope (wall, roof, windows, and basement ceilings) and refurbishing the lighting system, most commonly without adding a new ventilation system. This was also true for colder climates such as cz 6 and milder cz 4a and 3c. To achieve DER, the greater part of the buildings were renovated, including their thermal envelopes.
- Only the building in cz 4c achieved DER by adding minor glazing measures by refurbishing the heat and cooling recovery of the HVAC and installing a lighting control system, with no significant refurbishment measures on the thermal envelope.

- On the supply side, the DER measure bundles included, in most cases, a change of the fuel sources and an implementation of renewable energies such as photovoltaics, solar thermal, and heat pumps. In addition to improving the source energy balance, the switch in energy supply systems can—if framework conditions are appropriately considered—also improve the site energy balance. Other supply side considerations are that:
- The implementation of a ground-coupled heat pump reduced fuel consumption for heating demand considerably; however, it also increased electricity consumption. Note that the impact of the air source heat pump, which is implemented in the majority of cases in the United States, cannot be analyzed in detail since the fuel for heating reduction and electricity COP is not given separately.
- In total, the maximum energy saving reduction by a DER without the incorporation of solar energy system is found in Case No. 9 to be 92%, and the average energy savings for all cases was 63%.

The "Total Energy Savings after Solar Production" column in Table 1.3 lists the impact of the implementation of solar energy systems. The implementation of a solar heating system resulted in a heating fuel energy reduction, which indicates that the demand for oil and natural gas was not reduced by savings, but was instead partly replaced by renewables.

In some cases, as in Case Nos. 5 and 13, a photovoltaic system had a large impact that resulted savings in excess of 90% in energy demand reduction from the utility power grid. (Note, however, this does not mean that the overall energy consumption was reduced by 90%.)

The average energy savings achieved by the addition of a solar energy system was 68%. The added savings to the demand side measures that were achieved by the solar production and that were related to the avoided use of energy from the utility grid varied from 4 to 42% (Denmark).

Table 1.3 lists energy savings related to total energy consumption before and after renovation, the net energy consumption from the utility grid, the total energy savings after DER, and the solar energy contribution to that energy consumption.

Case No. 22 shows very high values because the figures show the original campus energy usage against the Architecturally Adjusted Baseline and the renovated campus energy usage. Note that all savings were measured from the Architecturally Adjusted Baseline due to the changes in campus square footage (resulting from buildings either having been demolished or newly constructed), changes in space use, and addition of new building facade(s).

Sometimes the change of a building's usage and purpose can itself lead to considerable energy savings. In one case (No. 18), the usage change from a post office building to a shelter building with a usage comparable to housing buildings, yielded energy savings of 90% despite the significantly increased usage density.

In two cases (Nos. 16 and 17), savings did not fully achieve the DER criterion (-50%) since the RUE measures were not sufficient and the energy supply had already been based on renewable energy sources (biomass) (Figs. 1.4 and 1.5).

Table 1.3 Energy before and after DER related to the energy saving

Case study	Before renovation	After renovation	Net energy consumption	Total% energy saving after DER (%)	Total% energy savings after solar production (%)
1. Social House Kapfenberg. AT	184.0	91.0	52.0	51	72
2. School Egedal. DK	148.3	101	39.0	32	74
3. OfficeVester Voldgade. DK	116.4	–	57.8	–	50
4. Kindergarten Valga. EE	280.0	36.0	36.0	87	87
5. Passivehaus LundMun. GE	250.0	48.5	16.0	81	94
6. Apartments Nuernberg. GE	229.6	43.1	34.8	81	85
7. Gym Ostildern. GE	181.1	99.5	99.5	45	45
8. School BaWue. GE	171.3	54.0	54.0	68	68
9. School Osnabrueck. GE	354.9	29.3	29.3	92	92
10. School Olbersdorf. GE	155.5	52.3	52.3	66	66
11. Passivehaus Office Darmstadt. GE	283.0	67.4	67.4	76	76
12. Town Hall—Baviera. GE	184.3	82.6	82.6	55	55
13. Passivehaus High school NordWest. GE	220.0	–	20.3	–	91
14. Social house Dún Laoghaire. IE	482.46	56.69	56.7	88	88
15. Apartments Riga. LV	351.30	162.53	162.5	54	54
16. Primary school Plevlja. MON	282.0	170.6	170.6	40	40
17. Student Dormitory Kontor. MON	199.0	–	132.2	–	34
18. Shelter home. Leeuwarden. NL	415.0	–	41.5	–	90
19. Mildmay Center London. UK	270.0	57.0	39.0	79	86
20. Federal building Grand Junction. USA	127.2	56.7	18.5	55	85
21. A. Office Silver Spring, MD, USA	372.2	197.8	197.8	47	47
21. B. Federal Building New Carrolton. MD. USA	382.2	162.9	146.4	57	62

(continued)

Table 1.3 (continued)

Case study	Before renovation	After renovation	Net energy consumption	Total% energy saving after DER (%)	Total% energy savings after solar production (%)
22. Intelligence Community MD. USA	935.9	–	492.1	–	47
23. Office Seattle WA. USA	224.0	102.1	91.5	54	59
24. Beardmore Priest River. USA	284.0	110.5	110.5	61	61
25. Office/Warehouse Indio. USA	231.1	–	77.3	–	67
26. Federal Building Denver-CO. USA	375.4	–	121.1	–	68

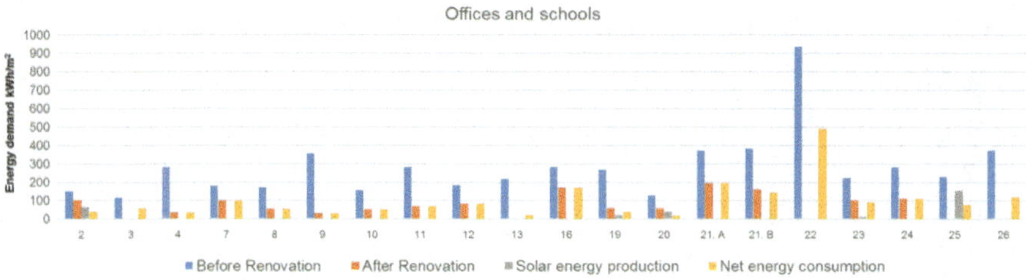

Fig. 1.4 Energy before and after retrofit comparison for public buildings

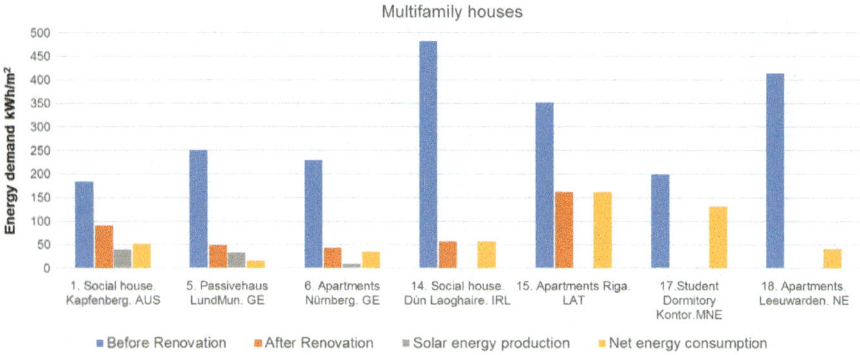

Fig. 1.5 Energy before and after retrofit comparison for family houses

Table 1.4 shows the energy savings for heating, electricity and the contribution of renewable energy to the energy balance of the project.

The data in Fig. 1.6 show that considerable energy savings were obtained in all climate zones. All cases show energy savings greater than 50% except the case study from Montenegro (No. 17), which achieved a total energy reduction of 34%. It can be seen that the countries located within Zones B (Humid climate) and Zone C (Marine climate) present average energy savings of 71% and 63%, respectively. Similarly, the case studies located within Zone 5 A (Dry climate) present an average energy saving of 72%, excluding the cases of Montenegro. Moreover, a general large energy reduction within Zone 4 (Mixed humidity climate) located in Europe can be seen with an average of 88%. However, the case studies located on the same climate zone in the United States (Case Nos. 21 and 22) present only an average of only 52%.

When the case studies are sorted by climate zone, it appears that the energy savings for these 26 cases does not show any climate zone dependency. The smallest variation is seen in Climate Zone 5B; however, since there were only three case studies in this climate zone, this cannot be considered statistically significant. The average net energy reduction obtained for all 26 case studies was 66.4%.

Table 1.4 Energy savings for heating, electricity and the contribution of renewable energy to the project's energy balance

	% Energy reduction		kWh/m^2 contribution		
	Heating	Electricity	Heat pump	PV production	Solar thermal
1. Social House Kapfenberg. AT (5A)	72				27 kWh/m^2
2. School Egedal. DK (5A)	34	100		62.2 kWh/m^2	
3. Office Vester Voldgade. DK (5A)	68	− 48	–		—
4. Kindergarten Valga. EE (6A)	87	–			—
5. Passive House LundMun. GE (5A)	94	–		32.5 kWh/m^2	
6. Apartments Nürnberg. GE (5A)	86	–			8.3kWh/m^2
7. Gym Ostildern. GE (5A)	51	2			
8. School BaWû. GE (5A)	72	0		8.5 kWh/m^{2a}	
9. School Osnabrueck. GE (5A)	96	17	45.5 kWh/m^2		
10. School Olbersdorf. GE (5A)	67	54	–		
11. Passive House Office Darmstadt. GE (5A)	78	70			
12. Town Hall—Baviera. GE (5A)	50	57			
13. Passive House High school NordWest. GE (5A)	90	–		–	
14. Social House Dún Laoghaire. IE (4A)	94	84			
15. Apartments Riga. LV (6A)	54				
16. Primary School Plevlja. MON (5A)	41	8			
17. Student Dormitory Kontor. MON (5A)	72	6			–
18. Shelter Home. Leeuwarden. NL (4A)	90				–
19. Mildmay Center London. UK (4A)	100	− 10		18 kWh/m^2	–
20. Federal Building Grand Junction. USA (5B)	100 (gas)	36	–	38 kWh/m^2	
21 A. Office Silver Spring. CO. USA (4A)	47				
21 B. New Carrollton Federal Building. MD. USA (4A)	61			8.6 kWh/m^2	7.7 kWh/m^2
22. Intelligence Community MD. USA (4A)	16	95			–

(continued)

Table 1.4 (continued)

	% Energy reduction		kWh/m^2 contribution		
	Heating	Electricity	Heat pump	PV production	Solar thermal
23. Office Seattle WA. USA (4C)	59				10.6 KWh/m^2
24. Beardmore Priest River. USA (5B)	61				
25. Office/Warehouse Indio. USA (3C)	67			148 kW	

[a] Calculated figure: 28.7 kW peak × 990 kWh/kW

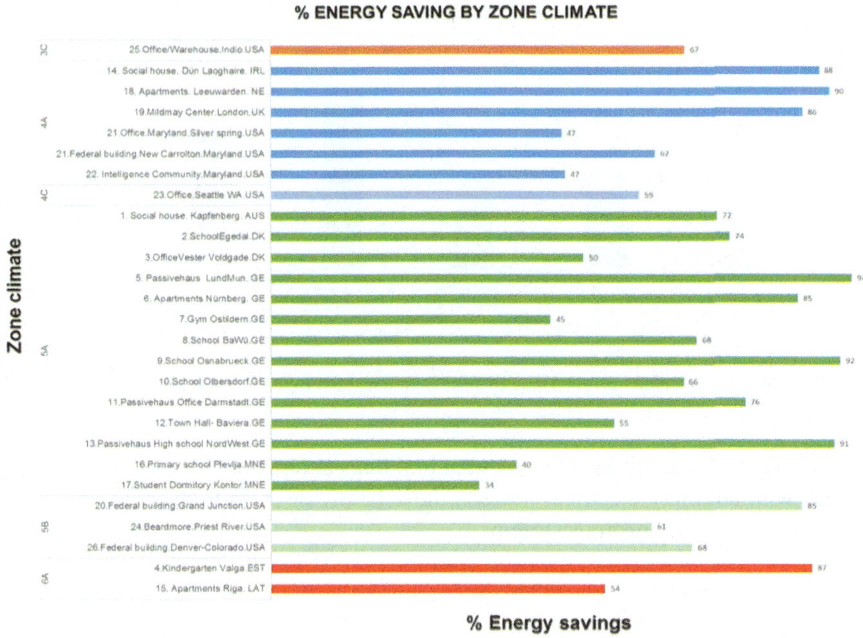

Fig. 1.6 Total energy savings distributed by climate zone

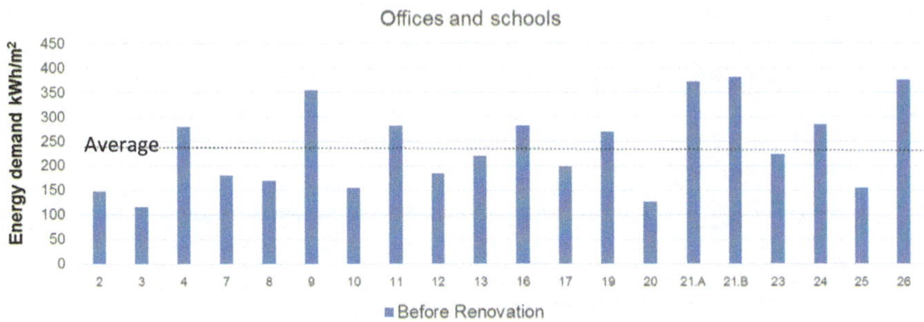

Fig. 1.7 Energy use intensity pre-DER for public buildings, average: 238 kWh/m^2; case 22 is not displayed

The Energy Use Intensity (EUI) of the case studies before Deep Energy Renovation was plotted for offices and schools (Fig. 1.7) and for multifamily houses (Fig. 1.8). The average EUI in the evaluated public buildings is 238 kWh/m^2. The EUI case study No. 22 resulted from the change in building footage. Some buildings such as 9, 21A, 21B, 28.

Figure 1.8 shows that the average ex-ante DER EUI in residential buildings is 319 kWh/m^2, which is 80 kWh/m^2 higher than that of the public buildings and may be related to the higher usage in residential buildings. The evaluation shows that

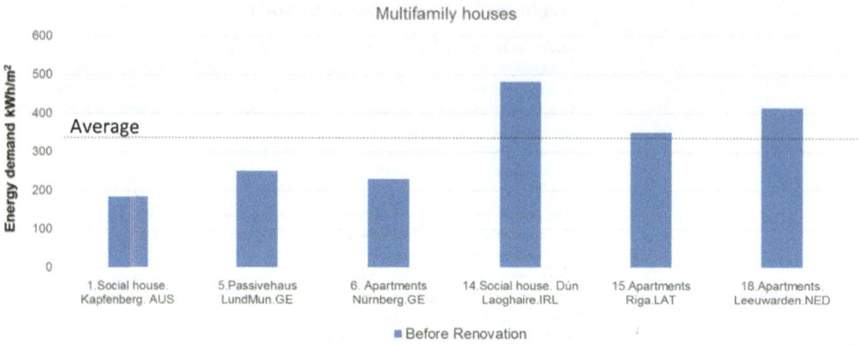

Fig. 1.8 Energy use intensity pre-DER of family houses; average: 319 kWh/m

over-average EUI are found at residential buildings equipped with individual oil/gas boilers with auxiliary electrical heaters; those below or close to average values are connected to district heating systems.

1.6 Decision Making Process

The observed reasons for undertaking a renovation were categorized in two main groups: energy related and non-energy related. In Fig. 1.9, the number to the right of each horizontal bar indicates the percentage of the case studies that undertook energy retrofits for the listed reason.

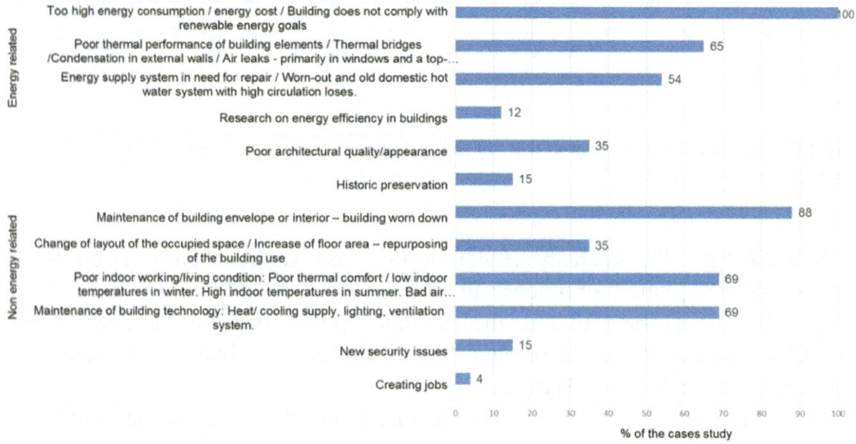

Fig. 1.9 Reasons for renovation (% of the cases studies); co-benefits of DER

Clearly, the greatest non-energy-related reason for carrying out retrofits (88%) is "general maintenance." This high percentage is of particular interest in that most of the case studies were not renovated solely to **save energy**, but rather to catch up with a backlog of maintenance and refurbishment, some of which were energy-related such as "poor thermal performance," "energy system in need for repair," etc. This indicates that many of the renovation measures had to be implemented to maintain the building and its functionality. Then, in a second stage of the decision-making process, the assessed buildings were considered as candidates for more ambitious projects (Deep Energy Retrofit, Passive House refurbishment, etc.) that could achieve energy targets that exceed national minimum requirements.

Consequently, the decision-making process regarding building refurbishment should consider not only the cost effectiveness of different energy standards, but the refurbishment measures that must be done anyway as well.

These reasons shown in Fig. 1.9 are obviously related to some extent. All 26 case studies were renovated for a combination of reasons, some energy related and some not. This supports a "rule of thumb" for energy renovation, that it always makes sense to consider DER when a building is undergoing renovation anyway.

The co-benefits of DER can also be categorized into energy- and non-energy related. The histogram in Fig. 1.10 clearly shows that three of the energy-related co-benefits (improvement of thermal comfort, improved green building image, and reduced dependency of fuel price variations) were identified in all case studies. Additionally, in 85% of the cases, an improved operational comfort of the building automation system was observed.

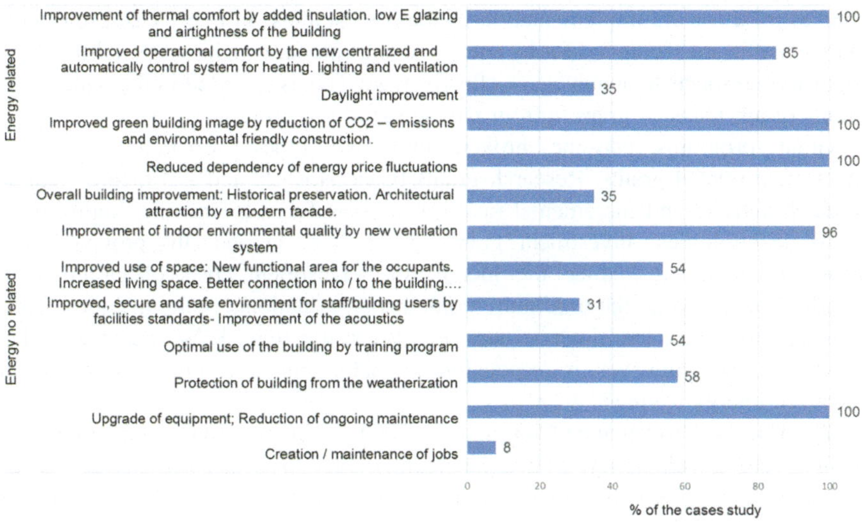

Fig. 1.10 Energy-related and non-energy-related co-benefits (% of the cases studies)

Of the non-energy-related co-benefits, "upgrade of equipment, reduction of ongoing maintenance," refers to the issue of maintenance backlog (as reasons for refurbishment) and also indicates a less clearly defined cost savings potential in all case studies.

The next most common co-benefit of DER (96% of the cases) is improved air quality due to an improved ventilation system. More than half of the case studies also cited "better weather protection" of the building. "Improved use of space" indicates an additional cost-saving potential (more available floor space). Again, some of the listed reasons may considered as being related. If one were to link these observations to the reasons for renovation and the relation between non-energy and energy-related renovation costs, it is tempting to say that energy savings are often a co-benefit of a renovation that would have been done "anyway," and that "reduction of ongoing maintenance" and "improved use of space" may be benefits that might also be quantified as additional monetary values.

1.7 Financing and Business Models

Financing DER is recognized as one of the major barriers for a large-scale implementation of DER projects. In spite of the above observations (regarding reasons for renovation), it is still most often a requirement that DER be "cost effective" by "paying itself off" in a reasonable amount of time.

In many cases, refurbishment budgets may be not sufficient to cover the entire cost of the projects, so renovation will only be started if the cost benefits cover at least a major part of the global (energy- and non-energy-related) investment costs. This requirement has a close relationship to the business models and funding sources used for the implementation of the case studies. In many cases, the monetary benefits are accounted for by energy savings alone, which makes it difficult for renovation projects to achieve cost effectiveness. This is reflected in the wide use of grants (grant loans or grant upfront payments) in the projects.

Grant programs can be broken into two types, research grants and investment-related grants. Research grants (Case Nos. 2a and 2b) mainly support research activity and incremental investment costs for very specific, highly innovative technologies. Investment grants (Case Nos. 3a and 3b) provide direct financial support to reduce the incremental investment costs of an energetic refurbishment beyond the minimum requirements of national building codes. Grant programs to reduce loan interest rates are also in use (e.g., KfW and others). The building owner, however, is responsible for achieving cost effectiveness and the predicted energy savings.

The Energy Performance Contracting (EPC) business model combines investment in energy efficiency with a financing model in which the energy service company (ESCO) provides the investment without any costs to the building owner. The revenue of the ESCO is related to the measured and verified energy cost savings after each year of the contract duration. The ESCO takes the responsibility

Table 1.5 Business models and funding sources

Business models and funding sources	Case study
Case 1: Self-financing Standard monthly "maintenance and improvement ***contribution***" by the tenants-funding model. Loan at low interest rates for Danish municipalities. Other loans —i.e., bank loans. In one case "private funding."	All cases except 4, 5, 10, 13, 14, 16, 17, 19, 20, 21, 25 (58%)
Case 2a: Research grants provided by national or international funding sources	2, 4, 6, 15 (15%)
Case 2b: National research program: American recovery and reinvestment act of 2009; agency provided funds (RWA); ARRA funding time-frame for completion	20, 22, 26 (12%)
Case 3a: National/regional/local investment grant program	1, 3, 4, 6, 8, 9, 10, 13, 14, 15, 16 (46%)
Case 3b: Subsidies: for implementation of ecological and sustainable measures; subsidized feed-in tariff for electricity generated by PV; subsidy loans for social housing companies—0.5%–25 years	1, 23, 24 (12%)
Case 4: EPC energy performance contracting; design-build business model including third party financing, planning and investment; the revenue payment is related to the savings performed	8, 25, 26 (12%)
Case 5: Public private public partnership: design build business model including third party financing, planning and investment; the revenue payment is related to a rental rate combined with other performance criteria	12

for achieving the predicted (and guaranteed) energy savings. The use of the ESCO business model enables building owners to reduce the upfront investment costs of a DER, which can significantly offset the scarcity of public funding sources. In the EPC business model, energy costs are turned into investment cost payback rates.

Table 1.5 lists the business models and funding sources, grouped into several main categories. Column 2 parenthetically lists the percentage of the case studies for which the indicated business model/funding source was used.

1.8 Cost Effectiveness of DER

Because the case studies are located in different countries, there was considerable variation in the way calculations of cost effectiveness were done. Some case studies focused on Net Present Value (NPV) of all current savings and expenditures including interest rate for capital costs; other case studies considered Simple Payback (SPB), which considers investment costs and the payback from energy and other cost benefits.

Besides different labor, material, and construction costs, financial parameters such as interest rate and price increase rate differ from case study to case study. Each study also provided a slightly different set of assumptions for what was considered an "energy related" or "non-energy-related" investment.

To generate an overall picture, NPV was calculated for each case study that could provide plausible investment costs, energy, and other cost savings. The calculation used the same discount rate for all the cases (2%). The expected economic and technical lifetime was set to 30 years. The calculation did not account for maintenance and replacement costs for some of the components implemented such as building automation, heating pumps, engines etc. Twelve of the 26 case studies presented sufficient information for this calculation (Fig. 1.11).

A positive NPV signals that cost effectiveness is achieved over the 30-year time period, while a negative NPV shows that cost effectiveness is not achieved in this period.

Figure 1.11 shows that eight of the 12 case studies represented here have a positive NPV of the energy investment costs. For four of them, the NPV was negative.

Three-fourths of the case studies are cost effective within the given framework conditions. Figure 1.11 also shows that there is a large variation between these 12 case studies. For Case Study No. 2, the NPV was as high as 538.80 Euro/m^2; for Case Study No. 26, it was 218.40 Euro/m^2. This reflects the fact that the investments costs vary widely. It may also reflect an issue already mentioned above, that

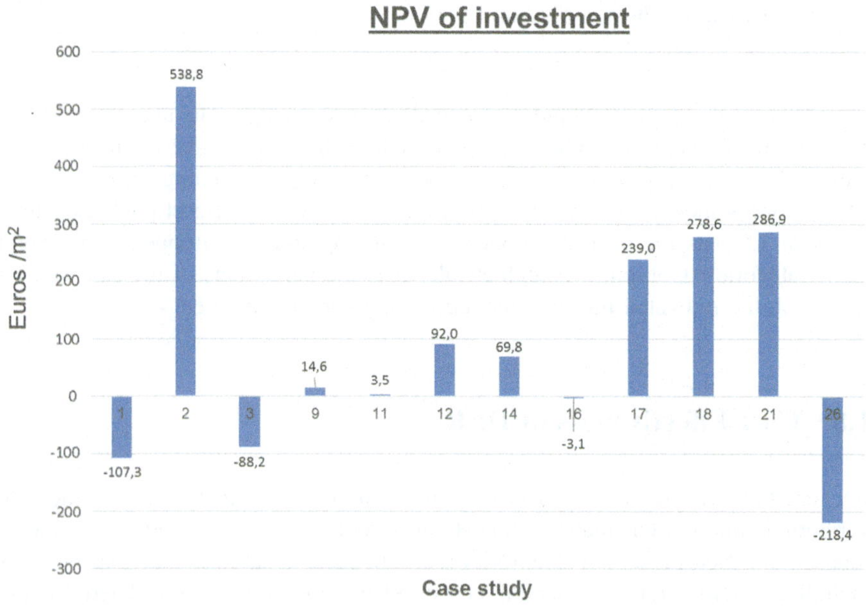

Fig. 1.11 Net present value of 12 case studies calculated with rates for discount and inflation equal to 2%

for those case studies that exhibit high, positive NPV (Case Study Nos. 2, 17, 18, and 21), a high fraction of the renovation costs has been assigned to non-energy measures, thereby leaving a smaller fraction to the energy-related measures.

1.9 Lessons Learned

Setting up a major renovation in the public building stock involves the allocation of a large amount of scarce public funding in one specific location. The decision-making process must consider whether it is more cost competitive to refurbish the existing building, or to construct a new building, and if the opportunity should be used to consider energy efficiency measures that strive to exceed national minimum requirements.

Since it can be difficult to accurately predict investment costs, cost savings from energy conservation measures, and other life cycle costs, any decision-making process used to determine whether (and how) to undertake refurbishment projects must draw information from recent experience with DER to overcome a great deal of uncertainty. Decision makers must consider the predicted key performance indicators of their specific case(s) (global and energy-related investment costs, energy and non-energy-related savings, cost efficiency, etc.), and the experiences and lessons learned from previous case studies (such as those listed in Table 1.6).

Table 1.6 Experience/lessons learned (energy related)

Energy
• To achieve 50% heating energy savings, the majority of case studies had to carry out refurbishment of major parts of the building's thermal envelope • DER measure bundle: To cut back heating energy up to 80–90%, use a holistic concept of combined building's thermal envelope, HVAC renovation, and a change of supply solutions
• Synergetic effects: The implementation of DER bundles contain multiple synergetic effects that help to decrease the overall investment costs compared to staged application of single refurbishment measures. This accounts for the reduction of investment costs (thermal envelope allows downsizing heating and cooling supply, LED lighting reduces internal cooling loads), and maintenance and energy costs
• DER to NZEB approach: It is possible to achieve NZEB and plus-energy standard for multi-story buildings when a DER with significant demand reductions is combined with renewable energy supply solutions
• Energy should also be reduced by means of demand side measures
• Energy exchange between buildings with different user/load profiles offer a potential for further energy reduction
• Especially in mid and northern European countries, a DER requires additional (new) ventilation systems, which will result in increased electricity consumption (often + 10–15 kWh/m^{2}year)
• To maintain the low energy consumption of a post-DER building it is required to conduct a continuous retro-commissioning/energy management of the implemented DER measures. Implementation of measures without stringent energy management may in many cases result in significant underperformance and lacking cost effectiveness of the DER concept

Table 1.7 Experience/lessons learned (use and comfort related)

DER benefits: Use and comfort
• The indoor air quality increased significantly: buildings with ventilation systems, which are still not very common in most of European cz. 4 and 5 countries, achieve a more stable humidity and a better fresh air quality with less ventilation heating losses than do buildings with window-based ventilation
• The implementation of building automation systems allows indoor temperature to be controlled more accurately; this improves indoor climate and energy efficiency
• Combining a DER with a major renovation allows energetic refurbishment to be combined with a new layout of the occupied space; this helps project designers to consider indoor climate, daylight usage, etc.
• A VOC sensor had to be installed in some classrooms to reduce high CO_2 levels
• In all case studies, the building users indicated that they perceived the impacts of the DER on the indoor climate either positively or very positively

The three first benefits listed in Table 1.7 clearly state that verified energy savings of beyond 50% have been achieved by implementing DER bundles of technologies that include synergistic effects among individual technologies; these effects may yield even greater positive impacts in terms of cost effectiveness and energy balance when the DER is carried out in neighborhoods, campuses etc., areas that allow an energy exchange between neighboring buildings with different user and load profiles. In such cases, the importance of a stringent commissioning and energy management increases.

Many evaluations and research work done over the past several decades provide evidence that, in addition to energy-related (cost-saving) benefits, major energetic renovations can improve occupant comfort and increase productivity. In schools and office buildings, especially, this could be one of the most important reasons for undertaking building renovations; improved productivity and learning in office, manufacturing, and education buildings may provide significant additional financial benefits.

One case study provided a simple calculation assuming 100 employees working in a 3000 m^2 office building with average energy costs of 20 €/m^2 (high energy costs) and a yearly salary of 50,000 €/employee. Assuming that a DER would reduce the energy costs by 70% and increase the productivity by 1%, the financial value of the improved working conditions would account for 50,000 €, while the energy savings of 70% would have a value of up to 45,000 €—over 1 year. Such quantifications of productivity may turn the NPV of a lagging renovation project into a positive value. So far, not many business models are able to monetize these obviously large potentials; recent approaches provided in the comfort.meter.org project show simple methods to evaluate and monetize the building comfort into productivity and revenue streams.

Table 1.8 Experience/lessons learned (user behavior and acceptance related)

User behaviour and acceptance
• The behavior of building users has a major influence on the performance of a DER project. Experiences from case studies shows that misbehavior of building users and facility management staff can lead to a significant underperformance of the DER project i.e., by inaccurate operation of the building. To ensure optimal performance, users and FM staff have in some cases been integrated in the preparation and planning phase of the project so they were able to provide valuable contributions to the design of the concept. Moreover, in such cases, the acceptance of the DER project will be improved. In addition, after the DER has been carried out, the implementation of **user training programs** is a necessary precondition for a good performance of the DER project. For the FM staff, it is important that the building automation system be accurately documented and that the functionality of the entire system and each sub-group of the building automation system be evaluated and verified
• Even after an initial training, users tend to revert to old habits. An annual update of FM staff and users is helpful to maintain a good performance of the DER over a long-time period. In some cases, especially in schools, DER projects have been carried out in combination with incentive systems for the users: energy managers in school classes are responsible for the "micro-management" of the building automation in a class room; if they perform well, the building authority gives the classes a small reward
• An additional, significant positive aspect of such a program is that **more users become aware of energy saving**
• In a few cases, the building owner and the users developed a "building user's guideline" that provides information on the DER concept and how it relates to the correct operation of office room equipment, lighting, ventilation, heating etc.
• An effective user-training program will improve the general level of care of the refurbished building (eliminate graffiti; ensure that damage is reported and repaired immediately, etc.)

Table 1.9 Recommendations to building owners

Recommendations to building owners
• Decisions made in early project stages have strong influence on energy performance and costs. To improve cost effectiveness of a DER project, energy efficiency (EE) measures must be combined with a general refurbishment of the building
• The DER project must begin by gathering all information concerning the energy consumption (energy baseline), building usage, and building construction and HVAC installation data. Next, a building model must be developed and calibrated against utility bills and other measured and verified data. The plausibility check of the modeling is an important quality management topic
• Communication between the building owner, designers, users, FM staff, and often financiers is a very important component to a successful DER project implementation
• Building systems should be commissioned and adjusted for optimal operation before the project can be handed over to the users/owners and commissioning should be an ongoing activity
• The cost effectiveness of DER projects can be improved by combining significant demand side reduction in a DER with renewable energy supply solutions in three stages: — Stage 1. Exploit non-investment energy savings by considering occupant engagement — Stage 2. DER implementation — Stage 3. After 1-year of post occupancy install renewable resources to offset tracked energy demand • The cost effectiveness of a DER concept can also be significantly improved by considering (and documenting/quantifying) non-energy-related benefits such as avoided maintenance costs, increased usable floor space, reduced insurance costs, improved productivity etc.
• To augment scarce public funding, use innovative business models for deep energy retrofit providing performance guarantees. For the building user, it is relevant that the energy service company take over the risk for the energy savings, and the overall cost effectiveness, availability, and functionality of the DER measure bundles over a time period of 10–20 years

Table 1.8 summarizes the experiences learned related to the user behavior and acceptance.

Table 1.9 lists recommendations that can help building owners achieve successful DER project(s).

Conclusion

The 26 case studies form an interesting collection of Deep Energy Retrocfit building projects from around the world. The interested reader will find valuable information about the actually implemented energy renovation technologies—often in terms of both technical parameters and costs.

It may be self-evident, but is nonetheless worth stating, that the overview of technologies that have been implemented clearly shows that, to reach DER, it is necessary to retrofit mechanical systems by implementing technology bundles in concert with a well-planned building envelope renovation. Implementation of extra insulation of the building facade significantly reduced heating energy consumption. The incorporation of a heat recovery ventilation system both reduces the energy consumption and enhances the indoor air quality, resulting in a positive occupant reaction. Moreover, the replacement of artificial lighting with low-energy lighting is often the easiest and most economic retrofit technology.

The investigation of the achieved energy savings shows that Deep energy renovation is quite possible. On average, these 26 case studies achieved 66.4% energy savings.

The analysis of reasons for renovation shows that the non-energy related reasons dominate. Buildings are renovated mainly to meet a need for maintenance. These reasons are often referred to as "anyway renovation." Anyway renovation is a term characterizing the renovation needed to maintain the building in good condition. It might also come from the fact that use of the building is to change, so the building will must be renovated/refurbished to some degree to accommodate the changes in use. These anyway measures may very well be the main reason for initiating the renovation process.

The building owners should be aware that costs of the "anyway" renovation must be established and documented to make sure that the energy part of the renovation is not required to "pay back" these elements of the renovation investment costs. The brief cost efficiency analysis carried out show a large variation in achieved NPV of the 12 case studies that could be further analyzed. Four of these have a relatively high NPV and it is assumed that the reason behind is that the

R. Lohse et al. (eds.), *Deep Energy Retrofit—Case Studies*,
https://doi.org/10.1007/978-3-031-17517-6

energy related costs presented in these case studies are really the net energy related investment costs, when the cost for the anyway renovation has been subtracted.

In this context, it is also worth noting that optimization is not always a straightforward financial optimization of the NPV of the energy saving measures. The optimization calculation depends on the situation of where and when it is carried out. It depends heavily on the parameters/assumptions used—energy prices, interest rates, etc. and might quickly change. Therefore, it is advisable to look also at what is cost-efficient and what uncertainty interval should be considered. The cost-efficient energy renovation may have a less advantageous NPV than the optimized renovation, but as long as the NPV is positive, the financial result will be better or equal to the outset situation and will result in higher energy savings than the optimized renovation.

Following this line of argument, when identifying the possible energy renovation measures, it is useful to consider the following:

1. In the long run, it may be advantageous to carry out the energy renovation to the fullest possible extent (a deep energy renovation), as each subsequent step will almost always be more costly.
2. The savings resulting from selecting the bundle of energy saving technologies should be calculated in energy and financial terms. As the individual measures influence each other, a separate calculation must be performed for each individual bundle investigated.
3. Co-benefits stemming from each energy saving measure should be noted and to the degree possible given an economical value. For example, improved comfort and indoor air quality have been shown to increase working efficiency and learning performance. Again, for each bundle investigated, the related co-benefits must be documented with the results of the financial analysis for each bundle.

The collection and analysis of the 26 case studies has proven a valuable activity to improve the understanding of the mechanisms behind deep energy renovation building projects and of how to advance the implementation of such projects. This knowledge has been used to promulgate guidelines under development of the IEA EBC Annex 61 project.

References

(1) Hermelink, A., and A. Muller. 2010. Economics of deep renovation. European Insulation Manufacturers Association. Project # PDEMDE 101646.

(2) NBI. New Buildings Institute. http://buildings.newbuildings.org

(3) RICS. 2013. Sustainable Construction: Realizing the opportunities for built environment professionals. RICS Europe. Brussels.

Appendix A: Case Study—Austria

A.1 Kapfenberg. Austria

A.1.1. Name of the Project, Location (City, Country)

Johann Böhmstrasse 34 + 36, Kapfenberg, Province of Styria, Austria.

These pictures show the building in its original and post-retrofit states, illustrating key features of the retrofit (Fig. A.1).

A.2 Project Summary

A.2.1. Project Objectives

The Austrian case study is a renovation of a multi-story housing block in the city of Kapfenberg. It was constructed in 1960–1961 with four floors and 24 flats with a size of 20–65 m^2. The total heated gross floor area is 2845 m^2. Due to the poor energetic, technical and architectural quality (overly small flats, outdated equipment) the housing company was forced to do a major renovation. To break new ground for ambitious concepts, renovation activities have been supported by the Austrian research program "Building of Tomorrow."

R. Lohse et al. (eds.), *Deep Energy Retrofit—Case Studies*,
https://doi.org/10.1007/978-3-031-17517-6

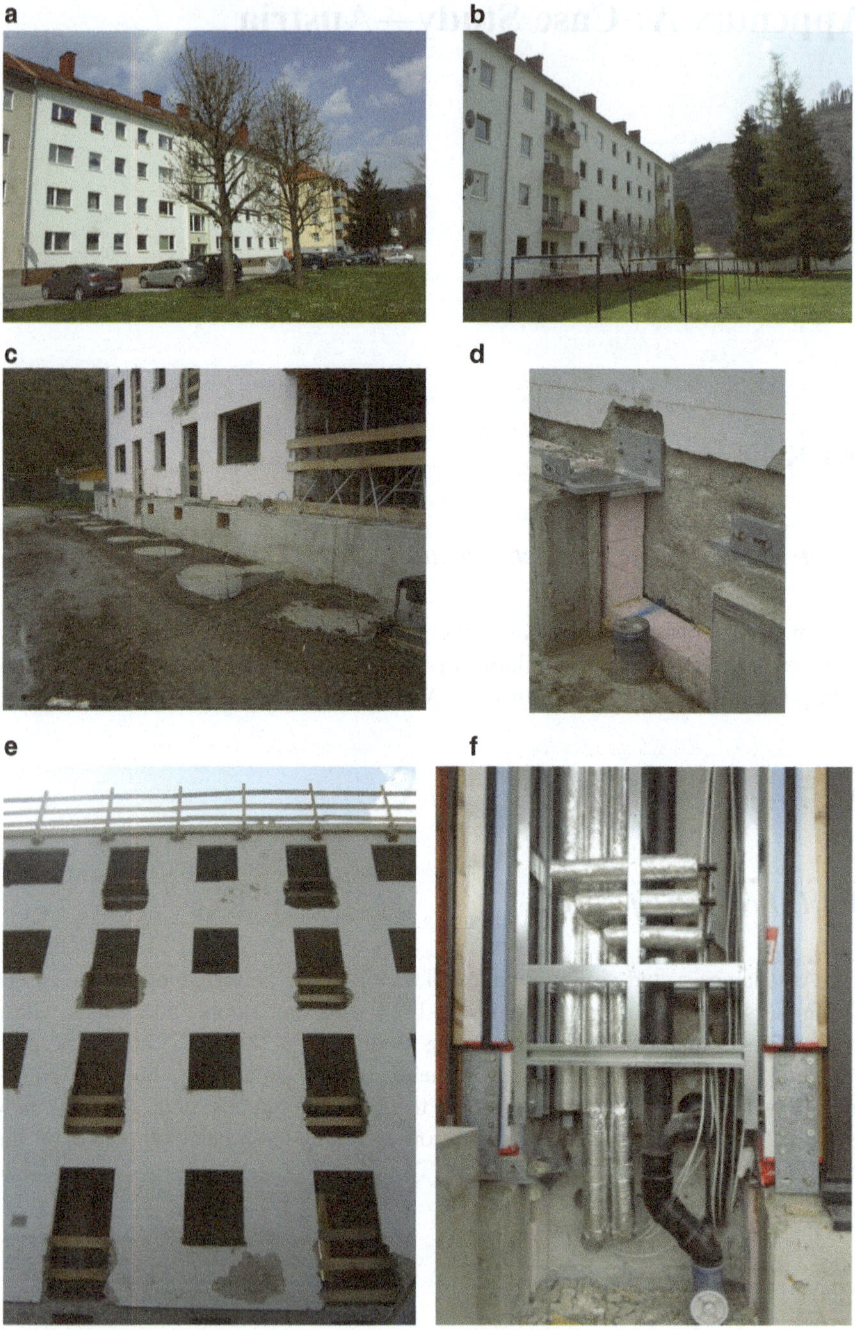
a
b
c
d
e
f

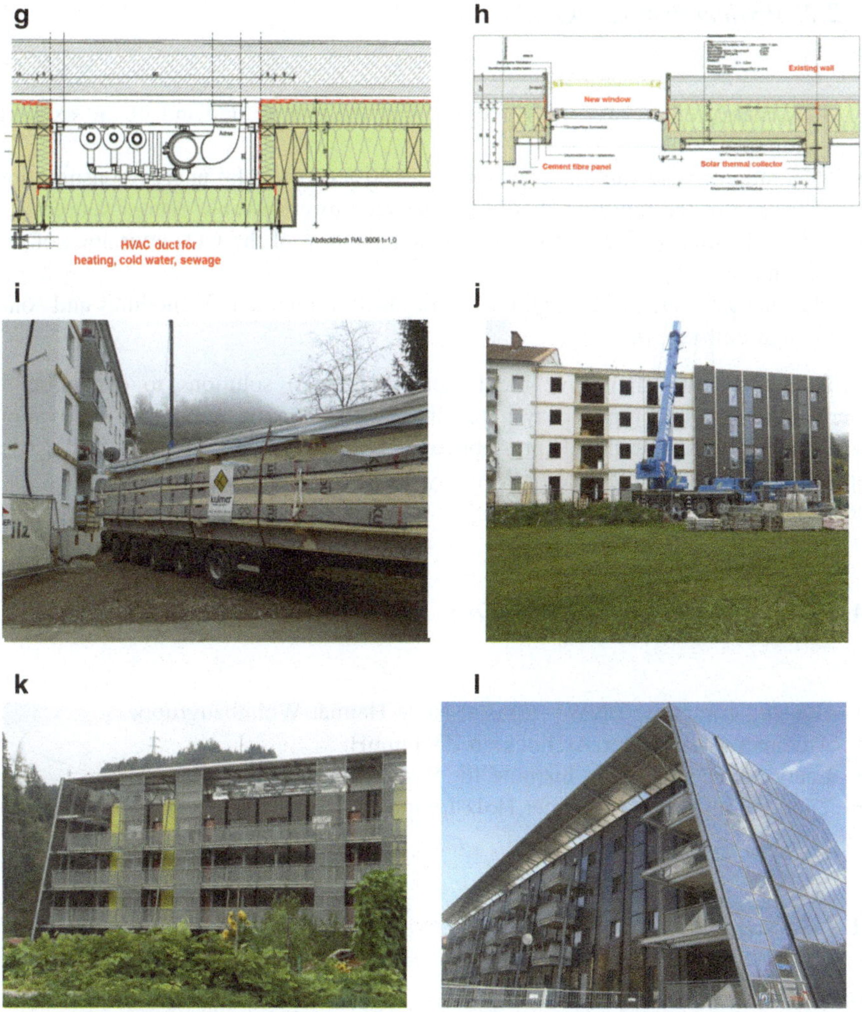

Fig. A.1 Renovation process of the Austrian case study: **a** existing building west facade; **b** existing building east facade; **c** building under construction; **d** foundation for the pre-fabricated element; **e** demolition works; **f** new installation duct in this figure: Renovation process of the Austrian case study: **a** existing building west facade; **b** existing building east facade; **c** building under construction; **d** foundation for the pre-fabricated element; **e** demolition works; **f** new installation duct in the facade; **g** horizontal section of the installation duct; **h** horizontal section of the facade element with integrated solar thermal collector; **i** transport of pre-fabricated facade elements; **j** mounting of facade elements; **k** renovated building from east; **l** west and south facade with solar thermal collector after renovation. *Source* g + h (Nussmüller Architekten ZT GmbH). Rest: AEE INTEC

A.2.2. Project Energy Goals

- 80% energy efficiency—80% reduction of the energy demand of the existing building.
- 80% ratio of renewable energy sources—80% of the total energy consumption of the renovated building should be provided by renewable energy sources.
- 80% reduction of CO_2 emissions—80% reduction of the CO_2 emissions of the existing building.
- Plus energy standard through energy production on site (PV modules and solar thermal collectors).

To demonstrate alternative (ecological optimized) solutions to conventional thermal insulation composite systems (like extruded polystyrene) the renovation was done with standardized, pre-fabricated wooden facade elements in Passive house standard with integrated HVAC systems (PV, solar thermal collectors, disposal systems).

A.2.3. Short Project Description

- Housing company: ENW—Ennstal Neue Heimat Wohnbaugruppe.
- Architect: Nussmüller Architekten ZT GmbH.
- Energy concept: AEE—Institute for Sustainable Technologies.
- Timber construction: Kulmer Holz-Leimbau GesmbH.

A.2.4. Stage of Construction

1960–1961.

A.2.5. Point of Contact (POC) Information

Heimo Staller.
AEE—Institute for Sustainable Technologies.
Feldgasse 19, A-8200.
Gleisdorf, Austria.
email: h.staller@aee.at.
Tel: +43 (0)3112 5886-364.

Table A.1 Design temperature

*Cooling Design Temperature—0.4% occurrence**	
Dry bulb temp C (F)	Mean coincident wet bulb temp C (F)
18.3 °C	No numbers
*Heating design temperature—99.6% occurrence***	
Dry bulb temp C (F)	
12	No numbers

A.2.6. Date of the Report

May 22, 2014.

A.2.7. Acknowledgement

A.3. Site

- Location: Johann Böhm Strasse 34 + 36, A- 8605 Kapfenberg, Austria.
- Latitude: 47.45°.
- Longitude:15.29°.
- Elevation: 502 m.
- Climate zone: Austrian climate zone S/SO, climate zone 5 A (ASHRAE 90.1-2004 Climate Zone).
- Cooling Degree Days (based on 65 F = 18.33 °C): 739 average (year 2009–2013) source: www.degreedays.net.
- Heating Degree Days (based on 65 F): 5619 average (year 2009–2013) source: www.degreedays.net (Table A.1).

A.4. Building Description/Typology

A.4.1. Typology/Age

Multi-story housing block/52 years.

A.4.2. Type (Office, Barracks, Etc.)

Social housing.

A.4.3. Typology/age

1950–1970.

A.4.4. General Information

- Year of construction: 1960–1961.
- Year of previous major retrofit—if known: no previous major retrofit.
- Year of renovation (as described here): 2013.
- Total floor area (m^2): 2845 m^2.
- Area of unconditioned space included above (m^2): 0 m^2.
- Other information as appropriate.

A.4.5. Architectural and Other Relevant Drawings

See Figs. A.2, A.3, A.4 and A.5.

A.4.6. National Energy Use Benchmarks and Goals for Building Type Described in the Case Study

A.4.6.1. Benchmark, According to Standard National Average, min, and Max

The baseline is the calculated site energy demand of the existing building and the target value is a reduction of 85%. Figure A.6 lists the values for the energy demand of the existing building, a renovation following the minimum requirements of the Austrian building code, and more ambitious scenarios. Scenario 5 will be the Austrian target value.

1. Existing building baseline.
2. Minimum requirements Austrian building code reduction of 47%.
3. Scenario e80^3 (improvement of u-values) reduction of 55%.
4. Scenario 3 + mechanical ventilation with heat recovery reduction of 62%.
5. Scenario 4 with PV modules and solar collectors reduction of 85% (target).

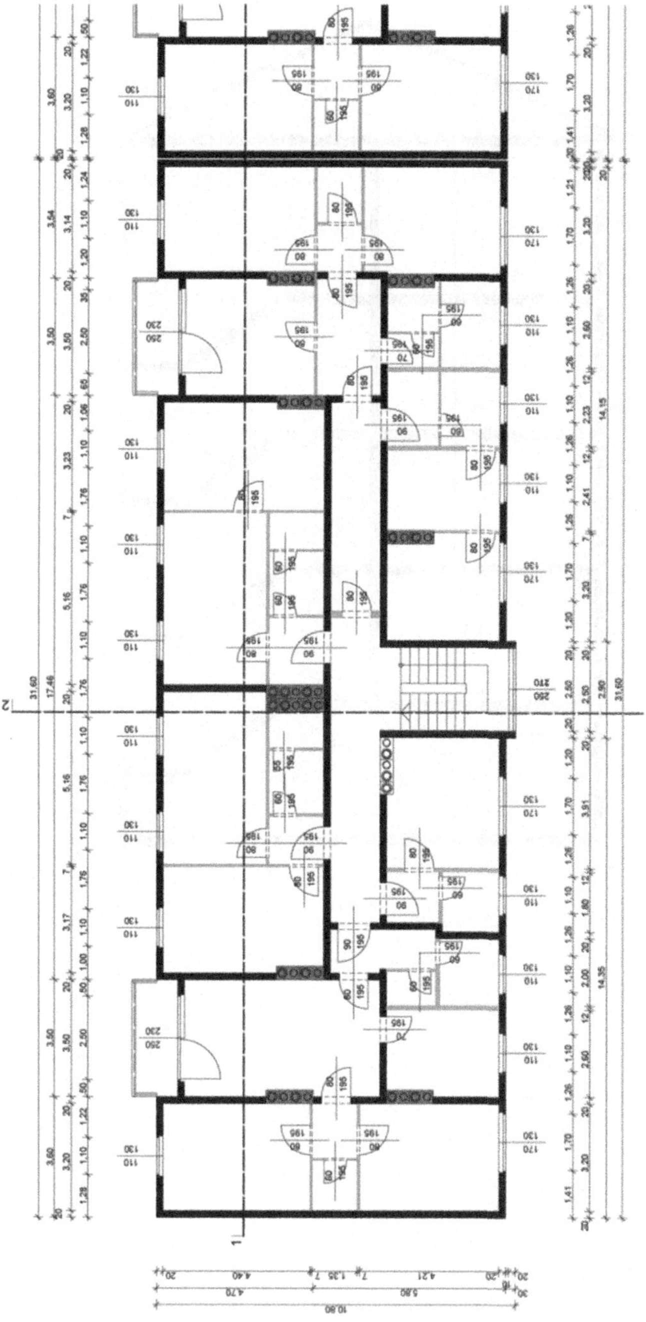

Fig. A.2 Detail of the ground floor plan of the existing building. *Source* Nussmüller Architekten ZT GmbH

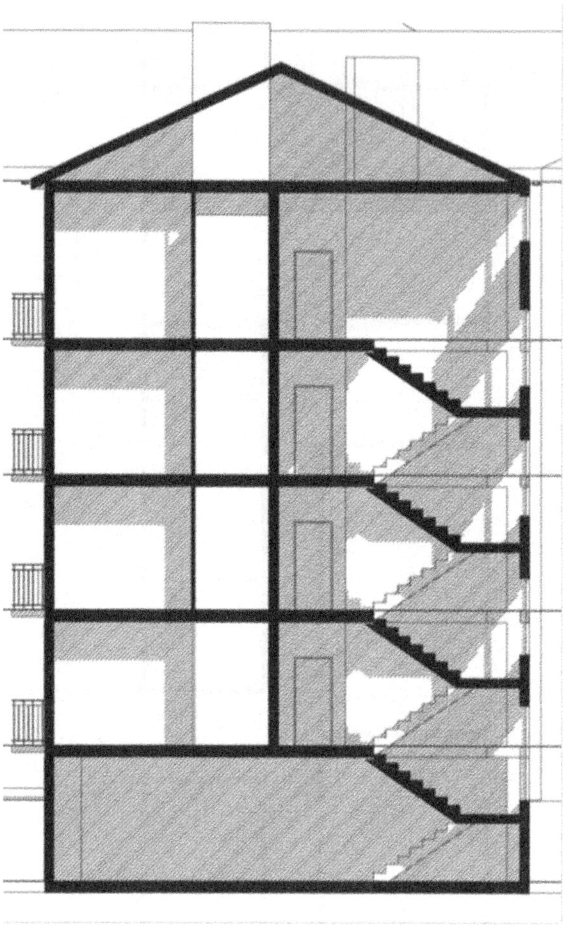

Fig. A.3 Cross section of the existing building. *Source* Nussmüller Architekten ZT GmbH

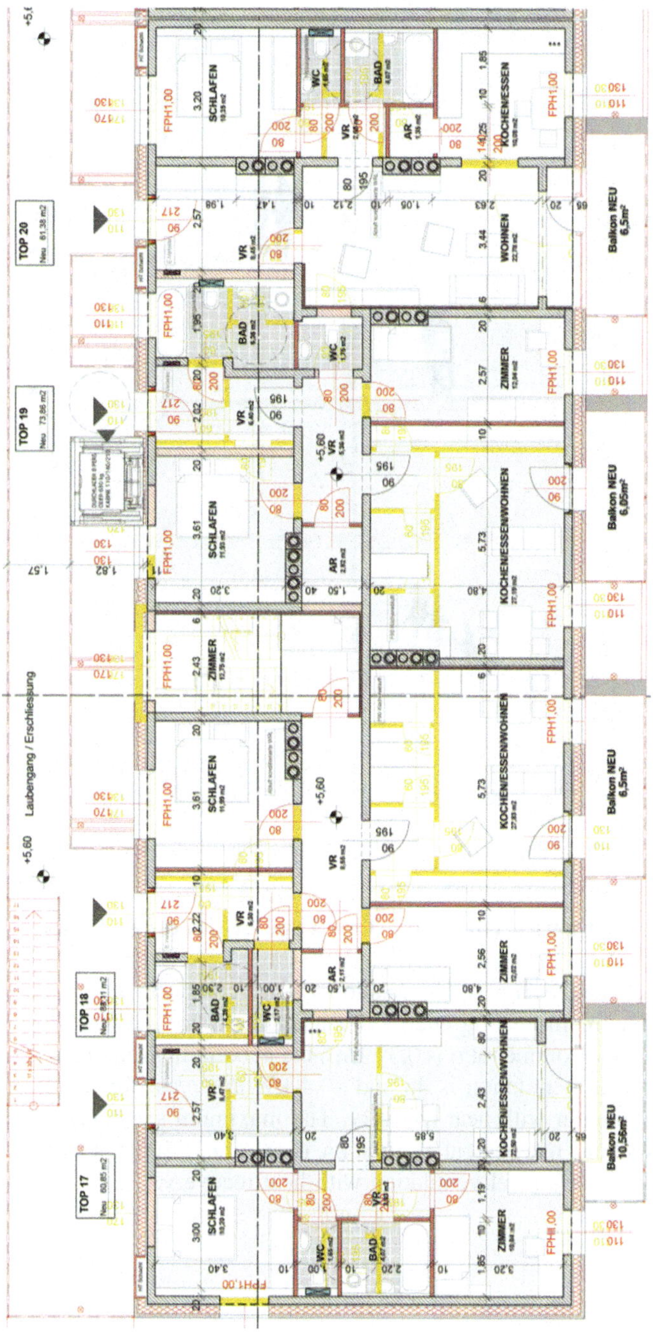

Fig. A.4 Ground floor plan of the renovated building. *Source* Nussmüller Architekten ZT GmbH

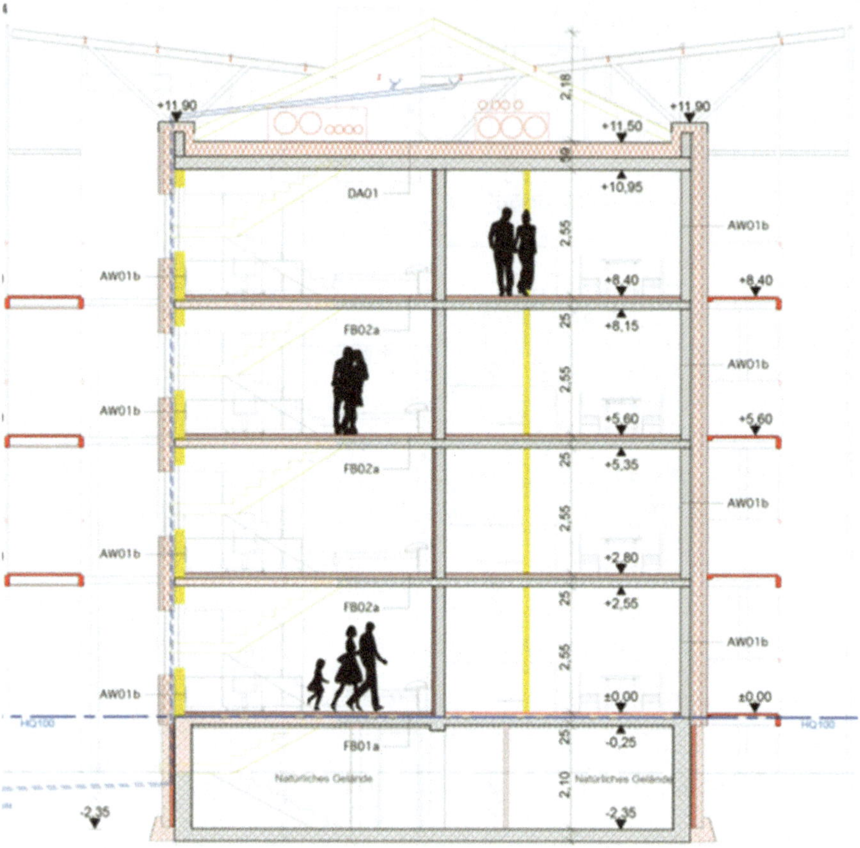

Fig. A.5 Cross section of the renovated building. *Source* Nussmüller Architekten ZT GmbH

A.4.6.2. National Energy Target for This Type of Building (if Any)

National energy targets are set in relation to the compactness of the building, whereby the maximum heating energy demand has to be 87.5 kWh/m^2 year. For this building (shape/volume ratio is 0.37 1/m) the maximum heating energy demand without mechanical ventilation is 48 and 40.4 kWh/m^2 year for buildings having mechanical ventilation with heat recovery. The maximum site energy demand for the building without mechanical ventilation is 99.1 and 105.8 kWh/m^2 year for buildings having mechanical ventilation with heat recovery.

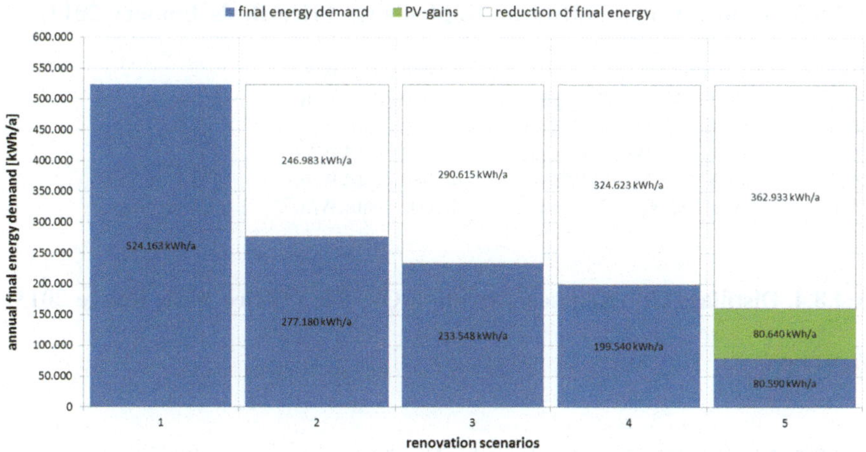

Fig. A.6 Renovation scenarios of the Austrian case study

A.4.7. Awards or Recognition

No awards or recognition.

A.4.8. Site Energy Cost Information (Electricity, \$/kWh; Gas, \$/m^3, Etc.)

In all values below, VAT is included.

A.4.8.1. Electricity (Stadtwerke Kapfenberg)

- Basic charge per month: €3.54.
- Energy price per kWh: €0.094103.

A.4.8.2. Renewable Energy Feed-In Tariff

PV OEMAG subsidy per kwh: €0.125.

A.4.8.3. Natural Gas (Stadtwerke Kapfenberg, Price Basis January 2013)

Zone 1	0–8000 kWh/year	46,920 cent/kW/h.
Zone 2	8001–15,000 kWh/year	46,920 cent/kW/h.
Zone 3	15,001–40,000 kWh/year	46,560 cent/kW/h.
Zone 4	40,001–80,000 kWh/year	46,320 cent/kW/h.
Zone 5	80,001–200,000 kWh/year	46,080 cent/kW/h.
Zone 6	over 200,000 kWh/year	45,600 cent/kW/h.

A.4.8.4. Distillate Oil (Kerosene, Heating Oil, Etc.) (Price Basis Range 2013)

€0.9–1.0/l.

A.4.8.5. District Heating (Stadtwerke Kapfenberg)

- Basic charge per year and per installed load kW: €23.467511.
- Energy price per kWh: 0.079865.

A.4.8.6. District Chilled Water

See Sect. 1.9.5.

A.4.8.7. Other (E.g., Biomass) (Price Basis January 2013)

- Wood chips: 3.51 cent/kWh.
- Pellet: 5.10 cent/kWh.

A.4.9. Pre-renovation Building Details

A.4.9.1. Envelope Details: Walls, Roof, Windows, Insulation Level

- Walls.
  - Sandwich concrete elements without an additional insulation. U-value: 0.87 W/m^{2}K.

- Roof.
  - Pitched concrete roof with no insulation. U-value: 0.87 W/m^{2}K.
- Basement ceiling.
  - Ceiling in concrete insulated with approx. 60 mm polystyrene.
- Windows.
  - Double-glazed wooden windows. U-value: 0.87 W/m^{2}K.

A.4.9.2. Heating, Ventilation, Cooling, and Lighting Systems

In the existing building a variety of different heating systems was installed: a central gas heating, electric furnaces, electric night storage heaters, oil heaters, wood-burning stoves and coal furnaces. There is no active cooling system and the lighting was mainly done with normal bulbs.

A.4.10. Description of the Problem: Reason for Renovation

Due to the poor energetic, technical and architectural quality (overly small flats, outdated equipment) the housing company was forced to do a major renovation. The enormous energy demand caused very high heating and operating costs.

A.4.11. Renovation SOW (Non-energy and Energy Related Reasons)

A high quality refurbishment of the building with a change in the layout of the apartments should make the building more attractive to new residents and young families.

A.4.12. Energy Saving/process Improvement Concept and Technologies Used

A.4.12.1. Building Envelope Improvement

- Walls (Fig. A.7).

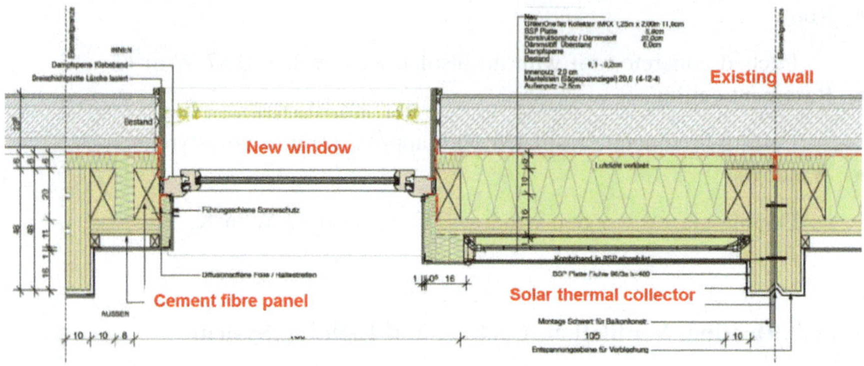

Fig. A.7 Renovated wall, structure from outside

Pre-fabricated wooden facade elements. U-value:	0.12 W/m^{2}K.
Cement fiber panel	0.8 cm.
Air space	5.0 cm.
Cross laminated timber	5.9 cm.
Wood construction with mineral wool insulation	20.0 cm.
Mineral wool	6.0 cm.
Vapor barrier.	
Exterior plaster existing wall	2.5 cm.
Existing wall: sandwich concrete elements	20.0 cm.
Interior plaster	2.0 cm.

- Roof

The existing old pitched roof was removed and a new flat roof was installed with 35–40 cm of polystyrene. U-value: 0.10 W/m^{2}K.

- Basement ceiling

Due to the low ceiling height, it was not possible to install more than 6 cm insulation even though a higher insulation would have been desirable to conserve energy.

The existing building has already been insulated with 6 cm insulation. This thermal insulation was renewed during the renovation but the energy performance of this component was not much improved (U-value: 0.30 W/m^{2}K) (Fig. A.8).

- Windows

The new windows are already integrated in the prefabricated facade modules and are of high thermal quality. An external shading device is also installed and already integrated in the facade module too. This external shading device helps to reduce the solar gains and therefore to avoid overheating of the rooms in the warm periods of the year.

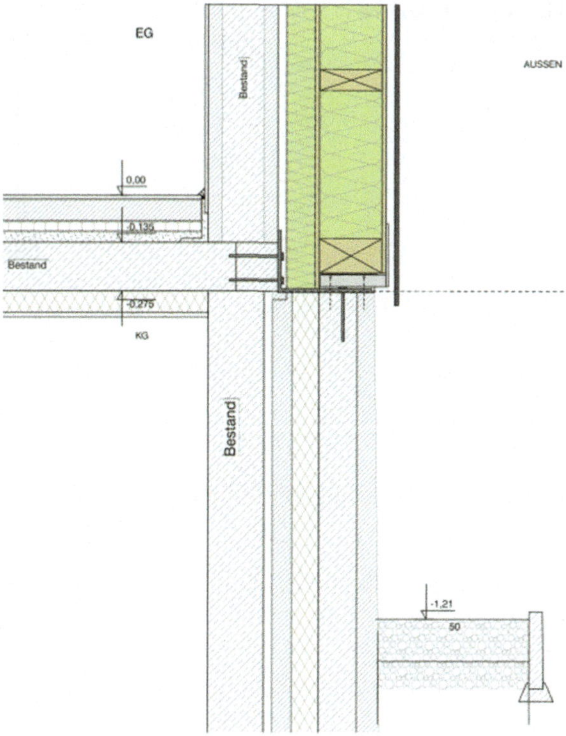

Fig. A.8 Detail of the basement

The integration of the external shading device in the prefabricated facade element had to be as thermal bridge-free as possible. So this point had to be considered in the planning stage.

U-values (Fig. A.9):

$$\text{Uglass} = 0.70\,\text{W/m}^2\text{K}.$$
$$\text{Uframe} = 1.17\,\text{W/m}^2\text{K}.$$
$$\text{Uwindow} = 0.97\,\text{W/m}^2\text{K}.$$
$$\text{g-value} = 0.60.$$

A.4.12.2. New HVAC System or Retrofits to Existing

General description

The basic heat supply of the renovated building is accomplished by the local district heating. The district heating grid of the city of Kapfenberg is largely supplied with waste heat (heat losses) of the local steel manufacturer (Böhler-Uddeholm).

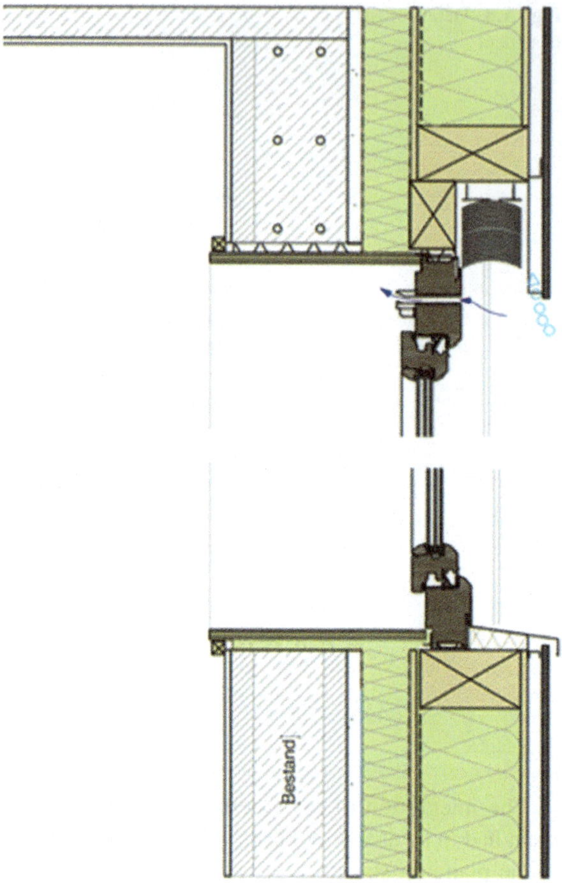

Fig. A.9 Detail of the window

In addition to the district heating, a solar thermal system with a collector surface of 144 m^2 was installed. A scaffold on the south facade was mounted to increase the area for the solar thermal panels and also to optimize the inclination of the panels (inclination of the solar thermal panels = 72°). An annual heat production from the solar thermal system of 39.5 MWh/a was calculated. Both district heating and solar thermal system store the produced heat in a 7500 l buffer storage, which is located right below the solar thermal panels. From the buffer storage a two-pipe-system (flow and return) brings the heat to the 32 flats where the heat for domestic hot water is stored in a small boiler Radiators emit the heat in the flats (Fig. A.10).

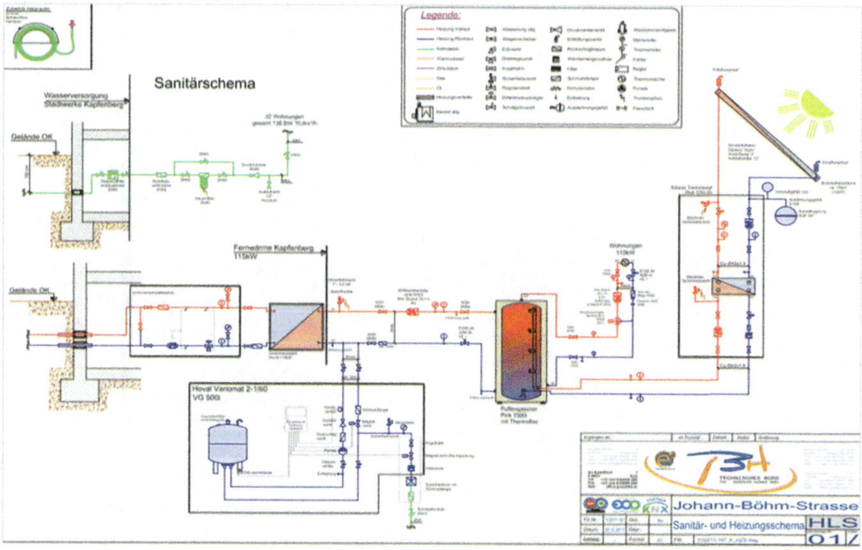

Fig. A.10 Plumbing and heating scheme

Domestic hot water

The boiler for the domestic hot water has a storage volume of 120 l and gets supplied with the necessary heat from the buffer storage via the two-pipe-system. Every apartment is equipped with one boiler. The small boiler is loaded twice a day from 06:00 a.m. to 07:00 a.m. and from 07:00 p.m. to 08:00 p.m. In this way the temperatures and therefore the heat losses of the two-pipe-system can be reduced because of the lower flow temperature of the heating system compared to the flow temperature for the domestic hot water preparation.

Ventilation

A new mechanical ventilation system with heat recovery is installed (heat recover efficiency = 65%/SFP = 0.45 Wh/m^3). The ventilation unit was positioned on the flat roof and the existing shafts of the building are used for the ventilation ducts. An additional benefit of using these existing shafts are the short ventilation ducts they accommodate.

In one half of the flats the ventilation system is controlled automatically based on the CO_2 concentration; in the other half of the flats, the residents can control the ventilation system manually by a three-stage switch (Fig. A.11).

A.4.12.3. New Lighting System

The lighting system is in the responsibility of the tenants.

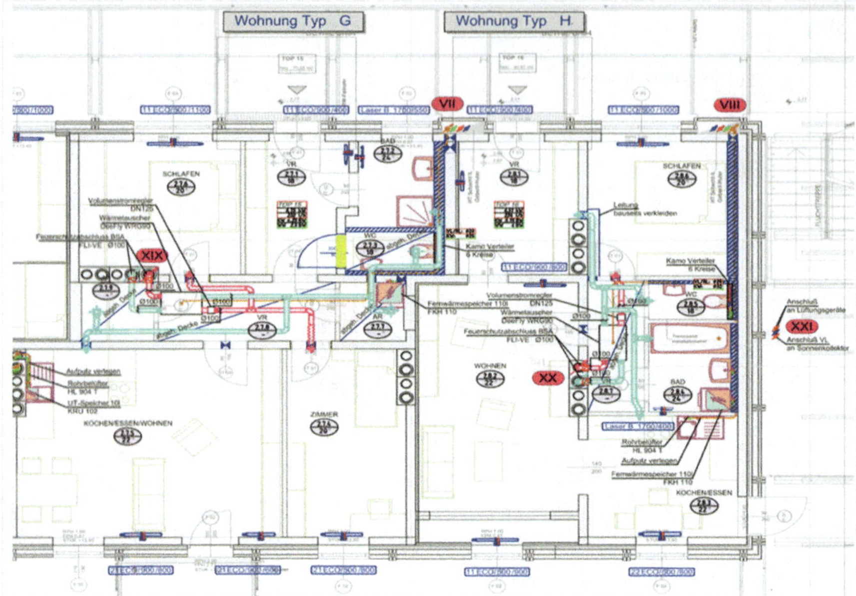

Fig. A.11 Ventilation scheme

A.4.12.4. New Generation/Distribution System

See "General description."

A.4.12.5. Renewable Energy

Photovoltaic

Photovoltaic panels with a size of 550 m^2 (80 kWp) are installed on the roof on an extra mounted scaffold that has the form of a wing. Additionally 80 m^2 (12 kWp) photovoltaic panels are installed on the south facade next to the solar thermal panels. In total, the calculated annual energy production of the 630 m^2 (92 kWp) photovoltaic panels is about 80 MWh/a.

Solar thermal system

A solar thermal system with a collector surface of 144 m^2 is installed. For this purpose a scaffold on the south facade was mounted to increase the area for the solar thermal panels and also to optimize the inclination of those (inclination of the solar thermal panels = 72°). An annual heat production from the solar thermal system of 39.5 MWh/a was calculated. Both district heating and solar thermal system store the produced heat in a 7500 l buffer storage located right below the solar thermal panels.

A.4.12.6. Daylighting Strategies

New windows and enlargement of old windows contributes to better daylight conditions.

A.4.13. Energy Consumption

A.4.13.1. Pre-renovation Energy Use (Total and per m²/year)

Values are based on the calculation required for the Austrian energy certificate.

Site energy demand	524,163 kWh/year	184 kWh/m^2/year

A.4.13.2. Predicted Energy Savings (Site, Source, GHG), Total and per m²/year

- Site energy

Savings total site energy demand	443,573 kWh/year	156 kWh/m^2/year
Savings heat energy demand	322,222 kWh/year	113 kWh/m^2/year
(generation of solar heat:	39,500 kWh/year	14 kWh/m^2/year)
Savings electricity demand:	121,351 kWh/year	43 kWh/m^2/year
(generation PV	80,640 kWh/year	28 kWh/m^2/year)

- CO_2

There are no CO_2 savings concerning electricity as the electricity of the company "Stadtwerke Kapfenberg" is totally renewable.

The district heating system of Kapfenberg has 54% renewable energy sources (waste heat of the local steel manufacturer, CO_2-equivalent is 0 g/kWh) and 46% natural gas (CO_2-equivalent is 236 g/kWh).

So the CO_2 savings are	31 to/year	0.0108 to/m^2/year

A.4.13.3. Measured Energy Savings (Thermal, Electrical), Total and per m²/year

Monitoring of the building has been done for 1 year from September 2014 to September 2015. The measured energy savings stated below are referenced to the calculated energy performance before renovation, as no detailed consumption data before renovation are available.

1. Savings total site energy demand	375,896 kWh/year	132 kWh/m²/year
2. Savings heat energy demand	292,060 kWh/year	103 kWh/m²/year
(generation of solar heat)	34,378 kWh/year	12 kWh/m²/year
3. Savings electricity demand	83,836 kWh/year	29 kWh/m²/year
(generation PV)	77,239 kWh/year	27 kWh/m²/year

A.4.14. Energy Cost Reduction

A.4.14.1. Split in All Energy Forms—Electricity, Oil, District Heating

- Electricity

Reduction of electricity

40,711 kWh/year × €0.094103	=	€3.831/year

Compensation for electricity fed into the grid

80,640 kWh/year × 0.125	=	€10,080/year
Total cost reduction	=	€13,911/year

- District heating

322,222 kWh/year × €0.079,865	=	€25,734/year

A.4.15. Non-energy Related Benefits Realized by the Project

Main non-energy related aspects are:

- Increased useful space.
- Increased indoor air quality.
- Reduced energy costs for tenants.
- Environmental friendly construction.
- Reduced maintenance.

A.4.16. Renovation Costs: Total and per m²

A.4.16.1. Total

Total renovation costs area €1717/m² gross floor area ~ €4.8 Mio.

A.4.16.2. Non-energy Related

€1280/m² gross floor area ~ €3.6 Mio

A.4.16.3. Energy Related

€438/m² gross floor area ~ €1.2 Mio

A.4.16.4. Cost for Each Measure

Prefabricated facade elements	€260/m² facade	€380,380
Windows (triple glazed wood windows)	€609/m² window	€215,586
Roof refurbishment (insulation)	€155/m² roof	€110,205
District heating		€14,950
Hot water system		€57,600
Costs for solar thermal system, including additional costs for larger storage system (with scaffolds)		€62.120
PV panels (with scaffolds)		€232,740
Ventilation system with heat recovery		€171,620

A.4.17. *Business Models and Funding Sources*

A.4.17.1. Decision Making Process Criteria for Funding and Business Models

The building is owned and managed by a social housing company, which is subjected by the Austrian Social Housing Law (Wohnungsgemeinnützigkeitsgesetz). To enable maintenance and renovation activities, tenants are obliged to pay a monthly "maintenance and improvement fee" (Erhaltungs- und Verbesserungsbeitrag). This amount is regulated by the Austrian Social Housing law and is €1.32/m^2 of useful area, per month (for flats with first time use more than 20 years ago), two-thirds of this amount for flats with first time use less than 20 years ago (but with a minimum of 10 years), and a fourth of this amount for other flats. The maintenance and improvement contribution is index-adjusted.

A.4.17.2. Description of the Funding Sources Chosen

Depending on the range of measures in the province of Styria, the following funding models for renovation are available:

- "Comprehensive energetic renovation" and "Small renovation."
- "Comprehensive renovation."

For the housing project in Kapfenberg, the "Comprehensive renovation" model was chosen. This model has the following requirements:

- Renovation of min. three flats at one time.
- Award of building license min. 30 years ago.
- Min. renovation cost of €30,000 per flat.
- 50% of measures have to contribute to the improvement of the flats.
- Max. heating demand for a shape/volume ratio > 0.8 is 75 kWh/m^2 year and for a shape/volume ratio < 0.2 is 35 kWh/m^2 year.

The maximum funding is €1130 per m^2 of living area. The amount of this funding increases for improved energetic standards:

- €40 per m^2 living area for a heating demand between 45 and 25 kWh/m^2.year (Low-energy standard).
- €70 per m^2 living area for a heating demand under 10 kWh/m^2 year (Passive house standard).
- €145 per m^2 living area for implementation of elevators.

For social housing companies, subsidy loans of the province of Styria (0.5% interest, 25 years runtime) are given.

In addition to the above mentioned funding, an extra nonrefundable subsidy of €7/m^2 and a year for the implementation of ecological and sustainable measures (use of renewable energy sources, mechanical ventilation with heat recovery, ecological building material, etc.) is given.

A national subsidized feed-in rate for electricity generated by PV is available in addition to the funding models of the province of Styria. For 2014 and 2015, the feed-in rate for PV facilities with a maximum capacity load over 5 kW peak is 12.5 cents/kWh. The production and sale of electrical energy is quite complicated for social housing companies (since it is fraught with legal and organizational problems); for the city of Kapfenberg, the energy company implemented PV panels using a contracting model.

To break new ground for ambitious renovation measures (Plus-energy standard), the Austrian case study was done as a pilot renovation within the Austrian research program "Building of Tomorrow." The Austrian Ministry for Transport, Innovation and Technology supported the project with a nonrefundable investment subsidy of 35% for the innovative cost (cost difference to standard renovation).

A.4.18. Cost Effectiveness of Energy Part of the Project (NPV, SIR, Etc.)

To get comparable numbers cost parameters in this chapter are calculated without funding models mentioned in Chapter 18. Compensation for electricity fed into the grid is accounted as cost reduction.

Simple pay-back time

Energy related investment costs	€1,245,201.00
Energy savings per year—electricity	€13,911.00
Energy savings per year—district heating	25,734.00
Energy Savings per year—total	€39,645.00
Simple pay-back time	31 years

Dynamic investment method

Results stated below are based on following assumptions:

Inflation rate per year:	2.2%
Interest rate:	3.75%
Interest rate inflation-adjusted:	1.52%
Price rise for electricity per year:	3%
Price rise for electricity per year inflation-adjusted:	0.78%
Price rise for district heating per year:	3%
Price rise for district heating per year inflation-adjusted:	0.78%
1. Internal interest rate:	1.52% per year
a. Cash value:	€254,362
b. Annuity method	
Annuity:	€14,266 per year
for a runtime of:	30 years
Annuity factor:	0.056
c. Dynamic amortization period	26 years

Investment costs of energy saved

To get an overview about energy related measures and their impact on energy savings, investment costs for energy saved in kWh for a certain operation period can be given. The investment costs of most important measures of energy saved for an operation period of 30 years (without maintenance and replacement costs) are:

1. Reduction of transmission losses (Improvement of the thermal building shell—windows, facade, roof): €0.08/kWh
2. Reduction of ventilation losses (mechanical ventilation with heat recovery): €0.25/kWh
3. Reduction through energy production on site—solar thermal panels: €0.05/kWh
4. Reduction through energy production on site—PV panels: €0.10/kWh

In the Austrian case study energy saved by implementation of mechanical ventilation with heat recovery shows the highest investment costs. The costs for reduction of transmission losses in this special case are quite high because the pre-fabricated facade elements (prototypes) have investment costa ~ 2.5 times higher than conventional thermal insulation composite systems.

Life cycle cost assessment (LCCA)

Within "Annex 56—Cost Effective Energy and Carbon Emissions Optimization in Building Renovation" a Life cycle cost assessment (LCCA) for the housing project in Kapfenberg was done. A first result is shown in Fig. A.12.

Life cycle cost assessment (LCCA)							
Component / Energy consumption				Investment cost		Investment cost [€/m²-GFA]	Annual cost [€/m²-GFA/y]
BITS	Heating			77.068	€	27,09	0,27
	DHW			57.600	€	20,25	0,20
	Cooling				€	0,00	0,00
	Auxiliaries			233.000	€	81,90	0,82
	Lighting				€	0,00	0,00
	Ventilation			171.616	€	60,32	0,30
	Common appl.				€	0,00	0,00
Envelope	Roof	711	m^2	155	€/m^2-element	38,74	0,00
	Facade	1463	m^2	260	€/m^2-element	133,70	0,00
	Win.	354	m^2	609	€/m^2-element	75,78	0,00
	Floor	711	m^2	-	€/m^2-element	0,00	0,00
Energy consumption	Heating						1,17
	DHW						2,54
	Cooling						-
	Electricity (misc.)						3,80
				1.245.455	Total	438	9,10

Fig. A.12 Life cycle cost assessment (LCCA) of the Austrian case study; reference study period 60 years; Methodology: LCIA. *Data source* Ecoinvent v2.2 (LCIA). *Source* Annex 56, econcept AG

A.4.19. User Evaluation

A.4.19.1. Description of User Training Program Within the Refurbishment

Before the refurbishment, tenant were requested to fill out a questionnaire on the following topics:

- Thermal comfort.
- Acoustic comfort.
- Indoor air quality.
- Malt problems.
- Expectation on refurbishment (kind of measures to be done, etc.).

After the renovation, two tenant surveys were taken (the first, 2 weeks after occupancy, and the second after the first heating period) on the following topics:

- Assessment/comparison of the living quality before and after renovation.
- Satisfaction with the mechanical ventilation with heat recovery.
- Thermal comfort in winter and summer.
- Satisfaction with the renovation process.
- Information strategy of the building owner.

The tenants were also provided with a short manual for the handling of the mechanical ventilation with heat recovery.

A.4.19.2. Integration of Users Demands in the Planning Process

No measures for this aspect.

A.4.20. Experiences/Lessons Learned

A.4.20.1. Energy Use

Within the research project monitoring/evaluation for 2 years will be done. The monitoring comprises following aspects:

- Energy consumption (heat, electricity) of flats and total building.
- Indoor and outdoor temperature and humidity.
- Solar radiation.
- CO_2 concentration in flats.
- Number of window openings by tenants.
- Parameters of the mechanical ventilation with heat recovery (air temperature, humidity, flow rate, electricity consumption).
- Household electricity (in one flat).

First monitoring results (monitoring period August 2013–January 2014) show following results:

- Analysis of energy consumption for space heating shows a big difference between flats (1.5–11 kWh/m^2 and month conditioned gross floor area). The summarized mean value from September to January is 16 kWh/m^2. The calculated energy demand for space heating is 16.9 kWh/m^2 per year, which is nearly equal to the monitored consumption. It has to be stated that the measured values are based on a monitoring period of 5 month, which from the scientific point of view is too small for verification.
- Analysis of energy consumption shows smaller differences between flats for hot water than for space heating (4 kWh/m^2 per month). The summarized mean value from September to January is 16.5 kWh/m^2. The calculated energy demand for domestic hot water heating is 12.8 kWh/m^2 and year, which is nearly equal to the monitored consumption (Figs. A.13 and A.14).

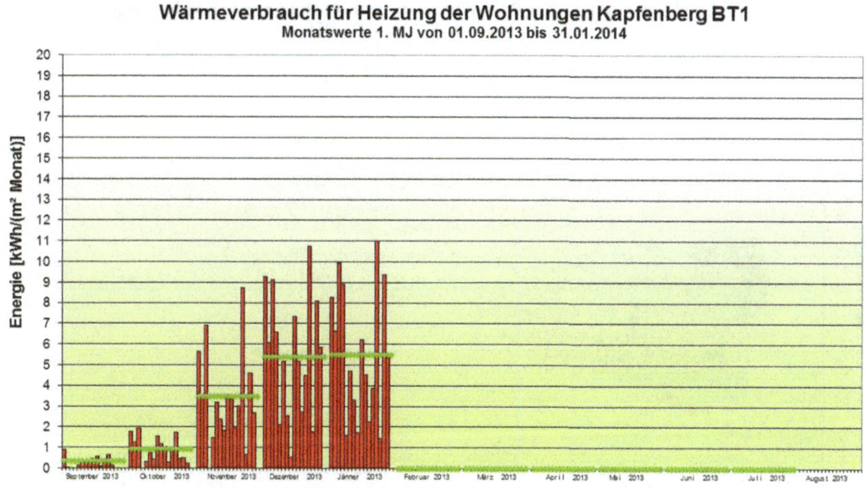

Fig. A.13 Energy consumption for space heating, month values of measured flats (monitoring period September 1, 2013–January 31, 2014)

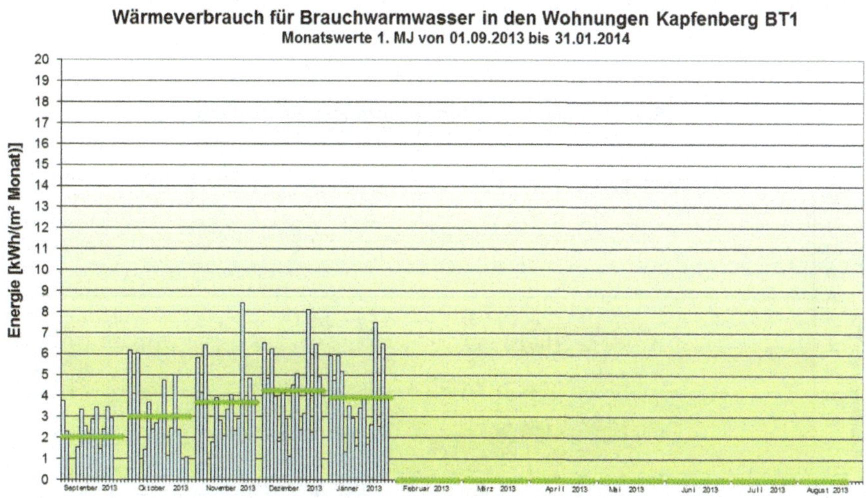

Fig. A.14 Energy consumption for domestic hot water, month values of measured flats, monitoring period 01.09.2013–31.01.2014

A.4.20.2. Impact on Indoor Air Quality

- CO_2 concentration in almost all flats is under 1000 ppm.
- Humidity in flats is between 20 and 60%.
- In all flats indoor temperature is in the range of comfort requirements (Austrian standards) (Figs. A.15 and A.16).

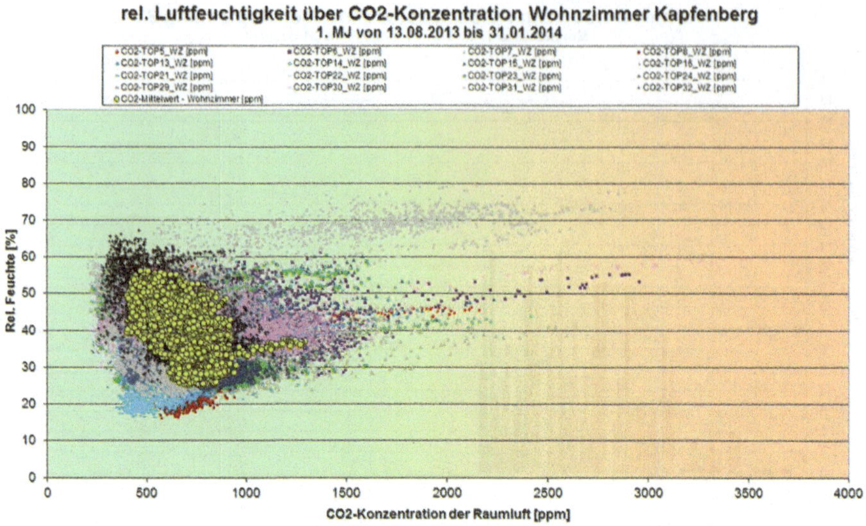

Fig. A.15 CO_2 concentration/humidity of measured flats, monitoring period 01.09.2013–31.01.2014

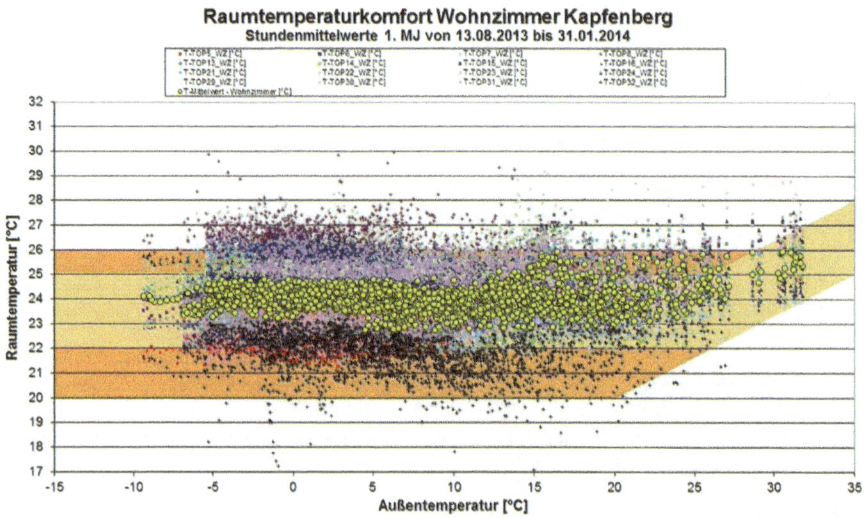

Fig. A.16 Indoor temperature/outdoor of measured flats (living rooms) (monitoring period September 1, 2013–January 31, 2014)

A.4.20.3. Practical Experiences of Interest to a Broader Audience

See below.

A.4.20.4. Resulting Design Guidance

Deep Energy Retrofit requires a multi-level approach. In a first step, energy should be reduced by means of demand side measures (reduction of transmission-, ventilation- and infiltration losses, minimizing the electrical energy demand of households, etc.). In a second step, energy efficient, smart HVAC-systems and energy production on site must be installed. In some cases energy exchange between buildings with different user and load profiles can offer further potential for energy reduction. Core aspect of all these steps is an optimized bundle of retrofit measures and the costs of energy saved. It is essential to find a compilation of cost-optimized solutions for different retrofit scenarios. Scenarios must be considered from early project stages (e.g., project development) to construction and operation stage. Because (especially for these early design stages) decisions made in early project stages have a strong influence on energy performance and costs (investment and operational costs), appropriate strategies and tools are required.

From technical point of view, ambitious projects like the Austrian case study Plus-Energy Renovation in Kapfenberg demonstrate that a plus-energy standard for multi-story housing can be achieved. However, a look at the financial aspect shows a different picture. A dynamic amortization period of 26 years differs greatly from common amortization periods (10–15 years) for investments on the Austrian market. To overcome these hurdles and to enable Deep Energy Retrofit, innovative business models must be developed.

A.4.26.3. Practical Experiences of Interest to a Broader Audience

See below.

A.4.26.4. Resulting Design Guidance

[illegible]

Appendix B: Case Study—Denmark

B.1. Stengårds School.DK

B.1.1. Renovation of Stengårds School

The school is located in the Municipality of Egedal of Copenhagen in Denmark.

B.1.2. Stengårds School Before and After Renovation

See Fig. B.1.

Key feature of the retrofit.

- New facade with improved insulation and new windows.
- Technical insulation of connections to hot water tank and the tank itself.
- Insulation of hot water pipes and circulation pipes incl. pump.
- Insulation of pipes of heating distribution system incl. pumps, connections and valves.
- New CO_2-controlled ventilation systems with heat recovery instead of old exhaust system.
- New low-energy motors for ventilation system in the big assembly hall.
- New electrical lighting system incl. control.
- Optimization of electrical lighting in the basement.
- New low-energy pumps on the heating distribution system.
- New BEMS.
- PV panels on the roof.

R. Lohse et al. (eds.), *Deep Energy Retrofit—Case Studies*,
https://doi.org/10.1007/978-3-031-17517-6

Fig. B.1 Stengårds School before energy retrofit (left), after retrofit (right). *Source* Ove Mørck, Cenergia

B.1.3. Project Summary

B.1.3.1. Project Objectives

Facade renovation of the school with the goal of giving a complete new impression from the outside reaching the energy target of the Danish building regulation 2008. An architectural company has designed the new facade and has been responsible for the implementation.

The renovation of Stengårds School was part of a project of Concerto- European Commission Initiative called Class 1, Cost-effective Low–energy advanced sustainable so1ution, (www.class1.dk), whose objectives are:

1. Optimize the integration of low-energy building technologies with supply (renewable and conventional) and distribution (heating and electricity) technologies.

2. Advance selected technologies within the three areas: low-energy building, renewable energy supply and distribution
3. Improve the design, checking and verification procedures. This relates directly to the implementation of the building energy performance directive -EPBD.
4. Integrate the European Eco label in the building project.
5. Demonstrate large scale implementation at close to market technical and economic conditions.

B.1.3.2. Project Energy Goals

The overall aim of the energy renovation process was that the renovated building ends with an energy consumption calculated according to the energy frame—as defined in the Danish Building Regulations (BR)—at the level of new buildings built according to the Danish BR08. The energy frame calculation includes heating energy consumption for space heating and hot water and the electricity consumption for ventilation and lighting. For this calculation, the electricity consumption is multiplied by a primary energy factor of 2.5.

The optimization procedure was performed based on the "before" situation of building modeled in the Danish calculation tool Be10, which is the official Danish building energy calculation tool used to verify that a building comprise with the requirements in the Danish Building regulation—the implemented EU EPBD requirements in Denmark.

B.1.3.3. Short Project Description

It has been calculated the energy savings of implementing for a number of different energy renovation measures, estimated the costs of the implementation of each measure and thereby the payback period for this measure. The following key improvements were implemented:

- High levels of insulation in building envelope.
- Energy efficiency windows.
- Energy efficient lighting.
- Heat recovery ventilation.
- BEMS systems.
- PV panels.

Architect:

Gottlieb Paludan Architects.
Kalvebod Brygge 30, 5. Sal.
1560 København V.
Energy concept:
Cenergia Energy Consulting.

www.cenergia.dk.
Construction Company/ Technical Supervision:
Cyril Olsen.
co@gottliebpaludan.com.
Project engineering:
Lars F. Kjems.
lfk@gottliebpaludan.com.

B.1.3.4. Stage of Construction

Finished construction.

B.1.3.5. Points of Contact

Project Acronym: Class 1.
REF EC: 038572.
Energi-og Projektleder:

Lisbeth Berg.
Lisbeth.Berg@egekom.dk.

Stenløse Rådhus Anneks.
Rådhustorvet 2.
3660 Stenløse.

Egedal Commune.
Mobil number: 7259 7223.

B.1.3.6. Date of the Report

01/09/2014.

B.1.3.7. Acknowledgement

The project is supported by CONCERTO. Co-funded by the European Commission.

B.1.4. Site

- Location: Stengårds Plads 2, 3650 Ølstykke in Denmark.
- Latitude: 55° 77′ 31″ N.
- Longitude: 12° 16′ 82″.
- Elevation: 21 m above water level.
- The climate zone corresponds to the zone 5 A.
- Cooling degree day: 0.
- Heating Degree Days (based on 17 °C): 2906 (Table B.1).

B.1.5. Building Description/Typology

B.1.5.1. Typology/Age

The school was constructed in 1970–1978 and has a total gross area of 9050 m^2. Stengårdsskolen is a compact plan building that consists of three buildings connected by cover corridor. The main building of 7000 m^2 on ground floor has a central yard and basement underground of 1300 m^2. The main building was extended with a second building of 750 m^2 where more classrooms are situated.

B.1.5.2. Type

School.

B.1.5.3. General Information

- Year of construction: 1970–1978.
- Year of previous major retrofit: No previous major retrofit.
- Year of renovation: spring of 2013.
- Total floor area (m^2): 9050.
- Area of unconditioned space included above (m^2): 213.

Table B.1 Design temperature

*Cooling design temperature—0.4% occurrence**	
Dry bulb temp C (F)	–
*Heating design temperature—99.6% occurrence***	
Dry bulb temp C (F)	− 12 °C

B.1.6. Architectural and Other Relevant Drawings.

Source: Gotlieb og Paludan Architects (Figs. B.2, B.3, B.4, B.5, B.6, B.7, B.8, B.9, B.10, B.11, B.12, B.13 and B.14).

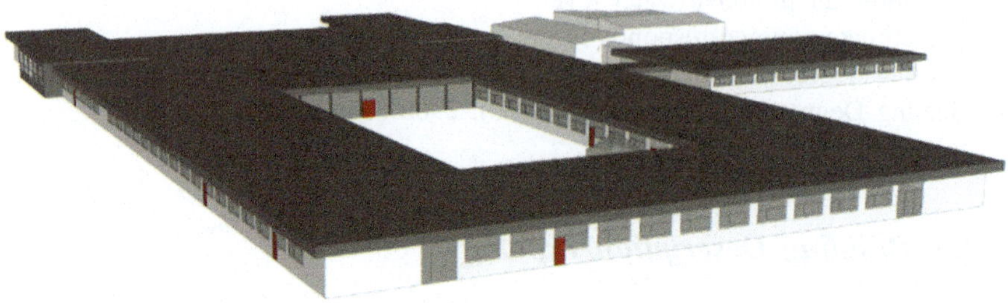

Fig. B.2 Rendering overview of Stengårds School

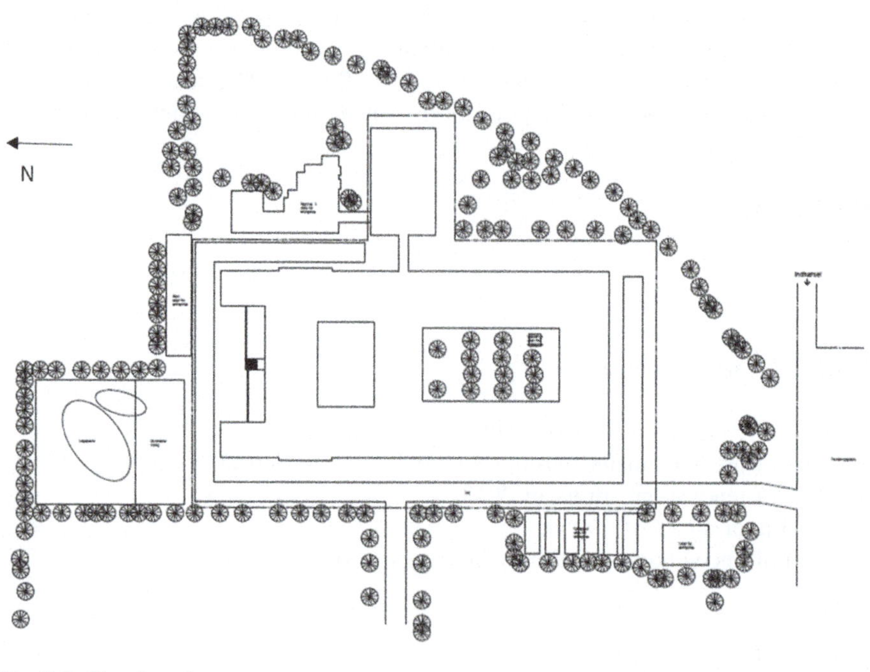

Fig. B.3 Situation plan

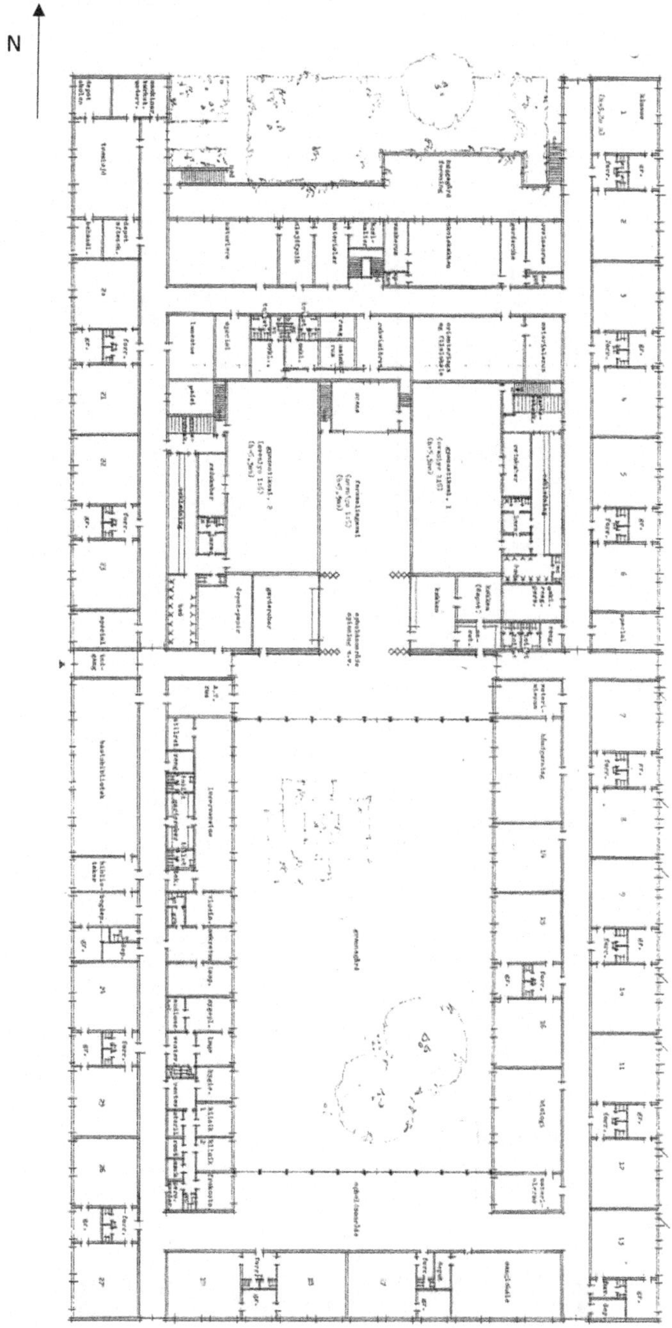

Fig. B.4 Distribution of ground floor of the main building

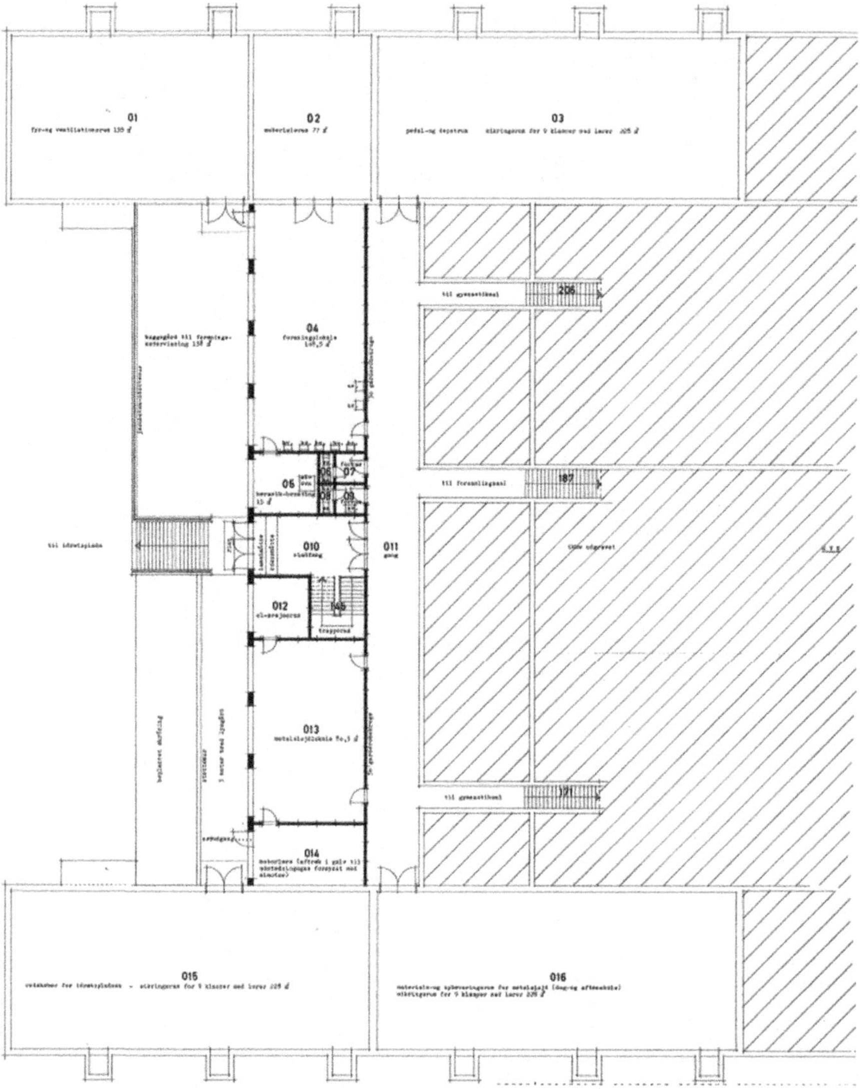

Fig. B.5 Basement floor facing north

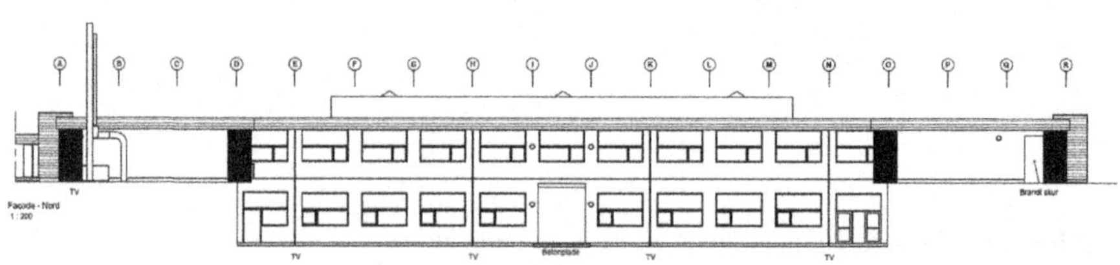

Fig. B.6 Cross section of Nord facade

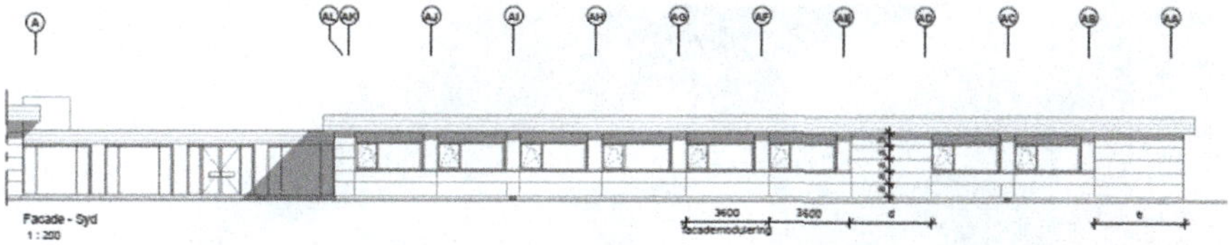

Fig. B.7 South facade, main facade

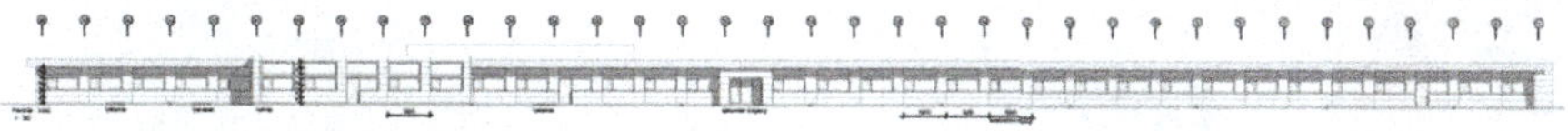

Fig. B.8 West facade

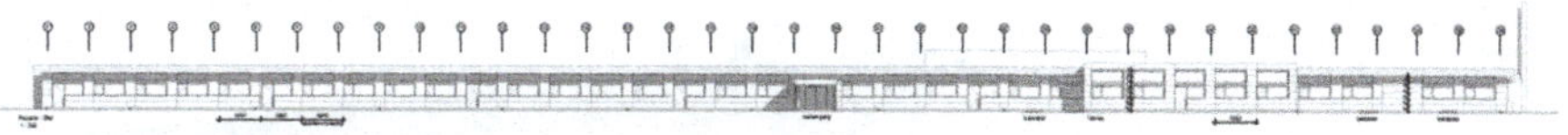

Fig. B.9 East facade

Fig. B.10 Rendering of renovated facade of the main entrance

Fig. B.11 Rendering of central back yard of renovated building

Fig. B.12 Photo of existing facade

Fig. B.13 Photo of renovated facade

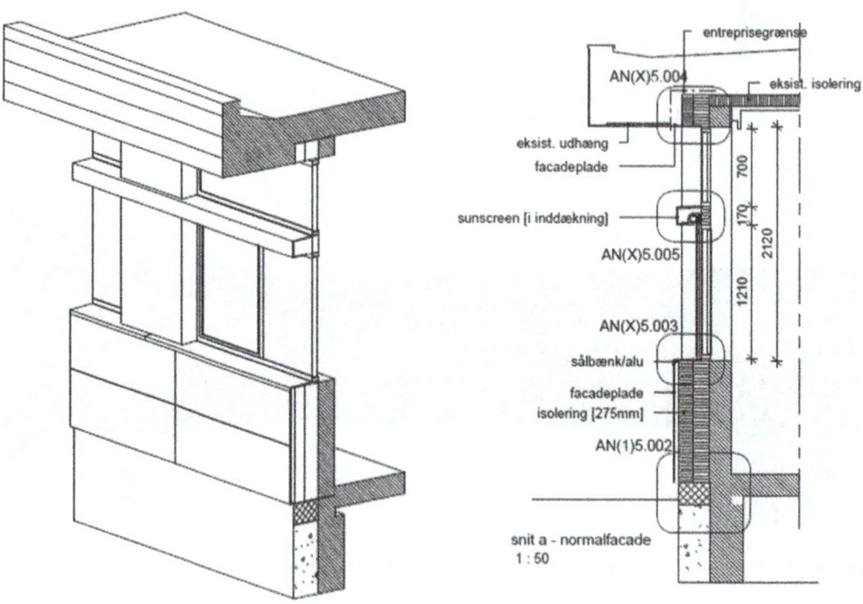

Fig. B.14 Cross section of renovated facade

B.1.7. National Energy Use Benchmarks and Goals for Building Type Described in the Case Study

B.1.7.1. Benchmark, According to the Standard National Average, Min, and Max.

See Table B.2.

B.1.7.2. National Energy Target for This Type of Building

The target values have been calculated on a purely theoretical basis, where the assumptions for the actual use a given building may vary considerably. The impact of different use/user behavior is to some extent balanced out by looking at the average energy frame consumption.

- Danish Building Regulation 2008: 95.2 kwh/m^2.
- Danish Building Regulation 2010: 71.5 kwh/m^2.
- Danish Building Regulation 2015: 41.1 kwh/m^2.
- Danish Building Regulation 2020: 25 kwh/m^2.

B.1.8. Awards or Recognition

No awards or recognition.

Table B.2 National energy use benchmarks (Germany)

Official annual average value for schools in DK		
Heating energy	Value	Unit
Heating energy	120	kWh/m^{2}a
Electricity	25	kWh/m^{2}a
Heating + electricity	145	kWh/m^{2}a
Water	260	l/m^{2}a

B.1.9. Site Energy Cost Information (Electricity, \$/kWh; Gas, \$/m^3, Etc.)

B.1.9.2. Electricity

2100 DKK/MWh.

B.1.9.3. Renewable Energy Feed-In Tariff

1.6 DKK/kWh.

B.1.9.4. Natural Gas

870 DKK/MWh.

B.1.9.5. Distillate Oil (Kerosene, Heating Oil, Etc.)

Heating oil: 800 DKK/MWh.

B.1.9.6. District Heating

Constant: 129.3 DKK/MWh, variable: 735 DKK/MWh.

B.1.10. Pre-renovation Building Details

B.1.10.1. Envelope Details: Walls, Roof, Windows, Insulation Levels

The building was construct according to Building regulation 1977.

- Walls: The external walls were double concrete sandwich type walls with an insulation thickness of 75 mm.
- Roof: Leca block with 150 mm of mineral wool.
- Basement floor: Concrete floor with 100 mm of Rockwool.
- Window: Double-glazed with wooden frame.
- The next table shows the u-values of the envelope building components before and after the renovation (Table B.3).

Table B.3 Envelope details

Envelope details	Before renovation	After renovation	Units
Hot water insulation of connectors for hot water tank	0.44	0.2	W/mK
Insulation of Domestic water and circulation pipe and pump boiler room in the basement	1.01	0.14	W/mK
Insulation of uninsulated heat distribution pipes and pumps, flanges and valves	1.23	0.16	–
Replacing the 2-glazing in windows energy windows			
U-value	2.8/1.4	1.1	W/m^{2}K
G	0.76/0.63	0.5	–
Replacement of 2 glazing in exterior doors with energy windows			
U-value	2.6	1.6	W/m^{2}K
G	0.76	0.5	–
Replacement of 2 glazing in patio doors onto the patio to energy windows			
U-value	2.7	1.8	W/m^{2}K
G	0.76	0.5	–
Modified U-value for skylight (skylight in sale). Existing windows: 2-layer energy windows			
U-value	2.7	1.4	W/m^2
G	0.76	0.65	–
Insulation in the roof of the classroom of 200–250 mm			
Area	7304	6581	m^2
U-value	0.15	0.19	W/m^{2}K
Insulation overhand entered separately			
Area	–	723	m^2
U-value	0.15	0.13	W/m^{2}K
Insulation of exterior wall sandwich with 175 mm and subtraction of the wall between windows			
Area	1437.8	1250	m^2
U-value	0.44	0.15	W/m^2
Adding wall between windows and re-insulation of 100 mm			
Area	0	187.8	m^2
U-value	0	0.20	W/m^2
Insulation of exterior wall light with 50 mm			
U-value	0.19	0.15	W/m^2

B.1.10.2. Heating, Ventilation, Cooling, and Lighting Systems

The existing building was heated by central district heating with the addition of an electrical heater. There is no a cooling system. The ventilation system is without heat recovery and the lighting was mainly done by normal bulbs. Table B.4 lists details the pre- renovation system.

Table B.4 Technological system details

Systems	Before renovation	After renovation	Units
Ventilation: replacing the fan motors in assembly hall			
Supply air temperature	0	18	°C
SEL	3.1	1.58	kJ/m^3
FO	0.6	0.55	–
Ventilation: changes in land for mechanical ventilation without heat recovery	5445	3762	m^2
Ventilation: FO modified for natural ventilation	1	0.65	–
Lighting in classrooms			
Area lighting is not replaced	4393	2196.5	m^2
Area lighting switch	-	2196.5	m^2
Effect	14.9	10	W/m^2
Lighting Level	200	300	lux
New lighting in the basement with motion detector			
Effect	9.6	7.5	W/m^2
Management	U	A	–
Lighting hallway			
Area		1844	m^2
Effect		7	W/m^2
Lighting level		50	Lux
Daylight factor		2	%
Fo		0.9	–
Mont. of new circulation pumps heating instead of Grundfos type UPE 25-45 (5 pcs)			
Effect	100	34	W
Reduction factor	0.6	0.4	–
Supply solar energy-220 kWp			
Area		1168	m^2
Peak power		0.15	kW/m^2
Efficiency		0.891	–
Orientation		194	Grader
Slope		20	Grader
Horizon cutting		5	Grader
Shadow left/right		0/0	Grader

B.1.11. Description of the Problem: Reason for Renovation

Due to the high energy consumption, technical and architectural quality of the building. The municipality was forced to renovate the facade and to change the technical systems. To give a complete new impression of the school from the outside and to reach the energy target of Danish Building regulation 2008.

B.1.12. Renovation SOW (Non-energy and Energy Related Reasons)

Although the school was in need for a general renovation, its primarily needs were improvements to the building envelope and ventilation system.

B.1.13. Energy Saving/Process Improvement Concept and Technologies Used

B.1.13.1. Building Envelope Improvement

New facade with improved insulation and new windows (Details are described in DER Measure Bundles).

B.1.13.2. New HVAC System or Retrofits to Existing (Details Are Described in Sect. B.2.10.2)

- New CO_2-controlled ventilation systems with heat recovery instead of old exhaust system.
- New low-energy motors for ventilation system in the big assembly hall.

B.1.13.3. New Lighting System (Details Are Described in Sect. B.2.13.3)

- New electrical lighting system incl. control.
- Optimization of electrical lighting in the basement.

B.1.13.4. New Generation/distribution System (Details Are Described in Sect. B.2.13.4)

- Technical insulation of connections to hot water tank and the tank itself.
- Insulation of hot water pipes and circulation pipes incl. pump.
- Insulation of pipes of heating distribution system incl. pumps, connections and valves.
- New low-energy pumps on the heating distribution system.
- New BEMS.

B.1.13.5. Renewable Energy. (Details Are Described in Sect. B.2.13.5)

- PV panels on the roof of 220kwp.

B.1.13.6. Daylighting Strategies. (Details Are Described in Sect. B.2.13.6)

- Control lighting system in order for the lighting system be controlled depended on the daylighting.

B.1.14. Energy Consumption

B.1.14.1. Pre-renovation Energy Use

The energy consumption was previously established based on the energy bills from the utility companies.

Before energy renovation, the average annual heating consumption was: 120.6 kWh/m^2 and the total electricity consumption was 27.7 kWh/m^2. The energy frame consumption (ventilation, pumps and lighting part of the electricity consumption is multiplied by a factor 2.5) was: 158.7 kWh/m^2 (Table B.5).

B.1.14.2. Predicted Energy Savings (Site, Source, GHG), Total and per m^2/year

The previous calculation was report by Jens-Peter Madsen aps Rådgivende ingeniører.

This memo states the total energy needs and scale value due to changes in XML file from the reported energy label of 28 October 2010 due to planned energy efficiency measures explained in detail above in Sect. 1.3, "DER Measure Bundles" and Sect. 1.5, "Impact of DER Measure Bundles." (Table B.6).

Table B.5 Pre-renovated energy use

Energy measured before renovation, kWh/m^2 per year	Heating	Total electricity	Energy frame
	120.6	27.7	158.7

Table B.6 Predicted energy saving

Predicted energy frame, kWh/m^2 per year	Before renovation	After renovation
	145	63

B.1.14.3. Measured Energy Savings (Thermal, Electrical), Total and per m²/year

Also as part of the Class1 project the public building was equipped with new monitoring equipment allowing for continuous online monitoring of the heating and electricity consumption.

A comparison of the energy consumptions before and after the energy renovation was carried out based on the monitoring over the winter 2013/2014. The energy renovation was carried out in the spring of 2013 and the monitoring equipment was also installed in the spring—in March 2013. Since experience shows that it always take some time before new installations and monitoring equipment is fine-tuned and working properly, it was decided that this analysis would be based on the data from August 2013 to March 2014—almost a full winter season—during the 8 months in Denmark that contain 80% of the average heating degree days in the year. Therefore to estimate the heating energy consumption for a full year the degree-day corrected monitored heating energy data for these 8 months was multiplied by a factor 1.25. The electricity consumption for the full year is established in a similar way by multiplying by a factor 1.5 (12/8).

The energy frame calculation includes heating energy consumption for space heating and hot water and the electricity consumption for ventilation and lighting. For this calculation, the electricity consumption is multiplied by a primary energy factor of 2.5 (Table B.7).

Besides the building energy renovation 220 kWp was incorporated.

The calculated result of the energy renovation, including PV of 220 kWp, was an energy frame of 95.2 kWh/m²—equal to the CONCERTO specification. The monitored data for the first year showed and energy frame of 48.4 kWh/m².

B.1.14.4. Annual Energy Use Reduction

See Table B.8.

Table B.7 Measured energy savings

Energy frame consumption BR08	2007–2012			After renovation		PV contribution	Sum	BR-08
	kWh	m²	kWh/m²	kWh	kWh/m²	kWh/m²	kWh/m²	kWh/m²
	1,402,586	8,837	158.7	977,352	110.6	62.2	48.4	95.2

Table B.8 Annual energy use reduction

Energy consumption [kWh/m²]	Before renovation	After renovation	PV contribution	Energy savings
Net heating consumption	120.6	80		40.6
Total electricity	27.7	21.5	62.2	68.3
Energy frame	158.7	48.4		110.3

B.1.15. Energy Cost Reduction

Total electricity is shown as a positive value indicating that the PV panels production generate a benefit cost by selling the exceed electricity back to the grid (Table B.9).

B.1.16. Non-energy Related Benefits Realized by the Project

- Architectural attraction.
- Improvement of acoustic and noise.

B.1.17. Renovation Costs: Total and per m^2

The main—most costly—energy renovation measure was the complete facade renovation with new external insulation and new windows. The total cost of this was 12,500,000. Table B.10 shows the Implemented energy saving measures at Stengaards School.

Table B.9 Energy cost reduction

Energy cost [euros]	Before renovation	After renovation
Heating consumption (district heating)	128,755	85,405
Total electricity	72,285	+ 105,905

Table B.10 Implemented energy saving measures at Stengaards School

Energy saving measure	Saving per year, Euros	Investment, Euros	Pay-back time, years
Technical insulation of ventilation ducts-25 mm insulation	818	12,967	15.9
New ventilation system with heat-recovery	7247	325,700	44.9
Exchange of ventilation motors on ventilation system for assembly hall	339	6667	19.6
New lighting system with control	2986	146,635	49.1
Optimization of the lighting in the basement	32	3520	110.0
New circulation pumps on the heating system	171	1933	11.3
New facade with improved insulation and new windows		1,023,400	
PV panels		451,500	
Non-energy related			
New facade		404,863	
Engineering cost		60,267	

- Energy related: 1,972,322 Euros or 217.9 Euros/m^2 of gross area.
- Non-energy related: 465,130 Euros or 51.4 Euros/m^2 of facade area.
- Total investment: 2,437,452 euros or 269.33 Euros/m^2 of gross area.

It is important to mention that the renovation of the facade is divided into two different measurements, energy-saving-related and non-energy-related, the latter of which is defined as a material that does not cause an impact on the reduction on the energy consumption, but is relevant on a functional or aesthetic basis alone.

B.1.18. Business Models and Funding Sources

B.1.18.1. Description of the Funding Sources Chosen

Danish municipalities have access to credit at low interest rates for renewables and energy efficiency programs to make energy improvements in an existing building or energy research. The model is based on actions that pay themselves back the loan over the course of their useful lifetimes. The lifetime value is specified in the Building Regulations.

B.1.18.2. Energy Management and Controlling in the Business Model

The funding of the renovation came primarily from the Municipality's own funds (and loans obtained at favorable rates). The renovation was also supported by EU (The EU CONCERTO Class1 project).

B.1.19. Cost Effectiveness of Energy Part of the Project (NPV, SIR, Etc.)

See Table B.11.

Table B.11 Cost effectiveness

Economical saving	kWh	Euro/year
Net heating saving	358,849	43,351
Electricity saving	603,418	178,191
Total saving		221,541
Total energy investment, Euros	2,437,452	
Simple payback time, years	11	
Heating cost, Euros/kWh	0.120805369	
Electricity cost, Euros/kWh	0.295302013	

B.1.20. User Evaluation

B.1.20.1. Description of User Evaluation Programs

Questionnaires were distributed to students in two classrooms, Room 40 and 58, to assess the following conditions in summer and winter, respectively:

- Temperature condition.
- Air quality.
- Daylight.
- Artificial light.
- Acoustics.
- Noise.

Note that the students in the classrooms before the renovation are different from the ones in the classrooms after the renovation (Table B.12).

B.1.21. Experiences/Lessons Learned

B.1.21.1. Energy Use

The overall aim of the energy renovation process was to ensure that the renovated building would reach an energy consumption level well below the target defined in the Danish Building Regulations (BR), equivalent to the level of "new built" construction according to the Danish BR08. In fact, the monitored energy consumption results are 23% below the predicted energy calculations. After the renovation, the total energy consumption was reduced by 70% to a level 50% below the target BR08.

The heating energy consumption was reduced by 34%. The electrical energy consumption decreased as well, but the electrical consumption reduction was relative smaller than reduction in heating consumption. This is because a large fraction of electricity consumption is directly related to the use within the buildings: Computers, photocopiers, refrigerators, and other electrical equipment. This contrasts with the fact that the implementation of PV panels contributed largely to the reduction in total energy consumption by satisfying the total electricity energy demand.

Table B.12 Number of completed questionnaires from the two classrooms

Room	Number of completed questionnaires
58	21
40	14

B.1.21.2. Impact on Indoor Air Quality

The indoor air quality was measured before the ventilation system was renovated (February 2013) and after the energy renovation (January 2014). During both periods, indoor air temperature, CO_2 concentration, and ventilation level were measured every 5 minutes using a passive gas technique ("PFT") technique. The resulting measurements were averaged during the measurement period over the normal working days, including the preceding weekend. Again, surveys were taken among the students.

The results of the measurements (Table B.13) show that classroom ventilation improved significantly. CO_2 concentration was reduced from over 3000 ppm in Room 58 and 1500 ppm in Room 40 to below 1000 ppm in both classrooms. The room air temperature in both classrooms decreased and became more stable; before the renovation, temperatures varied between 22 and 25 °C; after the renovation, temperature ranged between the stable temperatures of 20–21 °C. The results of the ventilation measurements using the tracer gas show that the ventilation rate before the renovation was 0.3 h^{-1} in Room 58 and over 0.8 h^{-1} in Room 40; after renovation, the average ventilation rate obtained was of 0.4 h^{-1}. This indicates the new ventilation system control conditions as intended (Figs. B.15 and B.16).

The student' assessment showed little or no differences between before the renovation and after the renovation on room air temperature and indoor air quality (and related acceptability) in summer conditions. In winter conditions, the students' assessment depended on the room. Regarding to the air temperature, students noted no particular shift in Room 40, but in Room 58 more students found the room air temperature after the renovation to be 'on the cold side' (60 → 80%) and more students found it acceptable (35 → 60%). Students found the indoor climate in Room 58 the students' had shifted to 'fresh'; however, at the same time, an increasing number of students find the air quality less acceptable. Students noted no significant changes in Room 40 from before to after the renovation.

Indoor climate measurements and occupants' interviews demonstrated that the occupants' behavior (as indicated by high measured indoor temperatures and wrong settings of the systems) had indeed formed the basis for these results. To achieve high energy savings, it is crucial to give occupants the appropriate information and training on the use of the systems.

Occupant satisfaction investigations and indoor climate measurements showed little correlation between low-energy buildings and poor indoor climate. On average, occupants are subjectively satisfied; the surveys indicated that they did not

Table B.13 Results of the ventilation measurement using passive tracer gas technique

Room	Measurement period	ARH [h^{-1}] %()
58	1st Feb–8th Feb 2013	0.29 ± 15
	17th Jan–24th Jan 2013	0.40 ± 11
40	1st Feb–8th Feb 2013	0.86 ± 15
	17th Jan–24th Jan 2013	0.45 ± 11

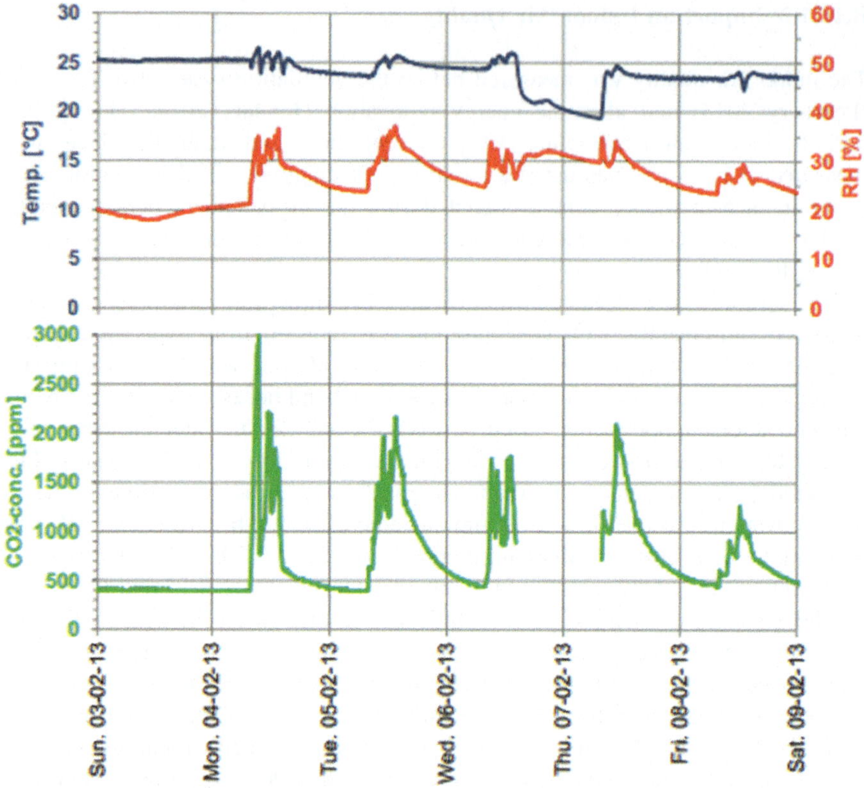

Fig. B.15 Room 58 before renovation

perceive poor indoor climate conditions. Clearly, the achieved energy savings did not reduce the IEQ standards set up in the design specification phase. It is hoped that satisfied occupants will be more collaborative and interactive occupants, who will be more disposed to develop energy-conserving habits that will lead to higher energy savings.

B.1.21.3. Impact on Light—Daylight and Artificial Light

The renovation of the ventilation systems is not expected to cause changes in the lighting conditions in the classrooms. Nevertheless, in both classrooms a shift is noted in the students' assessments towards more 'bright' and less 'dark' conditions. This applies to both daylight and artificial light.

The students indicated a high degree of satisfaction with the issues of 'operation of the artificial light', 'glare from the artificial light', and 'ability to see out of the windows'. Figures B.17 and B.18 show the Room 40 survey results before and after the renovation.

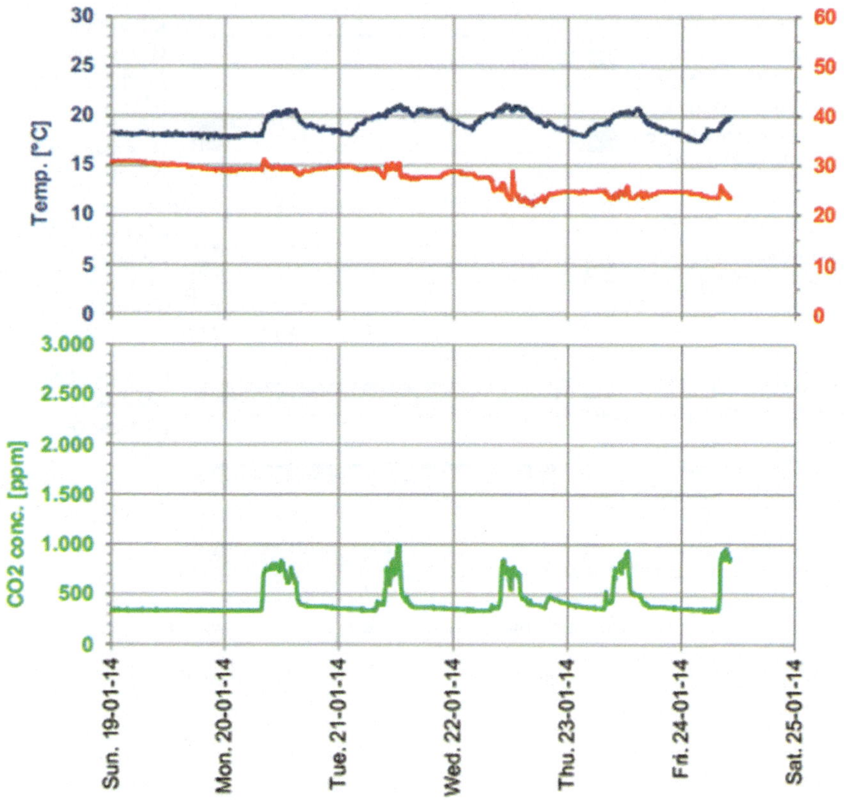

Fig. B.16 Room 58 after renovation

B.1.21.4. Impact on Acoustics and Noise

The renovation of the ventilation systems is not expected to cause changes in the acoustic conditions in the classrooms. However, after the renovation an increasing number of students in Room 58 find the acoustics neither 'hard' nor 'resonant'. Of the students not ticking 'neither' in the questionnaire (i.e., ticking either 'hard' or 'resonant'), 1/3 find the acoustics to be 'hard' and 2/3 find the acoustics 'resonant'. This is valid for both the "before situation" and the "after situation."

In Room 40, the students' assessments of the acoustics in the room show a slight shift towards 'resonant'. The students' assessments of acoustics and noise in terms of 'ability to hear instructions from the teacher', 'noise from ventilation, projector etc.', 'noise from outside (windows open)', 'noise from outside (windows closed) and 'noise from adjacent rooms' generally indicate improved conditions and generally a high degree of satisfaction, as it can be seen Figs. B.19 and B.20. One exception, though, is in Room 58 where 'noise from adjacent rooms' apparently has worsened significantly after the renovation. Before the renovation about 10% of the students found the conditions 'not ok' whereas after the renovation 60% finds the conditions 'not ok'.

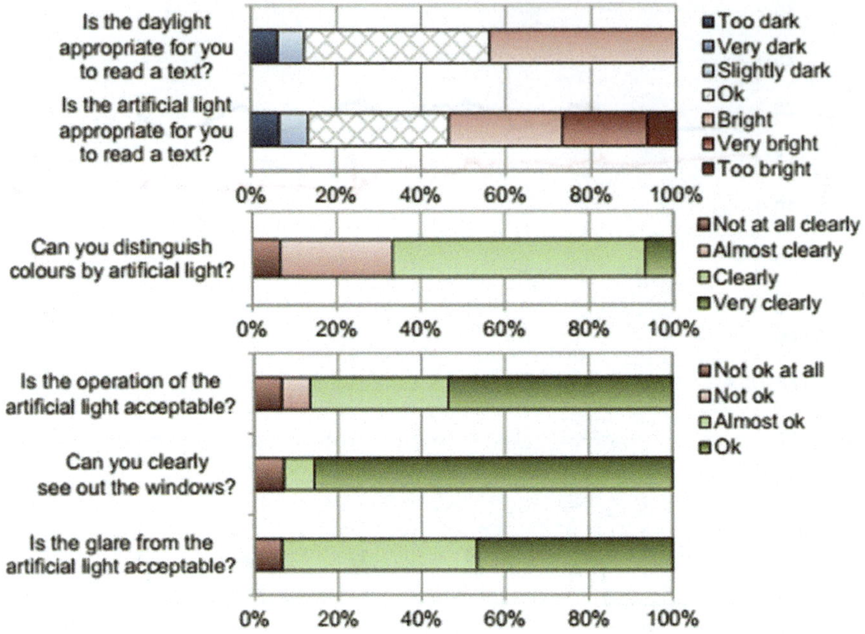

Fig. B.17 Room 40 before renovation, students' assessment of the daylight and artificial light conditions

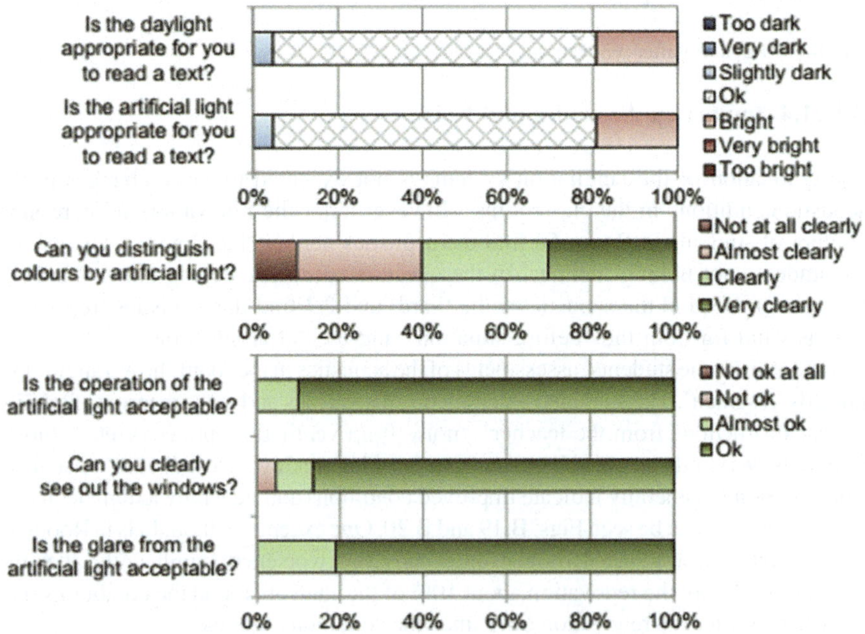

Fig. B.18 Room 40 after renovation, students' assessment of the daylight and artificial light conditions

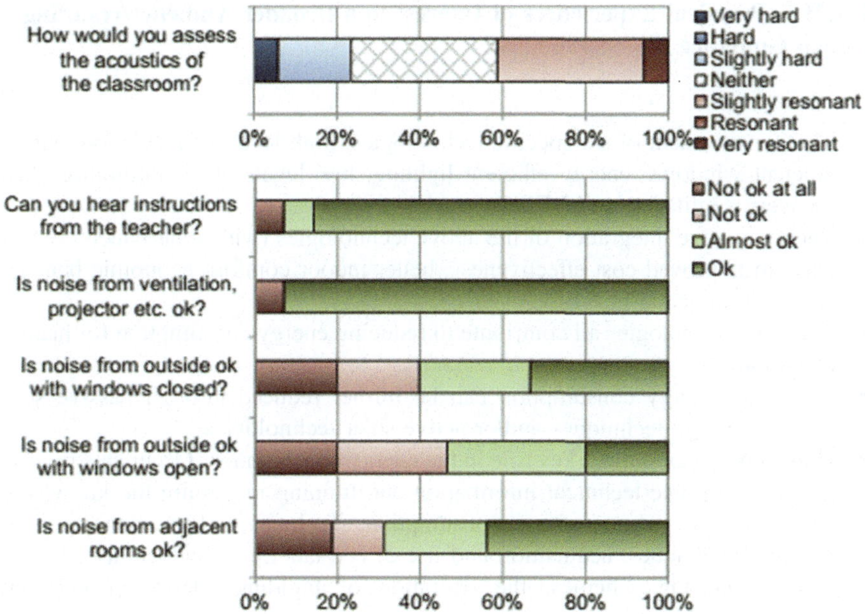

Fig. B.19 Room 40 before renovation, students' assessment of acoustics and noise in the classroom

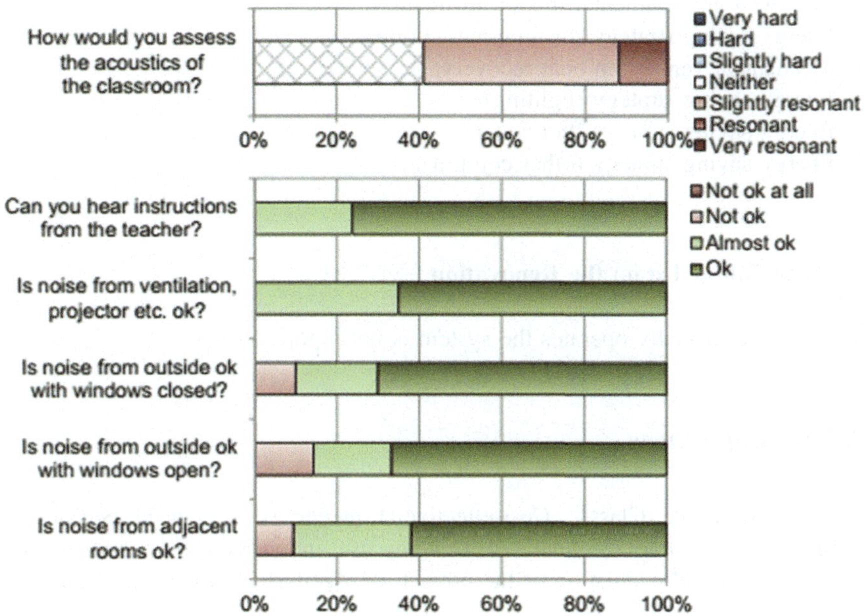

Fig. B.20 Room 40 after renovation, students' assessment of acoustics and noise in the classroom

B.1.21.5. Practical Experiences of Interest to a Broader Audience/resulting Design Guidance

- The improvement of six specific technologies: high levels of insulation, energy efficient windows, energy efficient lighting, low levels of air infiltration, heat recovery ventilation and BEMS.
- The innovative integration of the above technologies (with solar panels), which leads to improved cost effectiveness, better indoor comfort, economic benefits, and environmental benefits.
- The new technologies all contribute to reducing energy consumption for heating and cooling.
- Electricity energy consumption can be further reduced through passive solar building design techniques and/or active solar technologies.
- Human behavior plays a key role in the energy consumption. Occupants must be given appropriate technical information and training to acquire the knowledge and habits to reduce energy consumption. Technical information refers to building techniques, acquisition and use of eco-label products, choice of innovative financing schemes, the operation of building energy management systems.

The "human factor" mainly involve the following issues:

1. Environmental indoor air quality.
2. Improved environmental indoor air quality.
3. Energy saving strategy (heating and cooling).
4. Mechanical ventilation heat recovery.
5. Energy saving strategy (lighting).
6. Energy saving strategy (hot water).
7. Energy saving strategy (other equipment).

B.1.21.6. Follow Up on the Renovation

How the user actually operates the system is not reported yet.

B.1.22. References

Concerto Initiative Class1. Cost-effective Low-energy Advanced Sustainable So1utions Deliverable 23: Analysis of energy use, indoor climate, occupant satisfaction and overall evaluation of the project. 22 April 2014. www.class1.dk.

Concerto Initiative Class1. Cost-effective Low-energy Advanced Sustainable Solutions Deliverable 4: The completed low-energy building projects—new buildings and energy renovated public buildings. 22 April 2014. www.class1.dk.

Concerto Initiative Class1. Cost-effective Low-energy Advanced Sustainable Solutions Deliverable 7: The overall energy investments and total economic. 22 April 2014. www.class1.dk.

B.2. Vester Voldgade.DK

B.2.1. Name of the Project, Location (City, Country)

Vester Voldgade 123, Copenhagen, Denmark.

B.2.2. Pictures

Figures B.21, B.22, B.23, B.24, B.25, B.26, B.27 and B.28 show the building in its original and post-retrofit states, illustrating key features of the retrofit.

Fig. B.21 Facade before renovation

Fig. B.22 Facade before renovation

Fig. B.23 Facade after renovation

Fig. B.24 Facade after renovation

Fig. B.25 Windows fastened to the existing facade

Fig. B.26 Insulation and new windows installed

Fig. B.27 Solar heat panels on the roof

Fig. B.28 Solar shading, adjustable vertical/horizontal

B.2.3. Project Summary

B.2.3.1. Project Objectives

This Danish case study documents the renovation of a multi-story office building in Copenhagen. The building was constructed in 1938 and has five floors and a full basement. The total gross heated floor area is 5460 m^2 and the unheated basement is 1274 m^2. In general, the building needed upgrading to repair degradation of the facade and to improve the building's very poor energy performance and indoor climate, These were the primary reasons for carrying out an extensive energy renovation.

The original project consisted only of replacing the building's windows and corroded water supply system. However, calculations of alternatively performing a more in-depth energy renovation clearly indicated that such a general renovation was both energy efficient and profitable without even considering further savings related to the avoidance of future concrete repairs to the facade.

B.2.3.2. Project Energy Goals

The overall goal of the energy renovation was to reduce the energy consumption in the building to "Passivhaus" level (according to the German definition, i.e., maximum 15 kWh/m^2 for heating). At the same time, this would significantly improve the building's indoor climate.

The project also included energy savings for lighting, active solar shading, new ventilation system, and pre-cooling/heating of ventilation air via horizontal tubes and two slanted wells reaching a depth of 21 m below ground.

B.2.3.3. Short Project Description

- Building owner: The Danish Building and Property Agency.
- Architect: tnt Arkitekter A/S.
- Energy consultants: Ørtoft A/S—Rådgivende Ingeniørfirma Strunge—Rådgivende Ingeniørfirma.
- Contractor: G.V.L. Enterprise A/S.

B.2.3.4. Stage of Construction

1938.

B.2.3.5. Point of Contact Information

Jørgen Rose, Department for Energy and Environment, Danish Building Research Institute, Aalborg University, A C Meyers Vænge 15, 2450 Copenhagen SV, Denmark. E- Mail: jro@sbi.aau.dk, Tel: +45 9940 2226.

B.2.3.6. Date of the Report

20.5.2015.

B.2.3.7. Acknowledgement

The authors wish to thank The Danish Building and Property Agency and the Danish Energy Agency for sharing all the necessary information concerning the energy renovation needed to complete the case study.

B.2.4. Site

- Location: Vester Voldgade 123, 1552 Copenhagen V, Denmark.
- Latitude: 55.67°.
- Longitude: 12.58°.
- Elevation: 3 m.
- Climate zone: Denmark, KOEBENHAVN/KASTRUP 061800 (ASHRAE).
- Cooling Degree Days (based on 62.6 F = 17.00 °C): 0.
- Heating Degree Days (based on 62.6 F = 17.00 °C): 2906 (Table B.14).

Table B.14 Design temperature

Cooling design temperature—0.4% occurrence*	Heating design temperature—99.6% occurrence**
Dry bulb temp °C (F)	–
Heating design temperature—99.6% occurrence**	
Dry bulb temp °C (F)	− 12 °C

B.2.5. Building Description/Typology

B.2.5.1. Typology/Age

Multi-story office building/76 years.

B.2.5.2. Type (Office, Barracks, Etc.)

Office.

B.2.5.3. Typology/Age

1930–1940.

B.2.5.4. General Information

- Year of construction: 1938.
- Year of previous major retrofit—if known: no previous major retrofit.
- Year of renovation (as described here): 2011–2013.
- Total floor area (m^2): 6734 m^2.
- Area of unconditioned space included above (m^2): 1274 m^2.
- Other information as appropriate.

B.2.6. Architectural and Other Relevant Drawings

See Figs. B.29, B.30, B.31, B.32 and B.33.

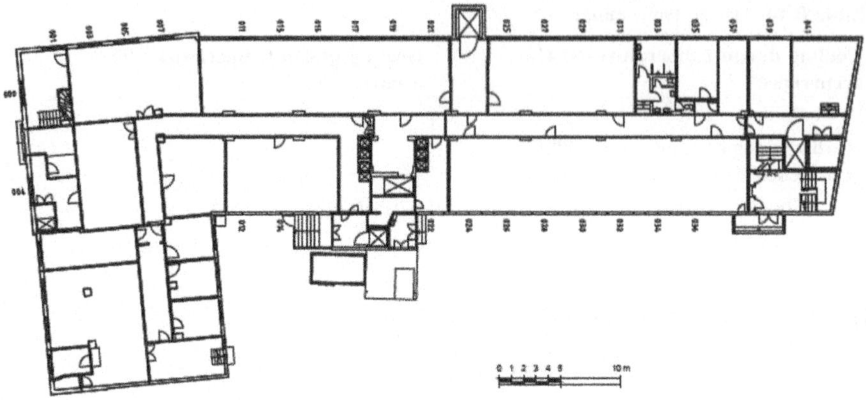

Fig. B.29 Floor plan, basement

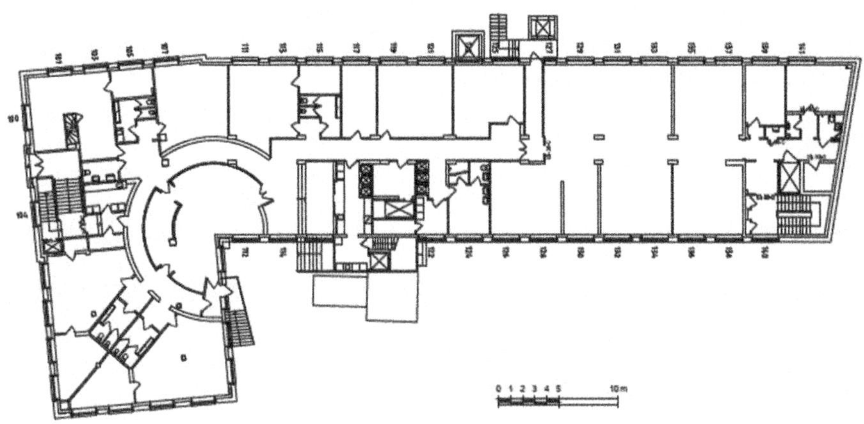

Fig. B.30 Floor plan, ground floor

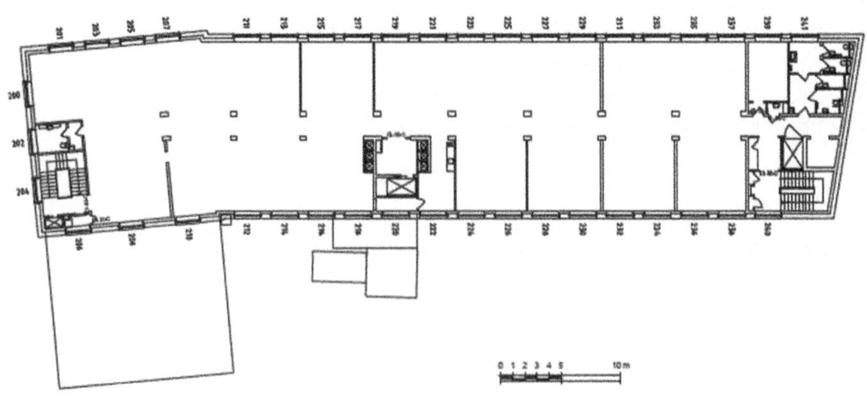

Fig. B.31 Floor plan, 1–4th floors

Fig. B.32 Cross section A-A

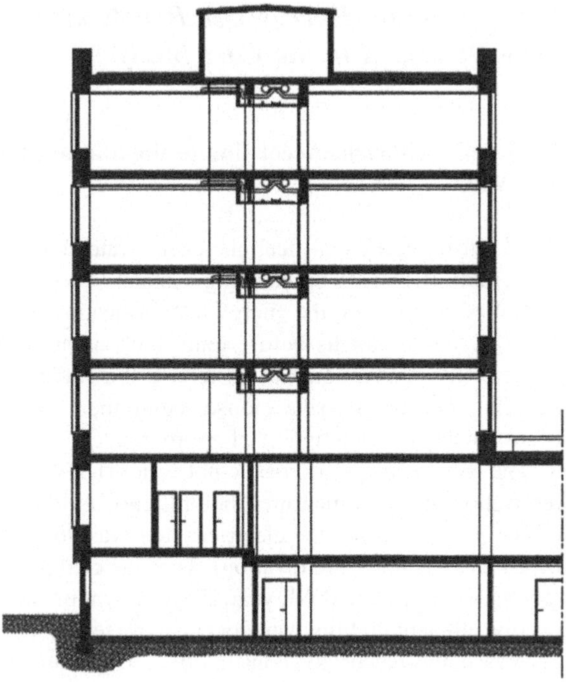

Fig. B.33 Cross section B-B

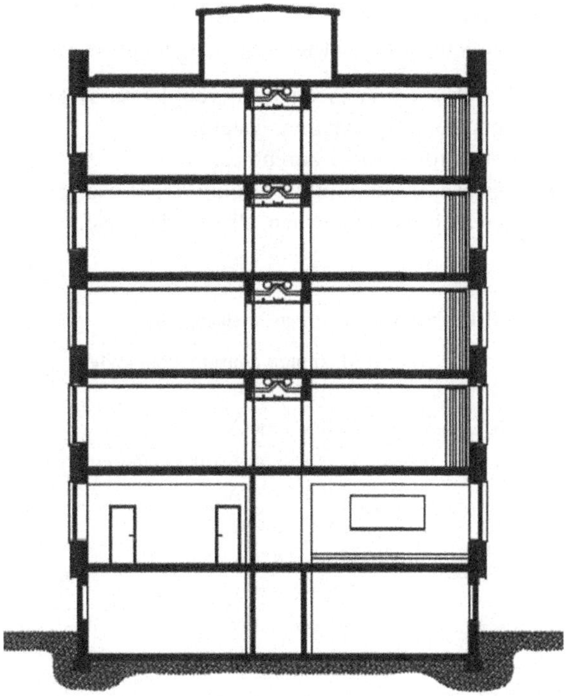

B.2.7. National Energy Use Benchmarks and Goals for Building Type Described in the Case Study

B.2.7.1. Benchmark, According to the Standard National Average, Min, and Max

The Danish energy certification scheme rates buildings by their energy consumption. Table B.15 lists the labeling system.

Before renovation, the energy use for heating was 519.3 MWh/year (based on 3 years of measurements before renovation adjusted for heating degree days). The target of the renovation was to reduce the energy consumption for heating to 15 kWh/m^2, i.e., the Passive House requirement. This would correspond to a total heating energy consumption of approximately 82.6 MWh/year for the building.

The energy use for domestic hot water (DHW) did not change as a result of the renovation and was measured before renovation to 25 MWh/year.

Before renovation, the electricity use was 96.5 MWh/year (based on 2 years of measurements before renovation). The expected (calculated) electricity use after the renovation is 141.5 MWh/year. The renovation included the installation of more energy efficient lighting and occupancy sensors; this change alone would have decreased the electricity consumption. However, after the renovation, the office building underwent a further change from individual offices to an open-plan so that the building now accommodates approximately 50% more occupants than before the reorganization.

Consequently, electricity use has increased due to:

- Two new larger mechanical ventilation systems.
- Pump for solar heating system.
- Pumps for vertical boreholes.
- Increase in the number of employees using two PC-displays.
- Increased use of kitchen, dishwashers etc.

Table B.15 National energy use benchmarks

Level	Maximum consumption kWh/m^2 per year (A is heated area)
A2020	25.0
A2015	41.0 + 1000/A
A2010	71.3 + 1650/A (minimum requirements for new buildings)
B	95.0 + 2200/A
C	135 + 3200/A
D	175 + 4200/A
E	215 + 5200/A
F	265 + 6500/A
G	> 265 + 6500/A

- Two new soda machines in the cafeteria.

To compare this usage with the energy labeling scheme, the electricity use must be multiplied by a factor of 2.5 to convert to primary energy.

B.2.7.2. National Energy Target for This Type of Building

Figure B.34 shows the ratings for this building along with actual measured data before renovation, expected consumption after renovation, and measured consumption after renovation. Before renovation, the building would have been labeled "D." The calculations suggested that, after renovation, the building would meet today's requirements for new buildings, i.e., an "A2010" label, but measurements show that was rated just above a "B."

B.2.8. Awards or Recognition

No award or recognition.

B.2.9. Site Energy Cost Information (Electricity, \$/kWh; Gas, \$/m^3, Etc.)

All values listed below include VAT.

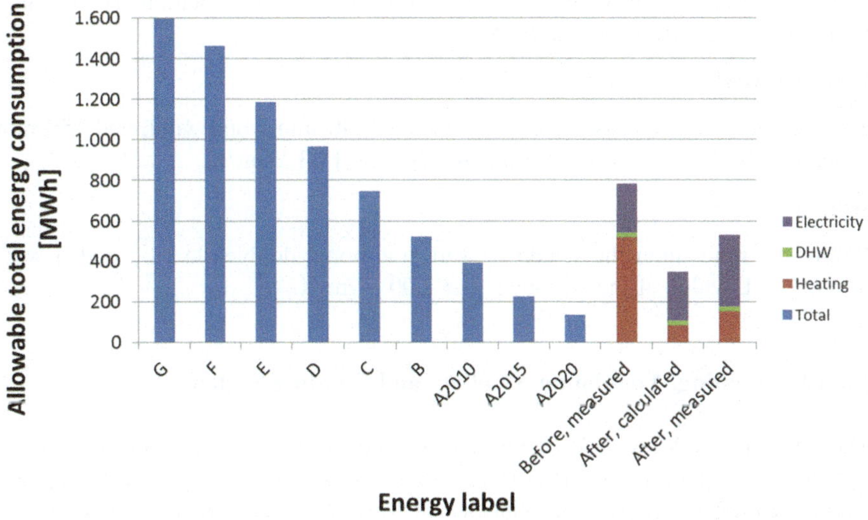

Fig. B.34 Energy labeling scheme and measured/calculated energy consumption for the office building

B.2.9.1. Electricity (Average for Copenhagen)

Basic charge per month: €1.33.
Energy price per kWh: €0.12.

B.2.9.2. District Heating (HOFOR—Hovedstadsområdets Forsyningsselskab)

Basic charge per year and per installed load kW: €28.2.
Energy price per m^3 (steam): €65.4, corresponding to €0.093 per kWh.

B.2.10. Pre-renovation Building Details

B.2.10.1. Envelope Details: Walls, Roof, Windows, Insulation Levels

Walls

The outer wall consisted of 10 mm screed, 40 mm wood wool, and 150 mm concrete. The U-value of the external wall was 1.32 W/m^2K.

Roof

The flat roof already had 200 mm insulation (added at a previous occasion). The U-value of the roof was 0.18 W/m^2K.

Basement ceiling

The ceiling in the unheated basement was concrete without insulation. The U-value of the basement ceiling was 2.00 W/m^2K.

Basement walls

The basement wall consisted of 10 mm screed, 40 mm wood wool, and 250 mm concrete. The U-value of the basement wall was 1.23 W/m^2K.

Windows

Before the renovation the windows were traditional double-glazed wooden windows. The U-value of the windows was 2.00 W/m^2K.

B.2.10.2. Heating, Ventilation, Cooling, and Lighting Systems

The existing building was connected to the old district heating system in Copenhagen, which is based on steam (rather than water). The steam-part of the district heating system in Copenhagen will be phased out over the next 15 years.

The heating installation in the building was quite complex; it was found that the installation actually covers the heating and hot water supply for two other buildings, i.e., a kindergarten and a gardeners lodge (both located on the property).

The ventilation was performed as balanced mechanical ventilation with heat recovery. The system was quite inefficient (62% recovery rate).

There was no cooling system in the building.

The lighting was primarily done with incandescent light bulbs and there were no specific energy-saving controls.

B.2.11. Description of the Problem: Reason for Renovation

The original project specified only the replacement of the building's windows and a corroded water supply system. However, calculations of alternatively performing a more in-depth energy renovation clearly indicated that this more general renovation was both energy efficient and profitable without even considering further savings related to the avoidance of future concrete repairs to the facade.

B.2.12. Renovation SOW (Non-energy and Energy Related Reasons)

Performing a deep energy renovation of the building has made the building more attractive and more easily rentable. The new tenant has asked for a change in the interior of the building from smaller offices for 3–4 persons to a new open office plan, i.e., making room for more people in the building.

B.2.13. Energy Saving/Process Improvement Concept and Technologies Used

B.2.13.1. Building Envelope Improvement

Walls

The walls were insulated on the outside with 300–430 mm insulation, which resulted in U-values ranging from 0.08 to 0.11 W/m^{2}K. Some parts of the exterior walls would not allow for large external insulation thicknesses and therefore these parts of the wall were insulated with 50 mm vacuum insulation ($\lambda = 0.007$ W/mK) instead. These parts would end up with a U-value of 0.38 W/m^{2}K.

Roof

No insulation was added to the roof, i.e., the U-value remained at 0.18 W/m^{2}K.

Basement ceiling

Even though the basement was unheated, the temperature in that area was quite high due to heat loss from the heating installation. Therefore no insulation was added to basement ceiling.

Basement walls

The basement walls were insulated with 260–290 mm insulation on the outside. The resulting U-values were between 0.12 and 0.13 W/m^{2}K.

Windows

The new windows were placed in the new facade of the building, i.e., flush with the insulation reducing the thermal bridges significantly. Automatic exterior shading with drop arm awnings with manual override was installed on the facades to the southeast and southwest. The external shading device help to reduce the solar gains and to thereby avoid building overheating.

The glass has a U-value of either 0.71 or 0.80 W/m^{2}K (the latter for special safety glass to avoid person injury) and a g-value of 0.49. The total U-value of the windows including the frame ranges from 0.92 to 1.32 W/m^{2}K and a mean value of 1.09 W/m^{2}K.

B.2.13.2. New HVAC System or Retrofits to Existing

General description

The basic heating system was not changed during the renovation of the building, i.e., the building still relies on the (steam) district heating provided by HOFOR.

Domestic hot and cold water

A rainwater harvesting system was installed to conserve water for toilets. Rainwater is gathered from the building's 1200 m^2 roof. Toilets were also changed for newer versions requiring less water.

In the domestic hot water system, the pipes throughout the building were replaced with new pipes that meet the most stringent technical insulation requirements.

Ventilation

To receive Passivhaus certification, the building needed to achieve an air tightness corresponding to n50 $\leq$ 0.6 l/s per m^2; therefore the air tightness of the building was a focal point. The overall insulation of the facade and the careful mounting of the new windows will help achieve this end.

The ventilation system in the building was changed during the renovation. The new system consists of two individual systems. The first part of the system is a "VE01" constant air volume (CAV) ventilation system that is used for offices to maintain an air change rate of approximately 1.5 h^{-1} during office hours. The second part of the system is a "VE02" variable air volume (VAV) ventilation system used for the intensive meeting rooms using a variable air flow between 0 and 5 h^{-1} during office hours, and controlled as needed using on temperature and CO_2-sensors.

In addition to this, another system was installed to take advantage of "free" pre-heating of ventilation air for the "VE01"-system by using horizontal ground collectors and two wells with oblique deep wells to approximately 21 m depth (top of the limestone layer). Groundwater is approximately 3.5 m below ground level. The COP is estimated approximately 50.

B.2.13.3. New Lighting System

The new lighting system in the building is established as a completely new LED lighting system in corridors and offices. In the offices the lighting is controlled as needed using PIR-sensors (motion detectors) and daylighting sensors.

B.2.13.4. New Generation/Distribution System

See "General description."

B.2.13.5. Renewable Energy

Solar thermal system

A 35 m^2 solar collector was installed on the roof and connected to a storage tank for solar thermal storage. The storage tank was placed beneath the basement floor to reduce the heat loss through the poorly insulated basement floor and thermal bridges around the rim foundations and concrete columns.

B.2.13.6. Daylighting Strategies

New windows and daylight-controlled LED lighting in offices contribute to better daylight conditions.

B.2.14. Energy Consumption

B.2.14.1. Pre-renovation Energy Use (Total and per m²/year)

Energy use values are based on degree day corrected measurements of energy for heating and domestic hot water (i.e., three heating seasons 2009–2010, 2010–2011 and 2011–2012) and electricity use (i.e., 2 years 2008 and 2009). Site energy use is:

Energy for heating	519,300 kWh/year	94.4 kWh/m²/year
Energy for DHW	25,000 kWh/year	4.5 kWh/m²/year
Energy for electricity	96,500 kWh/year	17.5 kWh/m²/year

B.2.14.2. Predicted Energy Savings (Site, Source, GHG), Total and per m²/year

Energy savings values are based on calculations performed with Be10 (the national Danish compliancy checker developed according to EN ISO 13790), which is based on monthly averages. The calculations were also performed using PHPP, which generated results similar to those of Be10. Site energy use is:

Energy for heating	376,700 kWh/year	68.5 kWh/m²/year
Energy for DHW	2400 kWh/year	0.4 kWh/m²/year
Energy for electricity	n-45,000 kWh/year	− 8.2 kWh/m²/year

CO_2

The mean CO_2-emission for electricity production in Copenhagen is 377 g/kWh (values for 2012–2013) and the mean CO_2-emission for district heating in Copenhagen is 104 g/kWh (value for 2013).

The total CO_2 savings are	21.0 to/year	3.82 kg/m²/year

B.2.14.3. Measured Energy Savings (Thermal, Electrical), Total and per m²/year

One year of measurements have been carried out since the renovation remodeling of the building was finished in 2013. Site energy use and measured energy savings are:

Energy for heating	368,100 kWh/year	66.9 kWh/m²/year
Energy for DHW	0 kWh/year	0.0 kWh/m²/year
Energy for electricity	− 45,900 kWh/year	− 8.3 kWh/m²/year

B.2.15. Energy Cost Reduction

B.2.15.1. Split in All Energy Forms—Electricity, Oil, District Heating

- Electricity. Increase in price for electricity is: 45,900 kWh/year €0.12 = €5508 per year.
- District heating. Reduction in price for heating and domestic hot water is: 368,100 kWh/year €0.093 = €34,233 per year.
- Total energy cost reduction = €28,725 per year.

B.2.16. Non-energy Related Benefits Realized by the Project, E.g., Improved Productivity, Increased Rent/Lease, Increased Useful Space, Reduced Maintenance, Etc.

Main non-energy related aspects are:

- Increased useful space.
- Increased indoor air quality.
- Reduced energy costs for tenants.
- Environmental friendly construction.
- Reduced maintenance.

B.2.17. Renovation Costs: Total and per m^2

B.2.17.1. Total

Total renovation costs are:	€565 per m^2 gross floor area	€3.09 mio

B.2.17.2. Non-energy Related

Non-energy related costs are:	€320 per m^2 gross floor area	€1.75 mio

B.2.17.3. Energy Related

Energy related costs are:	€245 per m^2 gross floor area	€1.34 mio

B.2.17.4. Cost for Each Measure

Windows (double-glazed windows)	€641,333
Extra cost (triple glazed windows)	€40,000
Facade insulation (including basement)	€297,760
Solar shading	€67,267
Heating system (pipes etc.)	€3333
Ventilation (reduced hours of operation)	€13,333
Solar heat storage + 30 m^2 solar panels on roof	€34,800
Mechanical ventilation + vertical wells for preheating	€195,547
Lighting system (controls and lighting)	€46,133

B.2.18. Business Models and Funding Sources

B.2.18.1. Decision Making Process Criteria for Funding and Business Models

The building is owned and managed by the Danish Building and Property Agency (Bygningsstyrelsen). The Danish Building and Property Agency is the state's property enterprise and developer. The agency is responsible for creating modern, functional, cost-effective frameworks for some of the country's most important government institutions, e.g., universities, the police, the courts, and government departments.

The total property portfolio consists of approximately 4 million m^2, which, when combined with current and planned construction projects, totals approximately DKK 14 billion; the agency is one of Denmark's largest public property enterprises and developers.

The agency plans all renovation for its building portfolio.

B.2.18.2. Description of the Funding Sources Chosen

The agency is funded by the government; its projects, as the one described here, are therefore partially financed by a general maintenance budget (€1.03 million in this particular case). The agency is additionally allowed to pass the part of the expenses that can be contributed to profitable energy savings in the buildings directly to the tenants (€190,333 in this particular case). The tenants, in turn, will receive all energy savings achieved in the renovation. The rest of the total cost is an investment made by the agency itself (€1.87 mio in this particular case).

B.2.19. Cost Effectiveness of Energy Part of the Project (NPV, SIR, Etc.)

Simple pay-back time

Energy related investment costs	€1,340,000
Energy savings per year—electricity	€− 5508
Energy savings per year—district heating	€34,233
Energy savings per year—total	€28,725
Simple pay-back time	46.65 years

Net present value

The results stated below are based on following assumptions:

Discount rate:	2.5%
Inflation of energy:	3.0%
Expected economic lifetime:	30 years
Which result in a NPV factor of	32.4
And a net present value of:	€− 409,310

B.2.20. User Evaluation

B.2.20.1. Description of User Training Program Within the Refurbishment

There has been no user training program within the refurbishment.

B.2.20.2. Integration of Users Demands in the Planning Process

User demands were not integrated.

B.2.21. Experiences/Lessons Learned

B.2.21.1. Energy Use

For this particular building, it was found that two separate buildings were heated by the same district heating installation. This was not discovered until after the renovation process when the first year measurements were scrutinized and the expected reductions in heating demand were not met. Therefore, an important lesson from the project is, if possible, to start off by gathering all information concerning the building, and to develop a building model that can be calibrated to actual measurements.

Another important lesson from this project is that it is necessary to require building systems, especially quite complex systems like those in this case, to be commissioned and adjusted for optimal operation before the project can be handed over to the users/owners. In this project, difficulties with the initial settings and calibration of systems meant that the energy consumption was significantly higher than expected for an extended period until systems were properly adjusted.

B.2.21.2. Practical Experiences of Interest to a Broader Audience

See below.

B.2.21.3. Space Utilization Changes

The office building has changed from offices for three to four persons to an open office solution.

Appendix C: Case Study—Estonia

C.1.1. Name of the Project, Location

Kindergarten in Valga, Estonia.

C.1.2. Pictures

Figures C.1, C.2, C.3, C.4, C.5, C.6, C.7, C.8, C.9, C.10, C.11 show the building in its original and post-retrofit states, illustrating key features of the retrofit.

The solar thermal system consists of:

- 28 pcs. SONNENKRAFT SK500N on the rooftop
- 2 × SONNENKRAFT PS2000 storage tank (in total 4 m^3)
- Stratified charging
- Insulation thickness 120 mm
- Calculated solar fraction hot water:
  - 54.4%; solar fraction space heat demand: 14.5%
  - Collector tilt angle: 50° from horizontal (Figs. C.12 and C13).

C.1.3. Project Summary

C.1.3.1. Project Objectives

- To modernize the existing kindergarten built in 1960s.
- The existing kindergarten was built as a two story building with an additional one story kitchen and dining area.

R. Lohse et al. (eds.), *Deep Energy Retrofit—Case Studies*,
https://doi.org/10.1007/978-3-031-17517-6

Fig. C.1 Before refurbishment

Fig. C.2 After refurbishment

Fig. C.3 Opening in Sept 2009

Fig. C.4 After refurbishment

Fig. C.5 Initial shell

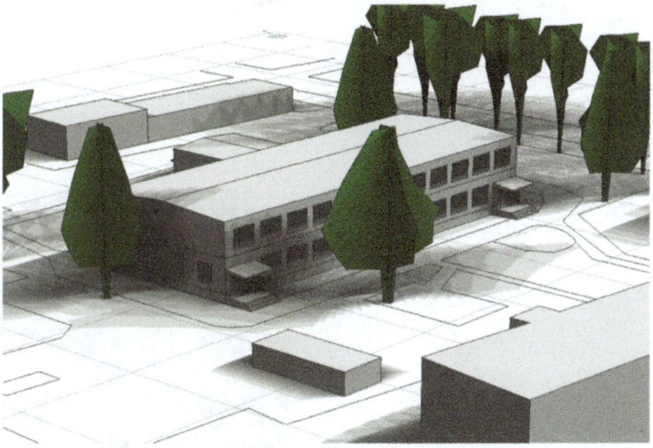

Fig. C.6 After refurbishment—overhangs for summer heat protection

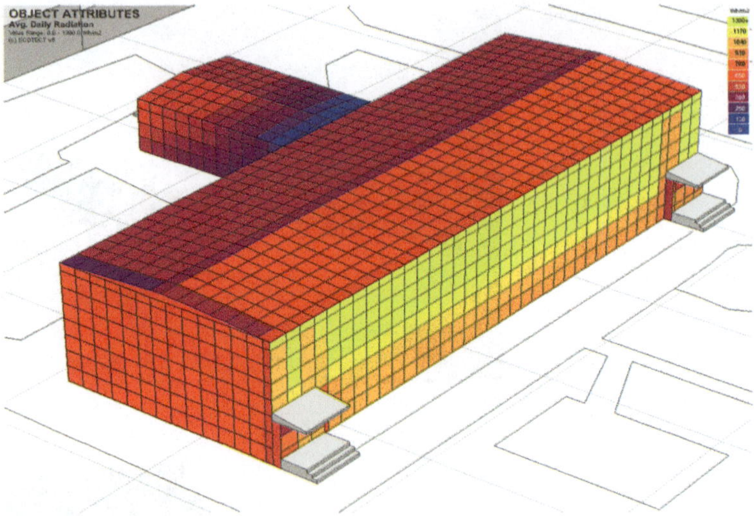

Fig. C.7 Average daily radiation in winter

Fig. C.8 Construction

Fig. C.9 Installed solar thermal system

Fig. C.10 Installed solar thermal system

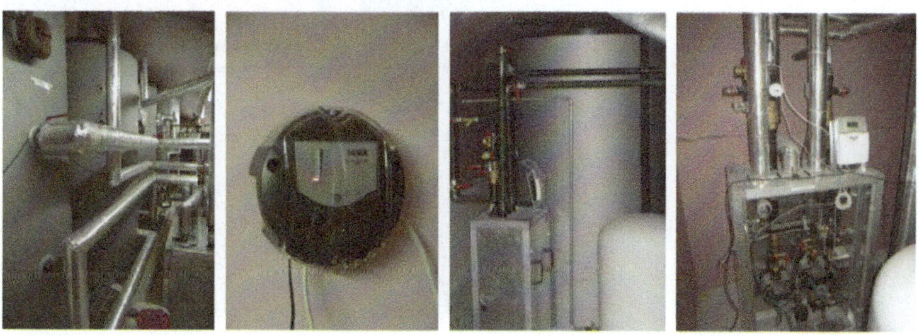

Fig. C.11 Installed solar thermal system

Fig. C.12 Ventilation with heat recovery

Fig. C.13 Building-time blower-door test, the value n50 = 0.47 1/h was reached

- The goal of the project was to increase the floor area of the house from the existing 901.4 to 1040 m^2 by adding a second floor to the one story building, to use passive house method for the refurbishment, and to install a solar heating system to use solar energy for the production of warm water and for room heating.

C.1.3.2. Project Energy Goals

The initial goal was to reach the level 20 kWh/m^2 a net space heat demand. After ventilation equipment replacement, the goal shifted to 40 kWh/m^2 a.

C.1.3.3. Short Project Description

The building shell and the HVAC systems were replaced completely. The building has an additional "new building" part connected to the old one (Fig. C.14).

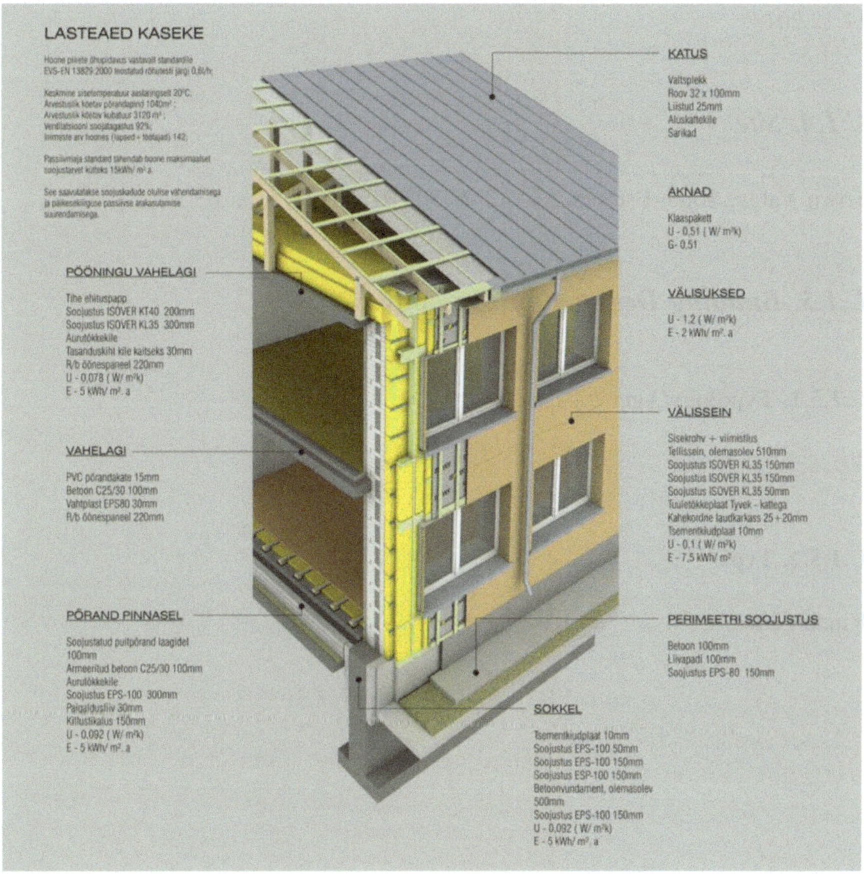

Fig. C.14 Basic refurbishment concept—well insulated, thermal-bridge free and airtight envelope

C.1.3.4. Stage of Construction

Finished Sept 2009.

C.1.3.5. Point of Contact Information

Valga town government.
Puiestee tn 8.
68203 VALGA.
Tel: +372 766 9900; +372 766 9910.
E-mail: valgalv@valgalv.ee.

C.1.3.6. Date of the Report

11.12.2014.

C.1.4. Site

Town Valga, South-Estonia, 57° 47′ N 26° 02′ E, elevation 65 m.

C.1.5. Building Description/Typology

C.1.5.1. Typology/Age

40.

C.1.5.2. Type

Kindergarten.

C.1.5.3. General Information

- Year of construction: 1960s.
- Year of renovation (as described here): 2008–2009.
- Total floor area (m^2): 1156.

C.1.6. Architectural and Other Relevant Drawings

See Figs. C.15, C.16, C.17, C.18, C.19, C.20 and C.21.

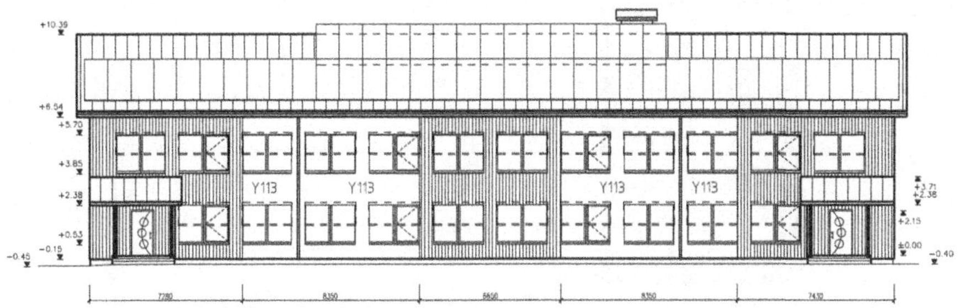

Fig. C.15 View from Southeast

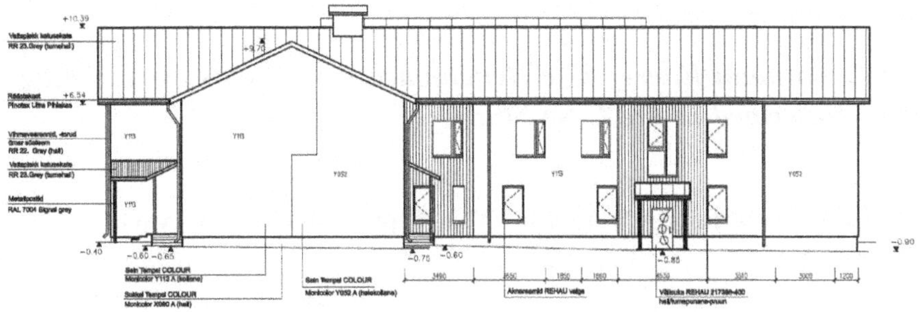

Fig. C.16 View from Northwest

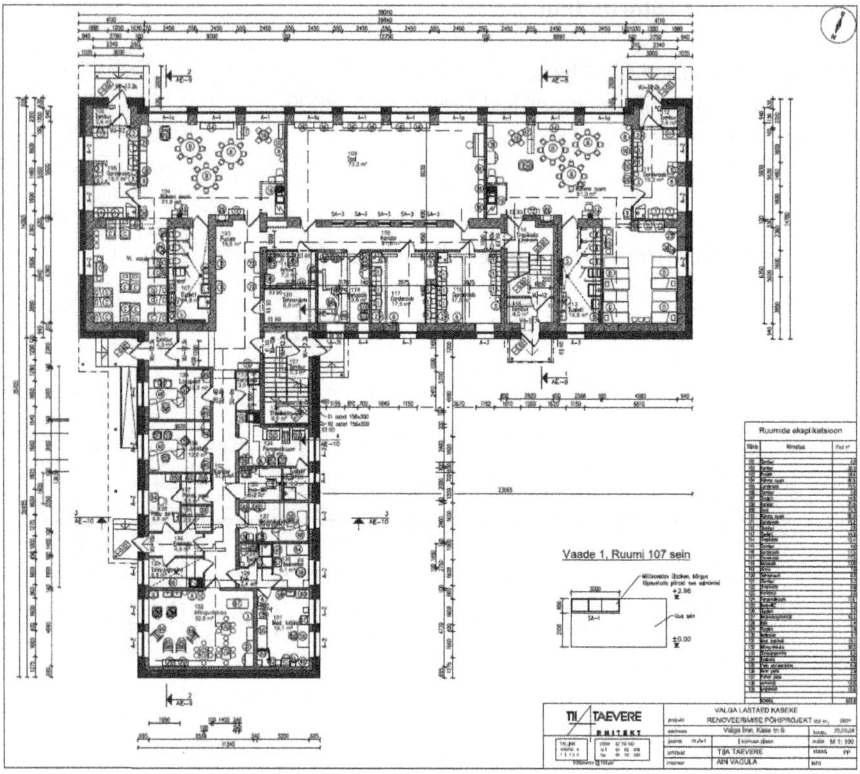

Fig. C.17 Floor plan

C.1.7. National Energy Use Benchmarks and Goals for Building Type Described in the Case Study

C.1.7.1 National Energy Target for This Type of Building (If Any)

None.

C.1.8. Awards or Recognition

None.

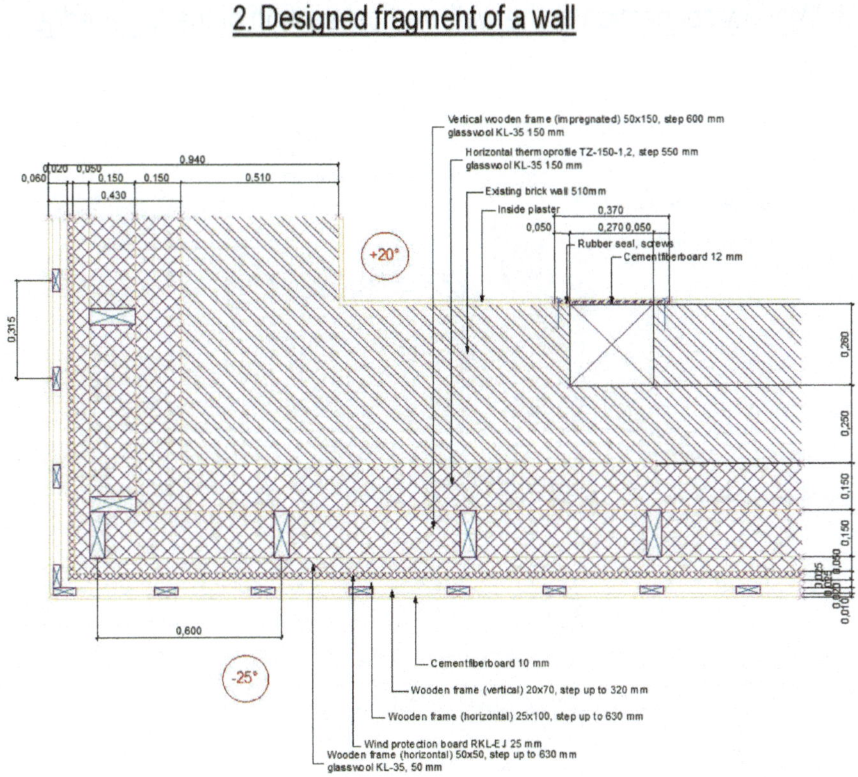

Fig. C.18 Designed wall

C.1.9. Site Energy Cost Information (Electricity, \$/kWh; Gas, \$/m^3, Etc.)

C.1.9.1. Electricity

0.11 Eur/kWh.

C.1.9.2. District Heating

0.06 Eur/kWh.

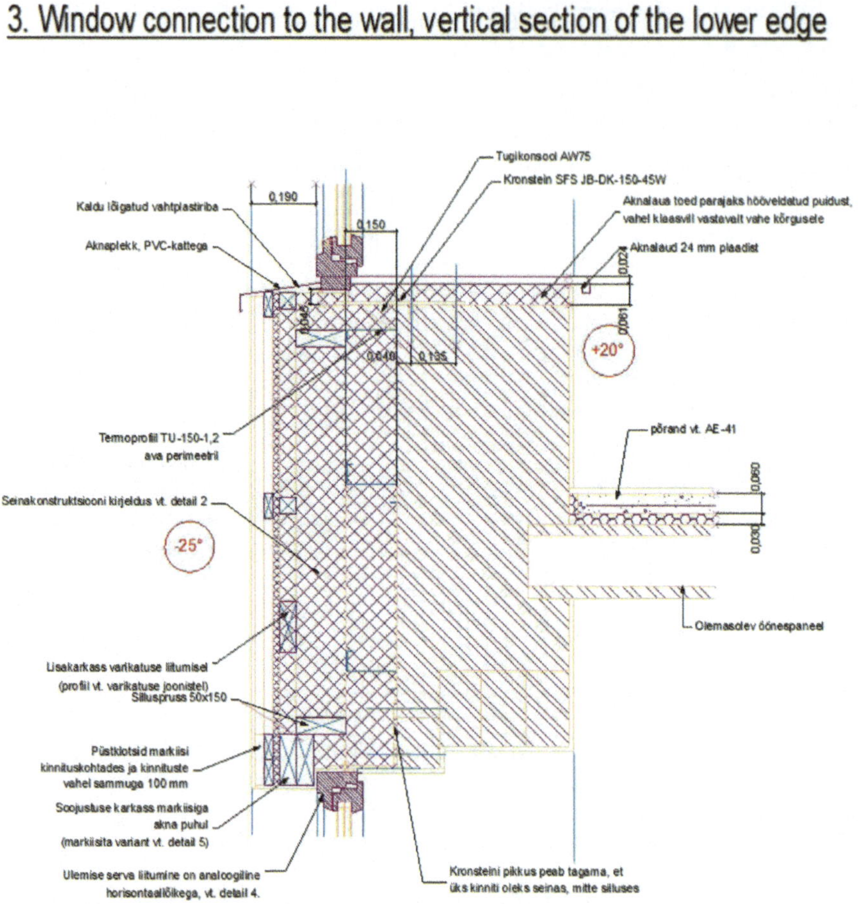

Fig. C.19 Designed window connection

C.1.10. Pre-renovation Building Details

C.1.10.1. Envelope Details: Walls, Roof, Windows, Insulation Levels

U wall = 1.0 W/(m^2 K).

C.1.10.2. Heating, Ventilation, Cooling, and Lighting Systems

Typical 1960s building, ventilation through windows, central heating, no cooling.

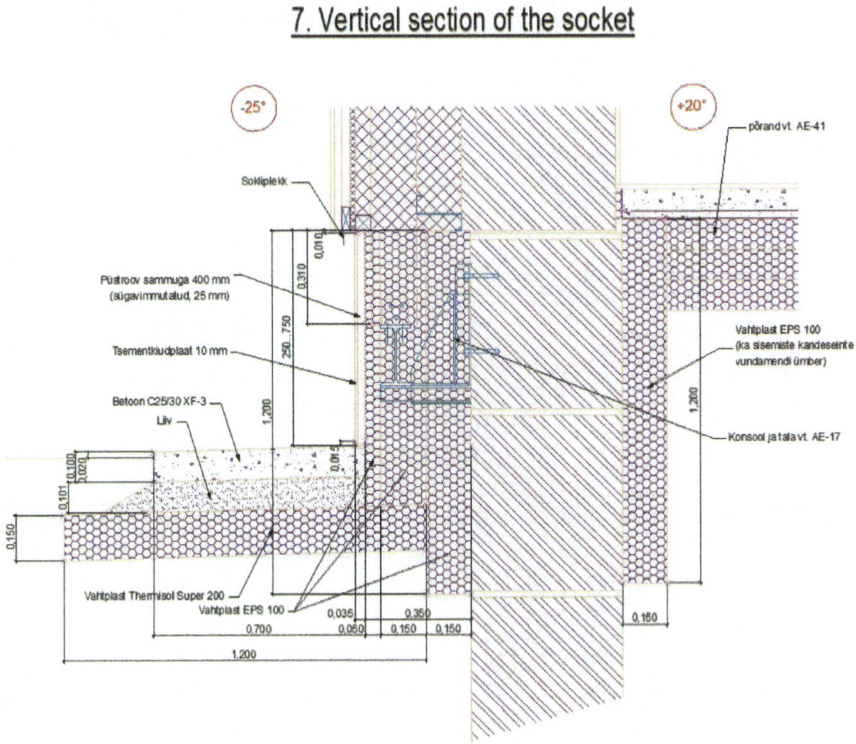

Fig. C.20 Designed socket

C.1.11. Description of the Problem: Reason for Renovation (Non-energy and Energy Related Reasons)

Not available.

C.1.12. Renovation SOW (Non-energy and Energy Related Reasons)

Not available.

C.1.13. Energy Saving/Process Improvement Concept and Technologies Used

None.

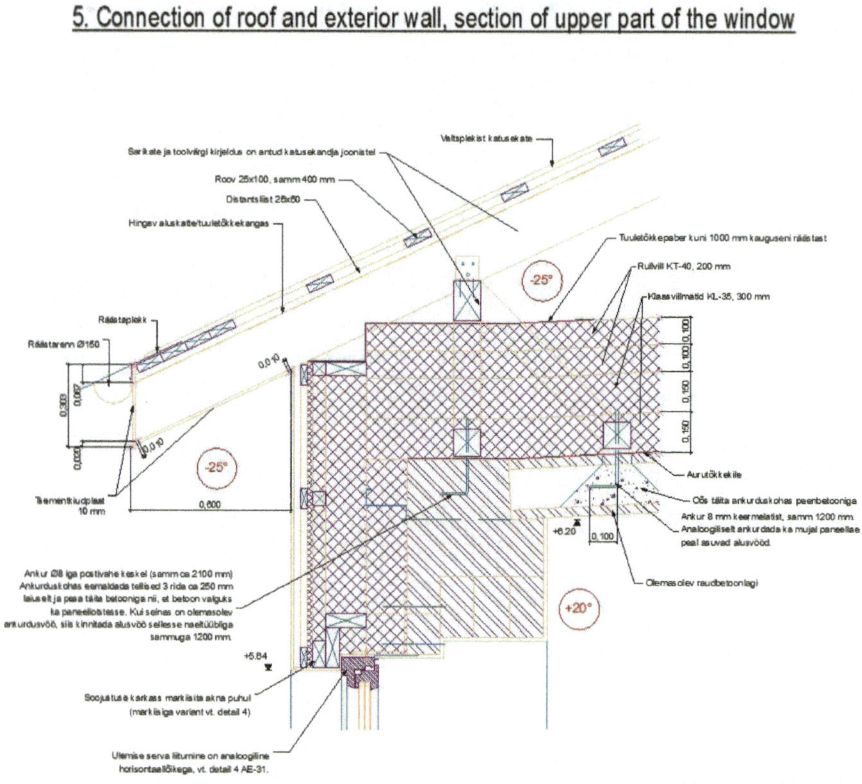

Fig. C.21 Designed wall-roof connection

C.1.14. Energy Consumption

C.1.14.1. Pre-renovation Energy Use (Total and per m²/year)

In the years 2003–2005, the average heat demand of the building was 253 MWh/year, which corresponds with an average specific heat demand of 280 kWh/m^2 per year.

C.1.14.2. Predicted Energy Savings (Site, Source, GHG), Total and per m²/year

PHPP calculated space heat demand: 36 kWh/(m^2 year).

C.1.15. Energy Cost Reduction

None.

C.1.16. Non-energy Related Benefits Realized by the Project, E.g., Improved Productivity, Increased Rent/Lease, Increased Useful Space, Reduced Maintenance, Etc.

None.

C.1.17. Renovation Costs: Total and per m^2

None.

C.1.18. Business Models and Funding Sources

C.1.18.1. Description of the Funding Sources Chosen

EU supported project.

C.1.18.2. Description of the Business Model Chosen (Option)

Initiated and financed by local government.

C.1.18.3. Operation Phase

Local government covers the costs.

C.1.19. Cost Effectiveness of Energy Part of the Project (NPV, SIR, Etc.)

None.

C.1.20. User Evaluation

None.

C.1.21. Experiences/Lessons Learned

C.1.21.1. Energy Use

Due to instalment of less effective heat recovery ventilation equipment and high ventilation rates with no possibility to reduce the ventilation rates comfortably the space heat demand was higher than expected. Envelope is high quality and meets all expectations.

C.1.21.2. Impact on Indoor Air Quality

Too dry air resulted from high ventilation rates.

There were no instructions for kindergarten personal. As air volumes and accordingly room temperatures rose, teachers opened windows on cold winter days, clearly illustrating a need for better instructions and for the occupants to have the ability to regulate air-volume/temp.

Appendix D: Case Study—Germany

D.1. PHI_GAG_Hoheloog_Ludwigshafen, GE

D.1.1. Name of the Project, Location (City, Country)

Passivhaus in Bestand, Ludwigshafen/Mundenheim (Germany).

D.1.2. Pictures

See Figs. D.1 and D.2.

D.1.3. Project Summary

D.1.3.1. Project Objectives

Achievement of the criteria for existing buildings, retrofitted to a passivehouse, with the least amount of monetary input.

D.1.3.2. Project Energy Goals

- Electricity: 18 kwh/m^2*a.
- Heating: < 15 kwh/m^2*a.
- Others, total: < 120 kwh/m^2*a.

R. Lohse et al. (eds.), *Deep Energy Retrofit—Case Studies*,
https://doi.org/10.1007/978-3-031-17517-6

Fig. D.1 Renovated building 2005

Fig. D.2 Original construction 1965

D.1.3.3. Short Project Description

The apartment building needed a renovation to create an appropriate living standard with very low energy consumption and a small amount of monetary input. So the achievement of the criteria for existing buildings, retrofitted to a passive house was the way to go.

D.1.3.4. Stage of Construction

Finished 2005.

D.1.3.5. Point of Contact Information

GAG Ludwigshafen am Rhein.
Wittelsbachstraße 32.
67061 Ludwigshafen am Rhein.

D.1.3.6. Date of the Report

December 2008.

D.1.3.7. Acknowledgement

Sponsored by ExWoSt (Rhineland-Palatinate).
KfW bank (German governmental housing bank).

D.1.4. Site

- Hoheloogstraße 1/3 67065 Ludwigshafen-Mundenheim.
- Coordinates: 49.4552785, 8.426207.
- Climate zone: ASHRAE 90.1-2007: 5/warm temperate/standard climate data for Mannheim (DIN 4108-6 Region 12).
- Cooling Degree Days (based on 65 F), Heating Degree Days (based on 65 F): 75 kKh/a = 3396 Kd/year (Table D.1).

Table D.1 Design temperature

*Cooling design temperature—0.4% occurrence**	
Dry bulb temp C (F)	Mean coincident wet bulb temp C (F)
28 °C	
*Heating design temperature—99.6% occurrence***	
Dry bulb temp C (F)	
15 °C	

D.1.5. Building Description/Typology

D.1.5.1. Age

49, solid construction.

D.1.5.2. Typology

Three-level Apartment House.

D.1.5.3. Typology/Age

1950–1970.

D.1.5.4. General Information

- Year of construction: 1965.
- Year of previous major retrofit—if known: unknown.
- Year of renovation (as described here): 2005–2006.
- Total floor area (m^2): 960 m^2 gross floor area/750 m^2 net floor area.
- Area of unconditioned space included above (m^2): none.

D.1.6. Architectural and Other Relevant Drawings

See Figs. D.3 and D.4.

D.1.7. National Energy Use Benchmarks and Goals for Building Type Described in the Case Study

D.1.7.1. Benchmark

To meet the Passive house standard for existing buildings.

Fig. D.3 Ground plan after renovation

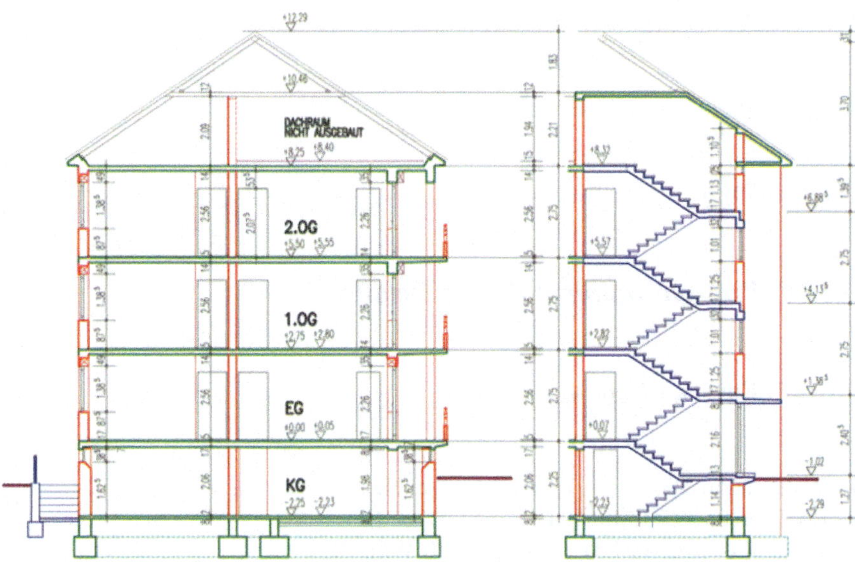

Fig. D.4 Section after renovation

D.1.7.2. National Energy Target for This Type of Building (If Any)

German KfW 40-criteria.

D.1.8. Awards or Recognition

None.

D.1.9. Site Energy Cost Information (Electricity, \$/kWh; Gas, \$/m^3, Etc.)

D.1.9.1. Electricity

- 30.4 kWh/m^2*a (First year).
- 28.5 kWh/m^2*a (second year).

D.1.9.2. Heating

18.2 kWh/m^2*a.

D.1.10. Pre-renovation Building Details

D.1.10.1. Envelope Details

- Walls (area: 500.17m^2; U = 1.33 W/m^2*K).
- Roof (area: 392.38 m^2 − U = 0.52 W/m^2*K).
- Windows (area: 194.21 m^2 − U = 2.8 W/m^2*K).

D.1.10.2. Heating, Ventilation, Cooling, and Lighting Systems

- Heating: via local heat and radiators (one per room).
- Ventilation: inexistent.

- Cooling: inexistent.
- Lighting systems: no daylight-usage.

D.1.11. Description of the Problem: Reason for Renovation

Heating costs for the inhabitants were extremely high. Since a refurbishment was necessary anyway, GAG Ludwigshafen decided to exceed the necessary *EnEV 2009* and *KfW 40* parameters for renovation and new buildings to build a passive house with an extremely low heating demand.

D.1.12. Renovation SOW (Non-energy and Energy Related Reasons)

- Redesign of ground plans of apartments.
- Better living comfort.

D.1.13. Energy Saving/Process Improvement Concept and Technologies Used

D.1.13.1. Building Envelope Improvement

- Walls U-Value: 0.1 W/(m^2*K), thickness of insulation: 30 mm.
- Windows: U-Value: 0.8 W/(m^2*K).
- Ceiling: U-Value: 0.1 W/(m^2*K), thickness of insulation: 30 mm.
- Floor: U-Value: 0.17 W/(m^2*K), thickness of insulation: 20 mm.

D.1.13.2. New HVAC System or Retrofits to Existing

- New Ventilation System.
- Heating: hydraulic damper registers (0.9 kW each) via ventilation.
- Ventilation: design flow rate: 1200m^3/h.
- Air-conditioning: inexistent.

D.1.13.3. New Lighting System

None.

D.1.13.4. Renewable Energy

Photovoltaic modules with an surface area of $105m^2$ and a installed power of 6.5 kW_{pea}, which results in a yearly gain of 16.5 kWh/(m^2*a).

D.1.13.5. Daylighting Strategies

None.

D.1.14. Energy Consumption

D.1.14.1. Pre-renovation Energy Use (Total and per m^2/year)

250 kWh/(a*m^2) (according to PHPP).

D.1.14.2. Predicted Energy Savings (Site, Source, GHG), Total and per m^2/year

94%, heat energy demand: 16 kWh/m^2*a.

D.1.14.3. Annual Energy Use Reduction.

(250 – 16) kWh/m^2a = 239 kWh/m2a.

D.1.15. Energy Cost Reduction

D.1.15.1. Split in All Energy Forms—Electricity, Oil, District Heating

Significant, but not quantitatively available as tariffs are not known.

D.1.16. Non-energy Related Benefits Realized by the Project

- Better air quality by mechanical ventilation system.
- Better thermal indoor temperatures and indoor climate.
- The upper floor was extended and therefore more living space was created.

D.1.17. Renovation Costs: Total and per m²

D.1.17.1. Total

920,319€ 1227 €/m^2.

D.1.17.2. Non-energy Related

- Facade: 26326 €;35 €/m^2.
- Installation: 20012 €;3 €/m^2.
- Scaffolding: 7850 €;10 €/m^2.
- Total: 615689 €;821 €/m^2.

D.1.17.3. Energy Related

- Roof: 18,702 €; 25 €/m^2.
- Air handling: 57,966 €; 77 €/m^2.
- Heating: 22,617 €; 30 €/m^2.
- Facade: 38,470 €; 51 €/m^2.
- Total: 304,630 €; 406 €/m^2.

D.1.17.4. Cost for Each Measure

Table D.2 lists total costs, the left part of table for the passive house building (EnerPHit),and the right part of table for low-energy buildings.

Table D.2 Total cost for each measure for passive house and low-energy buildings

	EnerPHit costs	No of dwellings	treated floor area (TFA)	component area		low energy building	No of dwellings	treated floor area (TFA)	component area		difference	
Cost Hoheloogstrasse LU		12	750 m²			costs	12	750 m²				
	€	€/WE	€/m² Wfl	[m²]	€/m² Bt.fl.	€	€/WE	€/m² Wfl	[m²]	€/m² Bt.fl.	[€/m² Wfl]	
Thermal insulation roof	10 358 €	863 €	14 €	346	30 €	0 €	0 €	0 €	346	0 €	14 €	
Thermal bridges at roof edge	8 344 €	695 €	11 €	346	24 €	0 €	0 €	0 €	346	0 €	11 €	
	18 702 €	1 558 €	25 €	346	54 €	0 €	0 €	0 €	346	0 €	25 €	
Thermal insulation cellar ceiling	11 358 €	946 €	15 €	346	33 €	5 078 €	423 €	7 €	346	15 €	8 €	124%
Thermal bridges at cellar edge	11 571 €	964 €	15 €	346	33 €	0 €	0 €	0 €	346	0 €	15 €	
Ground digging at perimeter	9 068 €	756 €	12 €	346	26 €	0 €	0 €	0 €	346	0 €	12 €	
	31 996 €	2 666 €	43 €	346	92 €	5 078 €	423 €	7 €	346	15 €	36 €	
	0 €	0 €	0 €	346	0 €	0 €	0 €	0 €	638	0 €		
Thermal insulation outside wall	26 284 €	2 190 €	35 €	638	41 €	13 012 €	1 084 €	17 €	638	20 €	18 €	102%
Thermal insulation fixings	7 796 €	650 €	10 €	638	12 €	7 796 €	650 €	10 €	638	12 €	0 €	0%
Thermal insulation connections	20 024 €	1 669 €	27 €	638	31 €	16 479 €	1 373 €	22 €	638	26 €	5 €	22%
outside wall rendering	10 691 €	891 €	14 €	638	17 €	9 874 €	823 €	13 €	638	16 €	1 €	8%
Sum	64 796 €	5 400 €	86 €	638	102 €	47 162 €	3 930 €	63 €	638	74 €	24 €	
anyway costs of above sum	26 326 €	2 194 €	35 €	638	41 €	26 666 €	2 222 €	36 €	638	42 €	0 €	
Windows	84 001 €	7 000 €	112 €	180	467 €	59 424 €	4 952 €	79 €	180	330 €	33 €	41%
Entrance doors	8 061 €	672 €	11 €	12	672 €	8 061 €	672 €	11 €	12	672 €	0 €	0%
Sum	92 062 €	7 672 €	123 €	192	479 €	67 486 €	5 624 €	90 €	12	5 624 €	33 €	
Ventilation system	57 966 €	4 831 €	77 €			9 585 €	799 €	13 €			65 €	505%
Ventilation second ceiling	3 046 €	254 €	4 €			0 €	0 €	0 €			4 €	
Sum	61 013 €	5 084 €	81 €			9 585 €	799 €	13 €			69 €	
Radiators	22 617 €	1 885 €	30 €			36 995 €	3 083 €	49 €			-19 €	-39%
Air tightness	3 582 €	298 €	5 €	1 510	2 €	0 €	0 €	0 €			5 €	
scaffolding	7 850 €	654 €	10 €	638	12 €	6 789 €	566 €	9 €	638	11 €	1 €	16%
Electric installation	2 012 €	168 €	3 €			0 €	0 €	0 €			3 €	
Sum of overall costs of Energy saving Actions	304 630 €	25 386 €	406 €			173 095 €	14 425 €	231 €			175 €	76%
the rest: not relevant for energetic performance	615 689 €	51 307 €	821 €			616 976 €	51 415 €	823 €			-1.72 €	0%
Total Renovation costs (due to billing, gros costs incl. VAT)	920 319 €	76 693 €	1 227 €			790 071 €	65 839 €	1 053 €			174 €	16%

D.1.18. Business Models and Funding Sources

None.

D.1.19. Cost Effectiveness of Energy Part of the Project (NPV, SIR, Etc.)

None.

D.1.20. User Evaluation

D.1.20.1. Description of User Training Programmes Within the Refurbishment

No user training programs were held, but the housing company does extended supervision of the building site.

D.1.20.2. Integration of Users Demands in the Planning Process

Since the building was not occupied during renovation, all tenants changed; the demands of the users were not integrated in the planning process.

D.1.21. Experiences/Lessons Learned

D.1.21.1. Energy Use

The energy consumption decreased significantly, but could be reduced even more with user training programs.

D.1.21.2. Impact on Indoor Air Quality

Indoor air quality improved dramatically, after the renovation, the air maintained a more stable humidity with much less pollution.

D.1.21.3. Follow Up on the Renovation

The users need a proper training to use the advantages of the building to its fullest.

D.1.22. References

- http://www.passiv.de/de/05_service/03_fachliteratur/030103_altbau_wohnungsbau/05_wirtschaftlichkeit_altbau/05_wirtschaftlichkeit_altbau.htm
- http://www.passiv.de/de/05_service/03_fachliteratur/030103_altbau_wohnungsbau/02_sanierung_phib/02_sanierung_phib.htm

D.2. Factor10_Nürnberg, GE

D.2.1. Name of the Project, Location (City, Country)

Jean-Paul-Platz 4, 90461 Nürnberg, Bavaria, Germany.

D.2.2. Pictures

See Fig. D.5.

Fig. D.5. Before and after retrofit. *Source* [Schulze-Darup, Burkhard. December 2011. Energetische Modernisierung: Modelprojekte. Wbg Nürnberg GmbH Immobilienunternehmen.]

D.2.3. Project Summary

D.2.3.1. Project Objectives

- To perform AN energy retrofit of an apartment building (six apartments) using passive house components within the EU-program ERDF (European Regional Development Fund) EU-Ziel-2 that supports the economic and social adjustment in areas facing structural difficulties.
- To perform a Factor 10 retrofit, specifically, to reduce the energy use by 90% (energy use = 10 times less than before retrofit).

D.2.3.2. Project Energy Goals

- Heat demand: 26 kWh/m^{2}a (3-L-house).
- Costs for constructional measures: 560 €/m^2 living area (Cost group 300/400 according to DIN 276 incl. VAT.

D.2.3.3. Short Project Description: Implemented Measures

- New boiler room in the attic, partial demolition of chimneys.
- Renovation of the base molding.
- Recovering of the roof, renewal of all sheet metal plates.
- Creation of a roof overhang.
- Implementation of a thermal insulation composite system.

- Insulation of the basement ceiling including entrance.
- Insulation of the attic including screed, flap tile and stairwell.
- Complete replacement of all windows including basement windows.
- New entrance door, attic and basement door, sealing of the apartment doors.
- Establishment of covered balconies for all flats.
- Modernizing the heating system.
- Solar thermal energy system with stratified buffer storage.
- Central hot water supply with new connections to all bathrooms.
- Local ventilation systems for all apartments.
- Modernization of the electricity system outside of the apartments.

D.2.3.4. Stage of Construction

Completed in winter 2002, operating phase in the 11th year.

D.2.3.5. Point of Contact Information

Wbg Nürnberg GmbH Immobilienunternehmen, Glogauer Straße 70, 90473 Nürnberg.

D.2.3.6. Date of the Report

December 2011.

D.2.4. Site

- Location: Jean-Paul-Platz 4, 90461 Nürnberg.
- Coordinates: 49.457833, 11.087018
- Climate zone: ASHRAE 90.1-2007: 5/warm temperate.
- Heating Degree Days (based on °C) GT20/15: 3688 Kd.

D.2.5. Building Description/Typology

D.2.5.1. Typology/Age

Year of construction 1930.

D.2.5.2. Type (Office, Barracks, Etc.)

Apartment building.

D.2.5.3. Typology/Age

1930–1950.

D.2.5.4. General Information

- Year of construction: 1930.
- Year of previous major retrofit—if known.
- Year of renovation (as described here): 2002.
- Total floor area (m^2): six apartments a 149 m^2 (net floor area: 894 m^2).
- Area of unconditioned space included above (m^2): 0.
- Other information as appropriate:
- Air tightness n_{50} = 4.9 h^{-1} (whole building).
- (Apartments: first floor: 4.2 h^{-1}; second floor: 6.2 h^{-1}; third floor: 9.9 h^{-1}).

D.2.6. Architectural and Other Relevant Drawings

None.

D.2.7. National Energy Use Benchmarks and Goals for Building Type Described in the Case Study

D.2.7.1. Benchmark, According to the Standard National Average, Min, and Max

Support of the Bavarian Ministry of Economics for refurbishment with passive house components linked to the EU-Ziel-2 program.

D.2.7.2. National Energy Target for This Type of Building (If Any)

The national objective, meeting the EnEV (energy saving regulations) requirements, could be far exceeded with the refurbishment with passive house components.

D.2.8. Awards or Recognition

Publications

Schulze Darup, Burkhard. (December 2011): Energetische Modernisierung Modelprojekte: Jean-Paul-Platz 4 in 90461 Nürnberg; Ingolstädter Straße 131-141, 90461 Nürnberg; Bernadottestraße 42-48 in 90439 Nürnberg; Kollwitzstraße 1-17 in 90439 Nürnberg. Projektberichte [p. 3–42]; Nürnberg: wbg Nürnberg GmbH Immobilienunternehmen.

D.2.9. Site Energy Cost Information (Electricity, \$/kWh; Gas, \$/m³, Etc.)

None.

D.2.10. Pre-renovation Building Details

D.2.10.1. Envelope Details: Walls, Roof, Windows, Insulation Levels

- Walls: solid brick wall (41 cm): U-value 1.4 W/m^2K.
- Ceiling above 2nd floor: U-value 0.87 W/m^2K.
- Basement ceiling: U-value 0.88 W/m^2K.
- Windows: origin box-type windows from 1930; single-glazed windows in side rooms and staircase.

D.2.10.2. Heating, Ventilation, Cooling, and Lighting Systems

- Heating: central gravitational heating system; in the 1960s the systems was changed to local single-story heating system by maintaining the radiators and voluminous heating cables.
- Ventilation, cooling: none.

D.2.11. Description of the Problem: Reason for Renovation

- Energy focused building refurbishment as main target in the context of the EU-Ziel-2 program. The company wbg Nürnberg GmbH as an innovative housing society could be won for a refurbishment with passive house components within this framework.

- Regular condensate loss in outer walls.
- Air leaks primarily in windows and top-floor ceiling.

D.2.12. Renovation SOW (Non-energy and Energy Related Reasons)

See Sect. D.2.3.3..

D.2.13. Energy Saving/Process Improvement Concept and Technologies Used

D.2.13.1. Building Envelope Improvement

- Outer walls: thermal insulation composite system, thickness 200 mm (λ = 0.035), U-value: 0.15 W/m^{2}K.
- Ceiling above 2nd floor: insulation thickness 250 mm (λ = 0.035), U-value: 0.12 W/m^{2}K.
- Basement ceiling: insulation with mineral wool, thickness 140 mm (λ = 0.035), U-value: 0.19 W/m^{2}K.
- Windows: passive house plastic windows (Rehau Clima Design).

D.2.13.2. New HVAC System or Retrofits to Existing

- Ventilation: passive house adequate ventilation system with heat recovery (Maico-Aerex, Reccobox); local installations for every single apartment; air volume per apartment: 140–150 m^3/h in normal position; quality assurance per measurement.
- Heating: gas condensing boiler 30 kW, calculated according to hot water demand plus solar panels with 17 m^2 flat-plate collectors and 1000 L buffer storage.

D.2.13.3. New Generation/Distribution System

New distribution system with new gas condensing boiler, dimensioned for the warm-water demand.

D.2.13.4. Renewable Energy

Solar panels with 17 m^2 flat-plate collectors and 1,000 liter buffer storage, for warm-water generation.

D.2.14. Energy Consumption

D.2.14.1. Pre-renovation Energy Use (Total and per m^2/year)

See Table D.3.
Electricity demand is not known.
Heating demand calculated: 204 kWh/m^2a (according to PHPP/EN 832).

D.2.14.2. Predicted Energy Savings (Site, Source, GHG), Total and per m^2/year

- Heating demand calculated: 27 kWh/m^2a.
- 84%, predicted reduction of site energy 177 kWh/m^2.

D.2.14.3. Measured Energy Use (Thermal, Electrical), Total and per m^2/year

- Heating: 26.9 kWh/m^2a (first heating period); 23.7 kWh/m^2a (second heating period).
- For warm water in total: 17.8, 8.3 kWh/m^2 are generated by solar panels.

D.2.14.4. Annual Energy Use Reduction

Retrofit factor 10 could be reached, energy use reduction by 90%.

Table D.3 Annual energy use reduction (Nürnberg)

	Heating		Warm water		Total	
Heating period	Final energy	Primary energy	Final energy	Primary energy	Final energy	Primary energy
	kWh/m^2a	kWh/m^2a	kWh/m^2a	kWh/m^2a	kWh/m^2a	kWh/m^2a
2001/2002	196.6	267.3	20.4	49.6	217.7	319
2002/2003	219.2	292.2	23	52.4	242.9	346.7

D.2.15. Energy Cost Reduction

None.

D.2.16. Non-energy Related Benefits Realized by the Project, E.g., Improved Productivity, Increased Rent/Lease, Increased Useful Space, Reduced Maintenance, Etc.

- Improved living comfort and indoor air quality.
- Increased rent by 1.20 €/m^2 per month (operation cost savings included).

D.2.17. Renovation Costs: Total and per m^2

D.2.17.1. Total (Including VAT)

450,515 €; 503 €/m^2.

D.2.17.2. Cost for Each Measure (Non-energy Related/Energy Related) Including VAT, Cleared Costs

See Table D.4.

Table D.4 Total cost for each measure (non-energy related/energy related)

Subsection	Costs	Energy related/non-energy related*
Scaffolding work	7225.46 €	Non-energy related
Excavation, bricklaying and concrete work	13,898.69 €	Non-energy related
Demolition work	4155.27 €	Non-energy related
Roofing work	23,606.48 €	Ca. 50% energy related
Locksmith	39,553.69 €	Non-energy related
Plumber works	6,172.70 €	Non-energy related
Plastering	111,901.79 €	48% energy related = 53,712.86 €
Screed work (attic)	9,968.12 €	Non-energy related
Tiling	2,772.33 €	Non-energy related
Carpentry work—windows	60,331.44 €	Energy related

(continued)

Table D.4 (continued)

Subsection	Costs	Energy related/non-energy related*
Carpentry work—doors	11,881.03 €	Non-energy related
Painting	5,884.85 €	Non-energy related
Cleaning	435.19 €	Non-energy related
Costs for own performance tenants	4,243.33 €	Non-energy related
Heating, sanitary, ventilation		
Establish construction side	423.40 €	Non-energy related
Dismounting	17,365.49 €	Non-energy related
Boilers with accessories	8,795.24 €	Non-energy related (maintenance if boiler is old/broken)
Solar system with accessories	15,000.42 €	Energy related
Pipes with accessories	11,446.54 €	Non-energy related
Gas supply	776.17 €	Non-energy related
Radiators with accessories	8118.78 €	Non-energy related
Sanitary installations	7316.60 €	Non-energy related
Insulation	5939.52 €	Energy related
Ventilation	37,980.43 €	Energy related
Others	19,894.72 €	Non-energy related
Labor hour work	6400.16 €	Non-energy related
Electrical installations	9027.66 €	Non-energy related
Total costs	450,515.50 €	
Costs per m^2 living space	503.37 €/m^2	

*Differentiation about energy and non-energy related cost was made by lessons learned

D.2.18. Business Models and Funding Sources

D.2.18.1. Decision Making Process Criteria for Funding and Business Models

- Program with the aim to refurbish the southern part of Nürnberg.
- Refurbishment while people keep living in their flats.

D.2.18.2. Description of the Funding Sources Chosen

Funding by the Bavarian Ministry of Economics in connection with the EU-Objective-2 program.

D.2.18.3. Description of the Business Model Chosen (Option)

Owner financed, partly with a grant.

D.2.19. Cost Effectiveness of Energy Part of the Project (NPV, SIR, Etc.)

Actual costs are lower than estimated: 560 €/m^2 calculated to 503 €/m^2 actual. There were almost no changes in the floor plan. The whole planning and refurbishment process maintained a strict and continuous cost reduction strategy.

D.2.20. User Evaluation

D.2.20.1. Description of User Training Programs Within the Refurbishment

Through a briefing in the beginning and about two following conversations, it was determined that good user energy-conserving behavior could be achieved in some apartments. Two or three parties still used window ventilation to a large degree, which highlights the advantages of improving the ventilation system.

D.2.21. Experiences/Lessons Learned

D.2.21.1. Energy Use

In three apartments, the heating energy use was higher than predicted; because of generously window ventilation, the other apartments do need more heating energy than predicted.

D.2.21.2. Impact on Indoor Air Quality

The relative humidity constantly remained in a comfortable range between 35 and 45%.

D.2.21.3. Practical Experiences of Interest to a Broader Audience

The result of the refurbishment in Jean-Paul-Platz 4 in Nürnberg shows that broad-scale refurbishment with passive house components to achieve a reduction up to Factor 10 of the former heating demand is a realistic option for the near future. Very good results are achieved in comfort, indoor air quality and subjective living atmosphere. The projected structural-physical and the energy parameters could be confirmed in practice and the energy consumption values are lower than the calculated demand.

Furthermore, if there is a further development and market penetration, the most economical standard to refurbish buildings would be the high efficient energy retrofit.

D.2.22. References

Source: Schulze Darup, Burkhard. (December 2011): Energetische Modernisierung: Modelprojekte: Jean-Paul-Platz 4 in 90461 Nürnberg; Ingolstädter Straße 131-141, 90461 Nürnberg; Bernadottestraße 42-48 in 90439 Nürnberg; Kollwitzstraße 1-17 in 90439 Nürnberg. Projektberichte [pp. 3–42]; Nürnberg: wbg Nürnberg GmbH Immobilienunternehmen.

D.3. Gym Ostfildern, GE

D.3.1. Name of the Project, Location (City, Country)

3-Feld-Sporthalle, Ostfildern, Germany.

D.3.2. Pictures that Show the Building in Its Original and Post-Retrofit States and that Illustrate

See Fig. D.6.

Fig. D.6 Before and after retrofit

D.3.3. Project Summary

D.3.3.1. Project Objectives

To refurbish the gymnasium to modern utilization and energy standards.

D.3.3.2. Project Energy Goals

- Electricity: 50,000 kWh.
- Heat: 175,000 kWh.

D.3.3.3. Short Project Description

This project undertook a complete energetic refurbishment of insulation done, excluded the floor (no floor height available). The existing low-temperature gas boiler could not be replaced with the other measures, but will be substituted by a pellet boiler in the future. (A storage room and an opening to fill in the Pellets is already installed.)

D.3.3.4. Stage of Construction

Construction complete.
Operation phase: 5th year.

D.3.3.5. Point of Contact Information

KEA Klimaschutz-und Energieagentur Baden-Württemberg.
info@kea-bw.de.

D.3.3.6. Date of the Report

April 2014.

D.3.4. Site

- Climate zone (ASHRAE 90.1-2007: 5/warm temperate.
- DWD Zone: Stuttgart (Leinfelden-Echterdingen).
- Cooling Degree Days (based on 65 F).
- Heating Degree Days (based on 65 F): GT20/15: 3688 Kd.

D.3.5. Building Description/Typology

D.3.5.1. Typology/Age

Three-Field Gymnasium.

D.3.5.2. Type (Office, Barracks, Etc.)

Gymnasium.

D.3.5.3. Typology/Age

Pre 1970.

D.3.5.4. General Information

- Year of construction: 1972.
- Year of renovation (as described here): 2008.
- Total floor area (m^2): 2440 m^2 heated gross floor area.
- Area of unconditioned space included above (m^2):
- Other information as appropriate:
  - Construction weight: middle weight.
  - Thermal bridge: middle.
  - Air tightness: New construction without tightness test.

D.3.6. Architectural and Other Relevant Drawings

None.

D.3.7. National Energy Use Benchmarks and Goals for Building Type Described in the Case Study

D.3.7.1. Benchmark, According to the Standard National Average, Min, and Max

KfW 218—This is a subsidy-program that gives very cheap credits for energetic renovation measures. After renovation, the U-Values must be about 12–15% better than the EnEV (energy saving regulations) requires.

D.3.8. Awards or Recognition

None.

D.3.9. Site Energy Cost Information (Electricity, \$/kWh; Gas, \$/m^3, Etc.)

None.

D.3.10. Pre-renovation Building Details

D.3.11. Envelope Details: Walls, Roof, Windows, Insulation Levels

- Outer walls: 1,365 m^2, U = 1.0 W/m^2*K.
- Windows: 101 m^2 + 46 m^2 (doors), Uw = 4.3 W/m^2*K.
- Roof light: 135 m^2, Uw = 4.3 W/m^2*K.
- Floor: 2,446 m^2, U = 0.6 W/m^2*K.
- Roof: 2,305 m^2, U = 0.6 W/m^2*K.

D.3.12. Heating, Ventilation, Cooling, and Lighting Systems

- Ventilation:
  - Constant intake and exhaust air system.
  - Construction year 1973.
  - Volume flow: 24,000 m^3/h.

- Lighting system:
  - Fluorescent tube t8 with low-loss ballasts.
  - Manual control.
- Heating:
  - Gas low-temperature boiler vissmann, paromat-triplex, elco-burner.
  - Construction year: 1992.
  - Nominal heat output: 460 kw.
  - Fuel: gas.

D.3.13. Description of the Problem: Reason for Renovation

The building was in a bad condition. It needed an overall refurbishment.

D.3.14. Renovation SOW (Non-energy and Energy Related Reasons)

Scope was to retain and reuse the gym.

D.3.15. Energy Saving/Process Improvement Concept and Technologies Used

D.3.15.1. Building Envelope Improvement

- Windows: Uw = 1.29 W/m^2*K (in average).
- Outer Wall: insulation thickness 140 mm (035), U = 0.21 W/m^2*K.
- Ceiling: insulation thickness 140 mm (030), U = 0.20 W/m^2*K.
- Rooflights: Uw = 1.4 W/m^2*K.
- Single rooflights: U = 1.6 W/m^2*K.

D.3.15.2. New HVAC System or Retrofits to Existing

- Air system with heat recovery, demand-driven.
- Construction year; 2008.
- Volume flow: 20,000 m^3/h.

- Fire protection requirements: yes.
- Sound insulation: yes.
- Hygiene: yes.

D.3.15.3. New Lighting System

Fluorescent tube T5 with electronic ballasts and present detector.

D.3.16. Energy Consumption

- Pre-renovation energy use (total and per m^2/year)
- Electricity: 52,000 kwh/a.
- Gas: 350,000 kwh/a.

D.3.16.1. Measured Energy Savings (Thermal, Electrical), Total and per m^2/year

- Electricity: 51,000 kwh/a.
- Gas: 170,000 kwh/a.

D.3.16.2. Annual Energy Use Reduction

- Electricity: 1000 kwh/a.
- Gas: 180,000 kwh/a.

D.3.17. Energy Cost Reduction

Not available.

D.3.18. Non-energy Related Benefits Realized by the Project

The whole building was given a face-lift. The impression of the renovated building is visually appealing. It is attractive to use again.

D.3.19. Renovation Costs: Total and per m²

D.3.19.1. Total

2,785,000 € for energy and non-energy related measures.

D.3.19.2. Cost for Each Measure

- Facade: 175,000 €; 71.72 €/m^2gross floor area.
- Windows: 190,000 €; 77.87 €/m^2gross floor area.
- Lighting: 130,000 €; 53.28 €/m^2gross floor area.
- Heating system: 3,000,000 €; 122.95 €/m^2gross floor area.
- Ventilation system: 220,000 €; 90.16 €/m^2gross floor area.
- Total: 1,015,000 €; 415.95 €/m^2gross floor area.

D.3.20. Business Models and Funding Sources

D.3.20.1. Decision Making Process Criteria for Funding and Business Models

The decision was to either continue using the building, or to demolishing it.

D.3.20.2. Description of the Funding Sources Chosen

- Self-financing.
- With KfW-credit: financing-part of energetic refurbishment measures: 47%.

D.3.21. Cost Effectiveness of Energy Part of the Project (NPV, SIR, Etc.)

None.

D.3.22. User Evaluation

None.

D.3.23. Experiences/Lessons Learned

None.

D.4. School BaWü, GE

D.4.1. Name of the Project, Location (City, Country)

School in Baden-Württemberg, Germany.

D.4.2. Pictures that Show the Building in Its Original and Post-Retrofit States and that Illustrate Key Features of the Retrofit

See Figs. D.7 and D.8.

Fig. D.7 The building in its original state

Fig. D.8 The building in its post-retrofit states. *Source* KEA Klimaschutz- und Energieagentur Baden-Württemberg, 2014

D.4.3. Project Summary

D.4.3.1. Project Objectives

The project objectives were to refurbish the technical facilities and insulate the building envelope.

D.4.3.2. Short Project Description

The project undertook to perform an energy retrofit of the Panoramaschool heat station (district heating connection).

D.4.3.3. Stage of Construction

Completed in 2010, operating phase: 3rd Year.

D.4.3.4. Point of Contact Information

KEA Klimaschutz- und Energieagentur Baden-Württemberg GmbH.
info@kea-bw.de.

D.4.3.5. Date of the Report

01/09/2014.

D.4.4. Site

- ASHRAE 90.1-2007: 5/warm temperate
- Coordinates: 48.687108, 9.401196
- Cooling Degree Days (based on 65 F), Heating Degree Days (based on 65 F) $GT_{20/15:}$ 3688 Kd.

D.4.5. Building Description/Typology

D.4.5.1. Typology/Age

1970.

D.4.5.2. Type (Office, Barracks, Etc.)

School.

D.4.5.3. General Information

- Year of construction: 1971.
- Year of previous major retrofit:2005: new roof with new insulation and Photovoltaic.
- Year of renovation (as described here): 2010.
- Total floor area (m^2): gross floor area: 1990 m^2.
- Heated net floor area: 1774 m^2.
- Area of unconditioned space included above (m^2): 0 m2.
- Other information as appropriate: A/V-ratio after: 0.33.

D.4.6. Architectural and Other Relevant Drawings

See Fig. D.9.

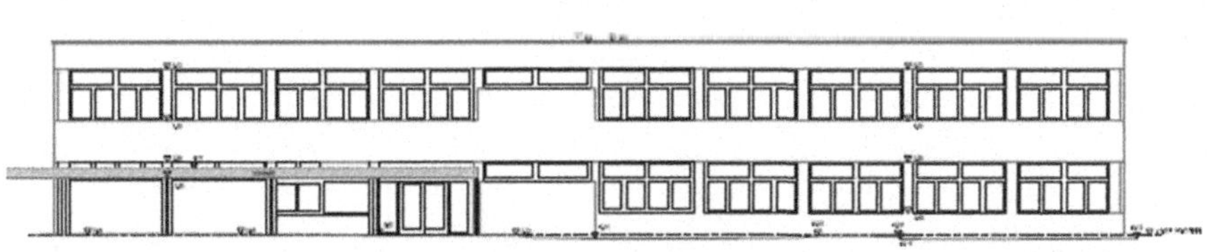

Fig. D.9 School in Baden-Württemberg—South Elevation

D.4.7. National Energy Use Benchmarks and Goals for Building Type Described in the Case Study

D.4.7.1. Benchmark, According to the Standard National Average, Min, and Max

EnEV 2009 for non-residential buildings.

D.4.7.2. National Energy Target for This Type of Building (If Any)

Stimulus package II (German Konjunturpaket II).

D.4.8. Awards or Recognition

None available.

D.4.9. Site Energy Cost Information (Electricity, \$/kWh; Gas, \$/m^3, Etc.)

None available.

D.4.10. Pre-renovation Building Details

D.4.10.1. Envelope Details: Walls, Roof, Windows, Insulation Levels

- Solid construction, strip structure facade, U-value 0.7 W/(m^{2}K), area: 1060.92 m^2.
- Windows U-value 2.5 W/(m^{2}K), area: 374 m^2.
- Flat roof: U-value 0.6 W/(m^{2}K), area: 1055 m^2.
- Floor: U-value 1 W/(m^{2}K), area: 1055 m^2.

D.4.10.2. Heating, Ventilation, Cooling, and Lighting Systems

- Heating: Gas boiler from 1980, 430 kW, natural gas.
- Ventilation: Exhaust air system in classrooms and rest rooms.
- Cooling: No cooling.
- Lighting:
  - Classrooms: fluorescent tubes, conventional ballasts for 300 lx.
  - Office: fluorescent tubes, conventional ballasts for 500 lx.
  - Other, sanitary: fluorescent tubes, conventional ballasts for 100/200 lx.

D.4.11. Description of the Problem: Reason for Renovation

Stimulus package II, energy retrofit, PCB pollution.

D.4.12. Renovation SOW (Non-energy and Energy Related Reasons)

The project was undertaken for energetic reasons.

D.4.13. Energy Saving/Process Improvement Concept and Technologies Used

D.4.13.1. Building Envelope Improvement

Walls: 16 cm insulation, U-value 0.18 W/(m^2K) (executed in 2005).

D.4.13.2. New HVAC System or Retrofits to Existing

- Heating: connection to district heating, 200 kW.
- Ventilation: exhaust air system only for rest rooms.

D.4.13.3. New Lighting System

- With electronic ballast, motion sensor, automatic switch-off.
- Classrooms: fluorescent tubes t5 (49 w or 35 w) for 300 lx.
- Office: fluorescent tubes for 500 lx.
- Other, sanitary: fluorescent tubes for 100/200 lx.

D.4.13.4. New Generation/Distribution System

Insulation of heating pipes, hydraulic balancing, new control valves.

D.4.13.5. Renewable Energy

Photovoltaic on the roof since 2005.

D.4.13.6. Daylighting Strategies

Manual.

D.4.14. Energy Consumption

See Tables D.5, D.6, D.7 and D.8.

Table D.5 Pre-renovation electrical energy use (total and per m^2/year)

Year	2007	2008	2009	2010
Total [kWh/a]	16,400	16,570	18,870	24,390

Table D.6 Natural Gas use before the insulation of the roof

Year	2007	2008	2009	2010
Total [kWh/a]			236,000	

Table D.7 Measured electrical energy demand after renovation (thermal, electrical), total and per m^2/year

Year	2007–2010	2011	2012
Total [kWh/a]		19,156	16,174

Table D.8 Measured electrical energy demand after renovation (thermal, electrical), total and per m^2/year (district heating)

Year	2007–2010	2011	2012
Total [kWh/a]		86,484	79,510

D.4.15. Energy Cost Reduction

None available.

D.4.16. Non-energy Related Benefits Realized by the Project

New ground plan design in ground floor and upper floor.

D.4.17. Renovation Costs: Total and per m^2

D.4.17.1. Total

About 215 €/m^2 heated net floor area: for new windows with blinds and external wall insulation system, some sheet metal work and improvement of the distribution system.

D.4.17.2. Energy Related and Non-energy Related

See Table D.9.

Table D.9 Energy-related and non-energy-related costs

Subsection	Costs*	Energy related/non-energy related
Insulation outer walls	83,784 €	27,912 € energy related
Plaster soffits, connections outer walls	55,872 €	non-energy related
Windows (aluminum)	285,850 €	energy related
Heating system (district heating connection)	29,750 €	energy related
Total costs	455,256 €	
Costs per m^2	256.60 €/m^2	

*The costs are estimated on consolidated basis

D.4.17.3. Cost for Each Measure

- External wall insulation system: about 70 €/m^2.
- New windows with blind: about 735 €/m^2.

D.4.18. Business Models and Funding Sources

D.4.18.1. Decision Making Process Criteria for Funding and Business Models

Town council.

D.4.18.2. Description of the Funding Sources Chosen

- Self-financing.
- Stimulus package II, bank loan + self-financing, Heating through EPC.

D.4.18.3. Description of the Business Model Chosen (Option)

Energy performance contracting.

D.4.18.4. Risk Allocation in the Business Model

Contractor/City.

D.4.18.5. Planning Process in the Business Model

Architect, engineering office.

D.4.18.6. Construction Phase in the Business Model

Architect, engineering office, contractor.

D.4.18.7. Operation Phase

Heating contractor.

D.4.18.8. Energy Management and Controlling in the Business Model

Yes.

D.4.19. Cost Effectiveness of Energy Part of the Project (NPV, SIR, Etc.)

Life cycle: envelope: 50 years, technic 25 years.

D.4.20. User Evaluation

D.4.20.1 Description of User Training Programs Within the Refurbishment

Training of caretaker.

D.4.20.2. Integration of Users Demands in the Planning Process

Yes.

D.4.21. Experiences/Lessons Learned

D.4.21.1. Energy Use

See above.

D.4.21.2. Space Utilization Changes

New ground floor design.

D.5. Angela School

D.5.1. Name of the Project, Location (City, Country)

- Angelaschule.
- Osnabrueck, Lower Saxony, Germany.

D.5.2. Picture

See Fig. D.10.

Fig. D.10 Angela School. *Source* [Final Report 2013] Fette, Max, Clausnitzer, Klaus-Dieter. Final report: Scientific monitoring and quality control of the energetic modernization of Angela School in Osnabrück

D.5.3. Project Summary

D.5.3.1. Project Objectives

Energy retrofit of a school block.

D.5.3.2. Project Energy Goals

- Energy demand reduction by more than 70%.
- Improvement of indoor climate.
- Energy results:
- Demand for electricity: 16.04 kWh/ (m^2*a).
- Demand for Heat (local heat + heat pump): 49, 16 kWh/ (m^2*a).

D.5.3.3. Short Project Description

- The building contains 18 classrooms for approximately 30 students each.
- It was built in 1965, the facade and the technical equipment is mainly > 40 years old.
- The building envelope is in poor energetic condition, double-glazed windows have high leakage rates.
- The project aimed to reduce the energy demand about 70% and to improve the air and light quality in the classrooms.

D.5.3.4. Stage of Construction

Completed in 2010.

D.5.3.5. Point of Contact Information

Mr. Fette and Mr. Clausnitzer from BremerEnergieInstitut.

D.5.3.6. Date of the Report

May 2013.

D.5.4. Site

- Location Bramstr. 41 a, 49090 Osnabrueck, Germany.
- Latitude, longitude: 52° 18′ 3.316″ N 8° 2′ 32.562″ E.
- Elevation: 64 m above sea.
- Climate zone (e.g., ASHRAE 90.1-2004 Climate Zone): ASHRAE 90.1-2007: 5/warm temperate.
- Cooling Degree Days (based on 65 F), Heating Degree Days (based on 65 F): $GT_{20/15:}$ 3688 kd.

D.5.5. Building Description/Typology

D.5.5.1. Typology/Age

- Concrete skeleton construction, 48 years old.
- 1950–1970.

D.5.5.2. Type (Office, Barracks, Etc.)

School.

D.5.5.3. General Information

- Year of construction: 1965.
- Year of renovation (as described here): 2010.
- Total floor area used for all calculations) (m^2):
  - Gross floor area: 2670
  - Net floor area: 2154
  - Heated net floor area 1850 m^2.
- Area of unconditioned space included above (m^2): in the net floor area of 2154 are 304 m^2 of unconditioned space included.

D.5.6. Architectural and Other Relevant Drawings

See Figs. D.11, D.12 and D.13.

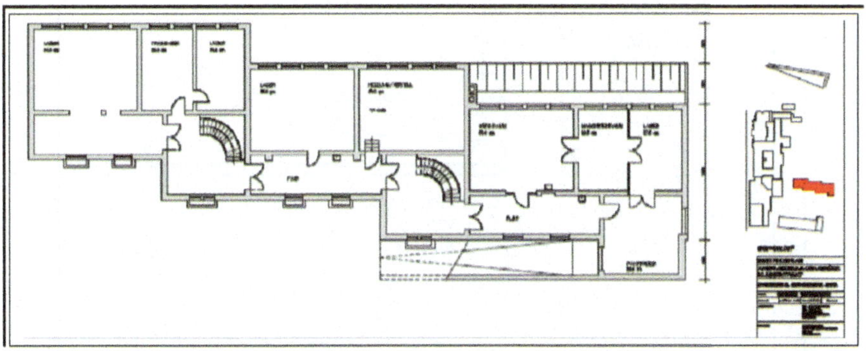

Fig. D.11 Ground floor plan. *Source* [Final Report 2013] Fette, Max, Clausnitzer, Klaus-Dieter. Final report: Scientific monitoring and quality control of the energetic modernization of Angela School in Osnabrück

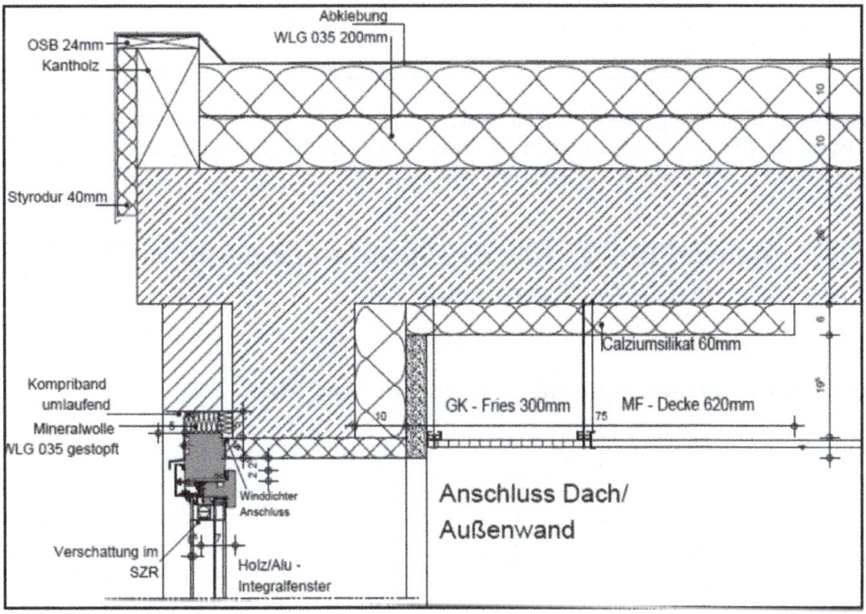

Fig. D.12 Connection roof/exterior wall. *Source* [Final Report 2013] Fette, Max, Clausnitzer, Klaus-Dieter. Final report: Scientific monitoring and quality control of the energetic modernization of Angela School in Osnabrück

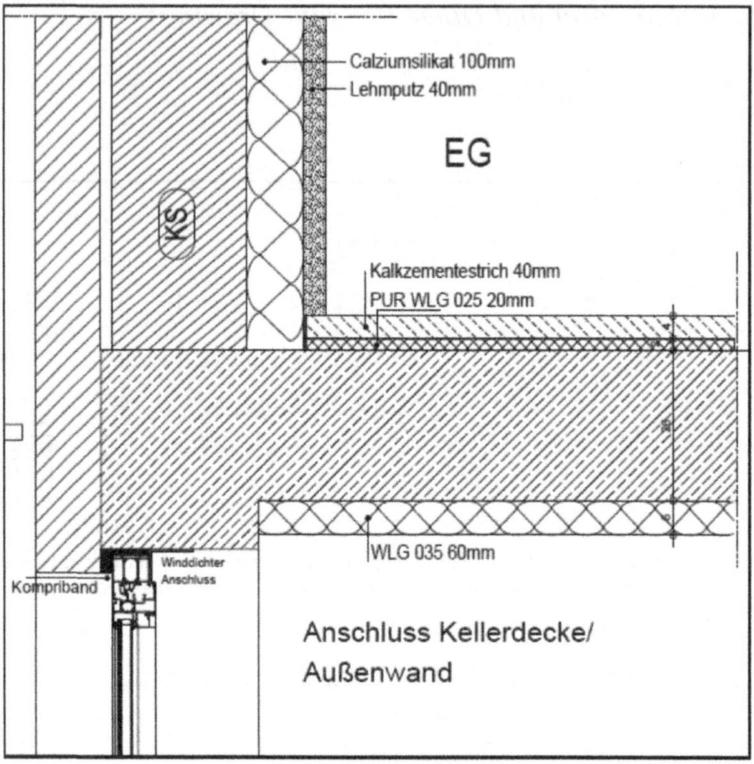

Fig. D.13 Connection basement ceiling/exterior wall. *Source* [Final Report 2013] Fette, Max, Clausnitzer, Klaus-Dieter. Final report: Scientific monitoring and quality control of the energetic modernization of Angela School in Osnabrück

D.5.7. National Energy Use Benchmarks and Goals for Building Type Described in the Case Study

D.5.7.1. Benchmark, National Average, Min, and Max

110 kWh/m^{2}a for heating and warm water and 10 kWh/m^2 a for electricity.

D.5.8. Awards or Recognition

None.

D.5.9. Site Energy Cost Information (Electricity, \$/kWh; Gas, \$/m³, Etc.)

Electricity: 0.177 €/kWh; Gas: 0.054 €/kWh upper heating value.

D.5.10. Pre-Renovation Building Details

D.5.10.1. Envelope Details: Walls, Roof, Windows, Insulation Levels

- Walls:
  - Curtain wall with concrete skeleton and infill of lime-sand bricks on the inner side, a 2 cm air layer and uninsulated clinker walls, U-value: 1.13 W/(m^2K).
  - Steal concrete structure with 4 cm insulation layer of wood-wool building boards on the inner side, U-value: 0.97 W/(m^2*K).
  - Heating slots with a light insulation layer of wood-wool building boards, U-value: 0.92 W/(m^2*K).
- Windows:
  - Most of the windows and the main doors were changed in 1983, U-value: about 3.0 W/(m^2K), PVC and aluminum frames.
  - Floors and toilets: wood frame windows, single glazing and glass bricks, U-value up to 4.7 W/(m^2K).
- Roof: Platform roof, concrete, poor insulated with 6 cm cork and bitumen waterproofing, U-value: 0.63 W/(m^2K), but insulation layer was moistened so that the real U-value was graded worse.

D.5.10.2. Heating, Ventilation, Cooling, and Lighting System

- Heating. The Micro District Heating Grid powered by:
  - Renewable oil CHP-Plant (base load) and gas condensing boiler
  - Heating medium: hot water (− 15 °C = 90 °C) distributed in two-pipe system, with three stage over dimensioned pumps, no hydraulic adjustment of the system, wrong dimensioned thermostats, heating in the rooms by cast iron radiators.
- Ventilation: No ventilation.
- Cooling: No cooling.
- Lighting systems:

- Classrooms: 16 × 58 W fluorescent tubes (T36)- conventional ballasts = 14.6 W el/m^2 for 300 lx.
- Hallways:16 × 40 W fluorescent tubes (T36) – conventional ballasts = 4.5 W/m^2 for 17 – 40 lx.
- Stairways:12 × 18 W fluorescent tubes (T26)- conventional ballasts = 5 W/m^2 for 17 – 40 lx.
- No active daylight- system.

D.5.11. Description of the Problem: Reason for Renovation

Problems included high energy costs, low indoor temperatures in winter, high indoor temperatures in summer, bad air-condition in classrooms.

D.5.12. Renovation SOW (Non-energy and Energy Related Reasons)

Renovations were undertaken to reduce energy costs, and to improve indoor climate in winter and summer.

D.5.13. Energy Saving/process Improvement Concept and Technologies Used

D.5.13.1. Building Envelope Improvement

- Roof: platform roof, old cork insulating board replaced by new insulation (20 cm, WLG 035).
- Outer walls (inside): 10 cm calcium-silicate boards plus 4 cm of clay-plaster. Additional cellulose fiber filling for the former radiator niches (radiators were replaced by wall heating radiators).
- Window reveal: calcium-silicate boards (4–6 cm).
- Windows: old windows and glass blocks were replaced by new ones with 2 + 1 glazing with blinds integrated in the building control system.
- Entrance: installation of vestibules.
- Cellar ceiling: new insulation underneath (6 cm) and above (2 cm).
- Story ceiling: 75 cm deep, 6 cm thick insulation next to the outer walls to minimize thermal bridges.

Table D.10 Construction building envelope characteristics (Angela School)

Building element	U-value before renovation	U-value after renovation	Thickness of new insulation
Windows	$\leq$ 4.7	1.1	
Outer wall	1.13	0.14–0.32	14 cm (10 + 4)
Roof	$\gg$ 0.63	0.19	20 cm
Cellar ceiling	0.96	0.23–0.27	8 cm (2 + 6)
Base plate	3.57	3.57	
Walls against earth		0.28	
Walls against unheated space		0.28	

- Inner walls: 75 cm deep, 6 cm thick insulation next to the outer walls to minimize thermal bridges (Table D.10).

D.5.13.2. New HVAC System or Retrofits to Existing

- Heating:
  - Base-load: water-water-heat-pump with 7 kW_{el}, figure of merit: 4.75, overall net heating capacity: 33 kW.
  - 2500 l heating water storage.
  - Peak load and back up: The school is still connected to the heating pipeline (backup).
- Ventilation:
  - One ventilation system with heat recovery for all restrooms.
  - Three ventilation systems (for six classrooms each): Every classroom has a volume flow control and a downstream heater battery. Hallways are ventilated by exhaust air from classrooms. The exhaust air is collected in the stairways and restrooms.
  - The systems aims for 90% heat recovery.
- Lighting system:
  - All conventional ballasts were replaced by electronic starters.
  - Classrooms: 6 × 35 W fluorescent tubes (T8).
  - 1 × 80 W fluorescent tube with 60° reflector for blackboard = 4.2 W/m^2 instead of 14.6 W/m^2.
  - Hallways:16 × fluorescent tubes 35 W (T16) = 3.5 W/m^2 instead of 4 W/m^2.
  - Stairways:12 × fluorescent tubes 18 W (T16) = 4.32 W/m^2 instead of 5 W/m^2.
  - Others (see Table D.11).

Table D.11 Other lighting

Zone	Type	Number	Total installed load (W)
Exterior lightning	Compact fluorescent tubes 26 W	6	190
Basement and others	Diverse fluorescent tubes T26, 35/58 W	58	3610
Workshop	diverse T26	18	1170
Restrooms	diverse T26	13	480

D.5.13.3. New Generation/Distribution System

- Heat is distributed by large sized wall heating (6.4 m^2 per classroom) and ventilation system with large heating battery.
- The flow/return temperatures were reduced from 90 °C/70 °C to 35 °C/30 °C.

D.5.14. Energy Consumption

D.5.14.1. Pre-renovation Energy Use (Total and per m^2/year)

Three different estimations were done of pre-renovation energy consumption and costs. Table D.12 lists Baseline 1 estimates, calculated according to DIN V 18599. There are no measured figures available.

D.5.14.2. Predicted Energy Savings (Site, Source, GHG), Total, and per m^2/year

See Table D.13.

Table D.12 Pre-renovation energy demand

	2008 calculated	
	kWh/a €/a	kWh/m^2/a €/m^2/a
Electricity (lighting + auxiliary energy)	0.177 €/kwh	
Consumption	35,695	19.29
Costs	6318	3.41
Heat (gas and vegetable oil)	0.054 €/kWh	
Consumption	620,846	335.59
Costs	33,526	18.12

Table D.13 Predicted energy demand (Angela School, Osnabruec, Germany)

	Average A + 1 and A + 2 (heat = weather-adjusted)		Conversion electricity for heat pump into thermal heat
	kWh/a	kWh/m^2/a	Merit factor = 4.75
Electricity consumption (ventilation)	11,699	6,32	
Electricity consumption (heat pump)	(17,705)	(9.57)	= 17,705 * 4.75 = 84,099 kWh/a thermal heat → 45.46 kWh/(m^2*a)
Electricity consumption heating (all pumps, controls, electricity)	1,952	1.05	
Electricity consumption class rooms	3,436	1.86	
Electricity consumption hot water, floor lighting, others	12,580	6.8	
Total electricity consumption (without heat pump)	47,372 – 17,705 = ***29,667***	16.04	
Heat (gas and vegetable oil)	6837	3.70	
Heat (heat pump)	84,099	45.46	
Total heat consumption (local heat + heat pump)	90,936	49.15	
Total costs	8754	4.73	

According to the energy certificate, consumption electricity (heating, ventilation and lighting) after retrofit is calculated as:

31.3 kWh/(m^2*a) * 1850 m^2 = 57,905 kWh/a  90% savings (only one value for electricity.

Note that it was not possible to separate the consumption of the heat pump).

D.5.14.3. Measured Energy Savings (Thermal, Electrical), Total and per m^2/year

- Site Energy.
- CO_2 reduction about 80%.

D.5.14.4. Annual Energy Use Reduction.

Annual energy use was reduced by more than 90% by application of insulation, combined with installation of a heat pump and a ventilation system with heat recovery.

D.5.15. Energy Cost Reduction

Pre-renovation energy costs were only estimated: between 40,000 and 58,000 Euro, for a cost reduction between 31,000 and 49,000 Euro, equivalent to 78–85%.

D.5.16. Non-energy Related Benefits Realized by the Project

- Better air-quality. Before the CO_2 concentration was often between 2000 and 3000 ppm; after refurbishment it is commonly under 1000 ppm.
- Higher comfort for the users.
- Contribution for the maintenance. Extended lifespan of the building.

D.5.17. Renovation Costs: Total and per m²

D.5.17.1. Total

2.24 Million € inclusive VAT = 1210 €/m^2 (heated net floor area).

D.5.17.2. Non-energy Related

1,058,194 €.

D.5.17.3. Energy Related

- 1,181,806 € = 638 €/m^2 heated net floor area (inclusive VAT).
- Without ventilation = 647,127 €.

D.5.17.4. Cost for Each Measure

See Table D.14.

Table D.14 Cost for energy-related and non-energy-related measures (Angela School, Osnabrueck, Germany)

Measure	Energy related costs including VAT and 18% incidental building costs
Building measures	
Insulation cellar ceiling	29,541 €
Inner insulation outer walls	152,882 €
Insulation ceilings	32,226 €
Insulation and screed new	53,135 €
Insulation roof	48,952 €
Windows, wood-aluminum	73,799 €
Steel structure for wind traps	8,029 €
Total	398,565 €
Other measures	
Heating system	69,982 €
Power installations	127,782 €
Control technology and building automation	50,798 €
Ventilation	517,873 €
Construction for ventilation system	16,806 €
Total other measures	783,241 €
Total	1,181,806 €
Without ventilation system	647,127

D.5.18. Funding Source

Owner and federal ministry for environment

D.5.19. Cost Effectiveness of Energy Part of the Project (NPV, SIR, Etc.)

Interest rate: 3.43% (government bond) (Table D.15)

D.5.20. User Evaluation

See Table D.16.

Table D.15 Cost effectiveness

Position	Present value (k€)
Investment costs	
Building measures	− 303
Technical measures without ventilation	− 279
Ventilation	− 600
Sum investment costs	− 1182
Maintenance costs	
Heat pump and building automation	− 59
Ventilation and heat recovery	− 119
Sum maintenance costs	− 178
Energy costs reduction	
Gas (and vegetable oil)	663
Electricity without ventilation	0.1
Electricity for ventilation	− 42
Sum energy costs reduction	621
Water	− 13
Total	− 752

Table D.16 Theoretical reduction of energy consumption through different measures

	Final energy (MWh/a)	Primary energy (MWh/a)
Insulation on the inside instead of on the outside	− 0.9	− 2.4
Insulation of walls	25	66
Insulation of roof	3	9
Tightness of windows and doors	57	153
Tight ventilation system	13	34
Ventilation system (not tight)	1	3

D.5.21. Experiences/Lessons Learned

D.5.21.1. Energy Use

Reduction by approximately 80% through insulation combined with heat pump.

D.5.21.2. Impact on Indoor Air Quality

Significant improvement through ventilation system.

D.5.22. References

Source: [Final Report 2013] Fette, Max, Clausnitzer, Klaus-Dieter. Final report: Scientific monitoring and quality control of the energetic modernization of Angela School in Osnabrück.

D.6. Friedrich-Fröbel-School Olbersdorf

D.6.1. Name of the Project, Location (City, Country)

Friedrich-Fröbel-Schule Olbersdorf, Olbersdorf, Germany.

D.6.2. Pictures

These pictures show the building in its original and post-retrofit states, illustrating key features of the retrofit.

	Original	**Post-retrofit**
View		

(continued)

(continued)

	Original	Post-retrofit
Heating		
Ventilation	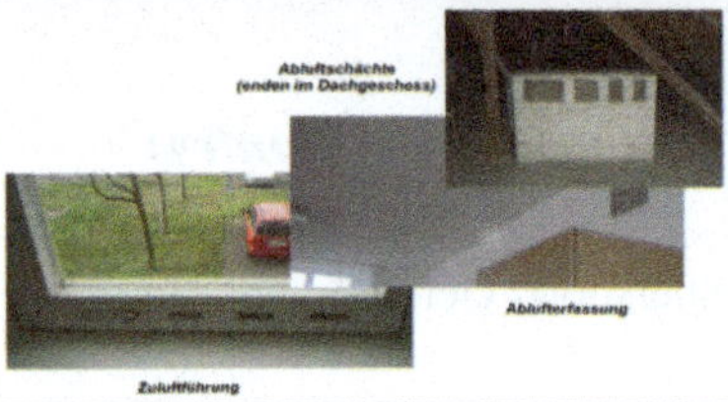	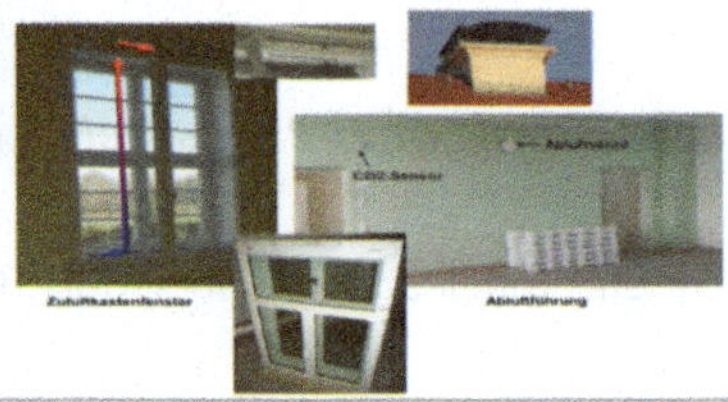

Source http://www.enob.info/en/refurbishment/projects/details/comprehensive-refurbishment-of-a-school-listed-as-a-historic-monument, http://www.eneff-schule.de/index.php/Demonstrationsobjekte/3-Liter-Haus-Schulen/3-liter-haus-schule-in-olbersdorf-landkreis-loebauzittau.html

D.6.3. Project Summary

D.6.3.1. Project Objectives

A model project for the research topic of "Energy-efficient schools." The building was built in 1927/28 with the gym and is today a cultural monument. The goal was to reduce the energy consumption to the "3-Litre-House-Standard": reduction of the annual primary energy demand for the heating and ventilation to max. 34 kWh/(m^2*a). The main target was to create a better climate in the building and to implement a best practice example of a successful compromise between historic conservation and energy-based refurbishment for this type of school.

D.6.3.2. Project Energy Goals

- Heating and ventilation: achieve a max primary energy value: 34 kWh/(m^2*a), with the renovation the heating demand must reduce more than 80%.
- Considerably reduce energy costs.
- Improve thermal insulation of the building envelope in accordance with listed building requirements.

D.6.3.3. Short Project Description

- The building contains 22 classrooms that accommodate 180 school students.
- The building was constructed in 1927/28.
- The building envelope is in poor energetic condition, the old box-type windows are energy inefficient.
- Some of the classrooms cannot use the daylight and must use artificial light; others suffer of the heat in summer because of the window orientation.
- The project aimed to reduce the energy demand by about 80% and to improve the air and light quality in the classrooms and at the same time to keep the original facade.

D.6.3.4. Stage of Construction

Completed in 2011.

D.6.3.5. Point of Contact Information

Administrative district Löbau/Zittau, Saxony, Germany.

D.6.3.6. Date of the Report

2013.

D.6.3.7. Acknowledgement

Subsidized by the Federal Ministry for Economic Affairs and Energy as part of its "Energy-Optimized Construction, EnEff:Schools" subsidy program.

D.6.4. Site

Location: Schulweg 13, 02785 Olbersdorf, Germany.
Latitude, longitude: 50° 52′ 24″ N and 14° 46′ 27″ E.
Elevation: 289,384 above the sea.
Climate zone (e.g., ASHRAE 90.1-2004 Climate Zone): ASHRAE 90.1-2007: 5/warm temperate.
Cooling Degree Days (based on 65 F).
Heating Degree Days (based on °C) GT20/15: 3688 Kd.

D.6.5. Building Description/Typology

D.6.5.1. Type (Office, Barracks, Etc.)

School.

D.6.5.2. General Information

- Year of construction: 1927/28.
- Year of renovation (as described here): 2009–2011.
- Total floor area (m^2): gross floor area: 5610; heated net floor area 4439 m^2.
- Area of unconditioned space included above (m^2):.
- Other information as appropriate: specific transmission-heat-loss 0.42 W/m^2K.

D.6.6. Architectural and Other Relevant Drawings

See Figs. D.14, D.15 and D.16.

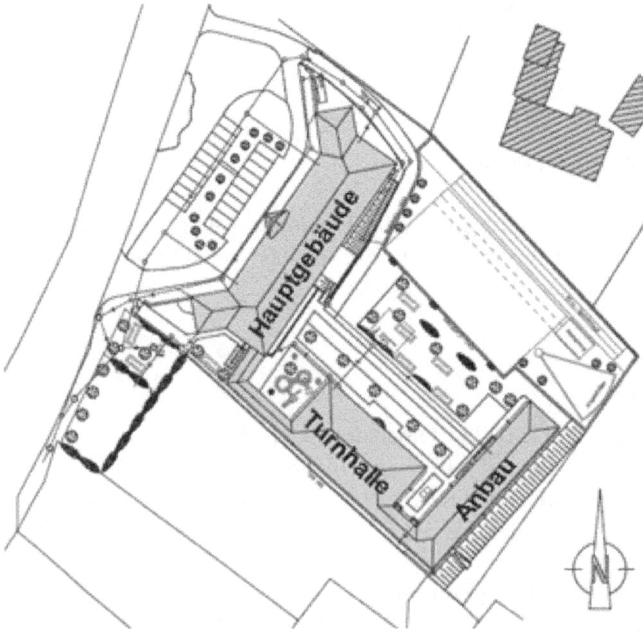

Fig. D.14 Plan of the whole site

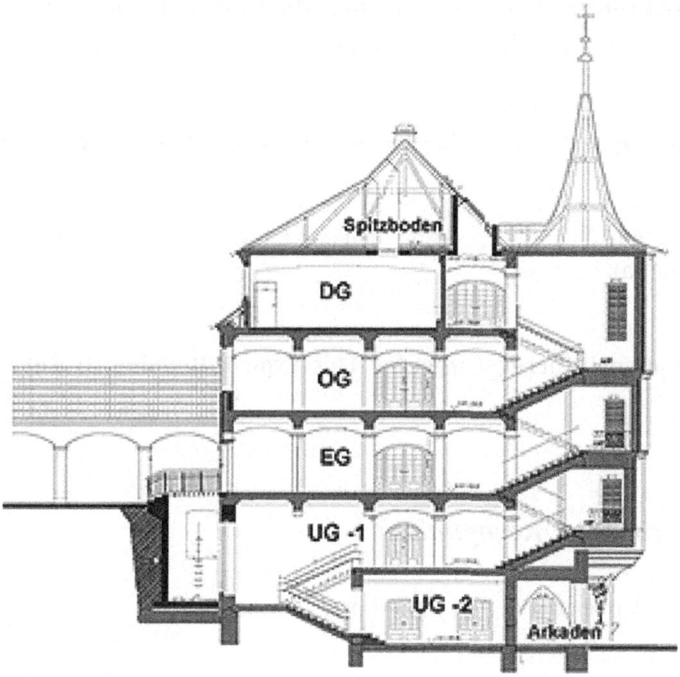

Fig. D.15 The main building

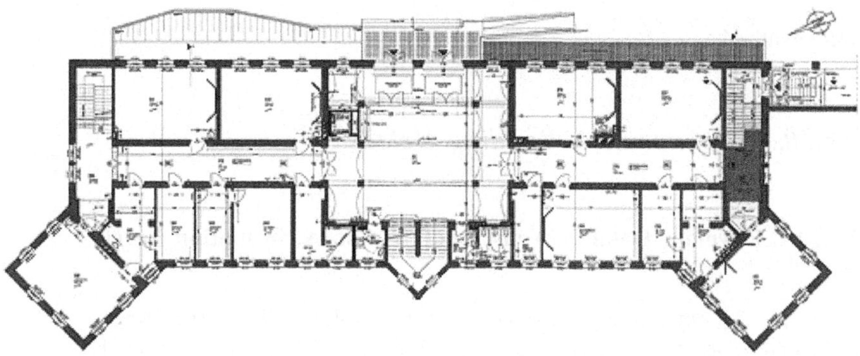

Fig. D.16 Ground floor plan of the main building. *Source* http://www.enob.info/en/refurbishment/projects/details/comprehensive-refurbishment-of-a-school-listed-as-a-historic-monument/

D.6.7. National Energy Use Benchmarks and Goals for Building Type Described in the Case Study

D.6.7.1. Benchmark, According to the Standard National Average, Min, and Max

3-l house standard = primary energy demand for a heating and ventilation max. of 34 kWh/(m^2*a).

D.6.7.2. National Energy Target for This Type of Building (If Any)

See Table D.17.

D.6.8. Awards or Recognition

Saxony state competition: "very good project solution."

D.6.9. Site Energy Cost Information (Electricity, \$/kWh; Gas, \$/m^3, Etc.)

None available.

D.6.10. Pre-renovation Building Details

D.6.10.1. Envelope Details: Walls, Roof, Windows, Insulation Levels

- Facade: 51 cm drilling masonry, plaster, U-value: 1.25 W/(m^2*K).
- Windows west: Twin pane isolation glazing, U-value: 1.70 W/(m^2*K).

Table D.17 Energy indices

Energy indices according to German regulation energy (kWh/m^2a)	Before refurbishment	After refurbishment
Heating energy demand	122.69	31.82
Overall primary energy requirement	174.23	48.89

- Windows east: box-type-windows with double glazing, U-value: 2.80 W/(m^2*K).
- Roof: 20 cm of concrete, 4 cm of screed, U-value: 1.70 W/(m^2*K).
- Floor: U-value: 1.70 W/(m^2*K).

D.6.10.2. Heating, Ventilation, Cooling, and Lighting Systems

- Heating: two gas boilers with a maximal performance of 250 and 283 kW.
- Ventilation: with ventilator, open windows and cooling air exhaust channel in the masonry.
- Cooling: no cooling.
- Lighting systems: No active daylight- system.

D.6.11. Description of the Problem: Reason for Renovation

High energy and ventilation costs, high indoor temperatures in summer, bad air-condition in classrooms.

D.6.12. Renovation SOW (Non-energy and Energy Related Reasons)

- Create a unique educational room climate under reduction of the energy consumption.
- Improve room air hygiene and acoustics.
- Improvement of indoor climate.
- Lowering the indoor temperatures in summer.
- Use low-maintenance technology.

D.6.13. Energy Saving/Process Improvement Concept and Technologies Used

D.6.13.1. Building Envelope Improvement

See Table D.18.

Table D.18 Construction building envelope characteristics (Friedrich-Fröbel-School Olbersdorf, Germany)

Building element	U-value before renovation	U-value after renovation	Description
Windows west	1.70	1.0	Supply air box-type windows, external windows: single glazing, intern windows: double glazing
Windows east	2.80	0.9	Supply air box-type windows, external windows with electro chromic glass: double glazing, intern windows: double glazing
Outer wall	1.25	0.34	51 cm masonry with 7 cm ESP-insulation
Top ceiling	1.70	0.22	Masonry, 5 cm screed, 15 cm mineral wool insulation, 2 cm OSB panels
Floor	3.09	0.36	10 cm concrete, 2 cm vacuum insulation panels, 4 cm screed
		0.32	10 cm XPS-insulation, 10 cm concrete, 4 cm screed

D.6.13.2. New HVAC System or Retrofits to Existing

The new energy concept reactivates existing ventilation and lighting systems.

Ventilation:

- Sanitary rooms: conventional central exhaust air system with presence control.

Classrooms:

- Customized ventilation that is substantially based on natural uplift and is only boosted with fans with low electrical consumption as appropriate ("hybrid ventilation").
- The existing double windows and the partly replaced windows are being rebuilt with thermal insulation glazing installed in double air-supply windows. Via an opening in the lower frame, the external air enters the cavity between the panes, warms up and enters the room via the upper window frame. Pre-heating the air and supplying it above the occupancy zone, reduces the risk of draughts. To prevent unwanted air currents, additional wind pressure reducers and check flaps are integrated in the windows. As soon as a specific external temperature is exceeded, the skylights in the inner panes of the windows are automatically opened, which increases the volume flow of the air change.
- Reactivation of the exhaust air ducts so that used air can be removed by means of natural uplift. Should a sensor measure increased CO_2 concentrations, an exhaust fan will be switched on to provide support. If the respective teaching space is not or only partly occupied, the fan remains switched off. Compared to a standard system, this considerably saves on electricity for powering the vent.

Cooling: employ a natural cooling system during summer by using effective night cooling (which is neutral in terms of the primary energy use).

Heating: A ground-coupled gas absorption heat pump with a 35 kW performance was installed, with peak load compensation provided by the gas boiler.

D.6.13.3. New Lighting System

Classrooms: pendant lamp, performance: 35 and 80 W for the panel. Artificial light is adjusted with solar shading.

D.6.13.4. Daylighting Strategies

- Reactivation or supplementation of old light shafts.
- Integration of louvre blinds in the cavities in the double windows to provide shading, glare protection and to redirect light.
- Daylight-controlled system for the need of artificial light.
- Light shaft to control the direction of lightning with aluminum reflection panels.

D.6.14. Energy Consumption

D.6.14.1. Pre-renovation Energy Use (Total and per m^2/year)

See Table D.19.

D.6.14.2. Predicted Energy Use/Savings (Site, Source, GHG), Total and per m^2/year

See Table D.20.

Table D.19 Consumption 2001–2003

Energy component	Final energy [kWh/m^{2}year]	Primary energy [kWh/m^{2}year]
Heting and drinking water	144.0	158.4
Lighting	11.5	34.4
Total	155.5	192.8

Table D.20 Calculations based on DIN V 18599 for Friedrich-Fröbel-School Olbersdorf, Germany

Energy component	Net energy [kWh/m²a]	Final energy [kWh/m²a]	Primary energy [kWh/m²a]
Heating	36.6	34.9	34.7
Drinking water	2.7	2.7	7.0
Ventilation	0	0.1	0.3
Cooling	0	0	0.0
Lighting	3.3	3.3	8.6
Total	42.5	41.9	50.4

D.6.14.3. Measured Energy Use/savings (Thermal, Electrical), Total and per m²/year

See Table D.21.

D.6.15. Energy Cost Reduction

None available.

D.6.16. Non-energy Related Benefits Realized by the Project, E.g., Improved Productivity, Increased Rent/lease, Increased Useful Space, Reduced Maintenance, Etc.

None.

Table D.21 Energy use after retrofit

	2011 (ab März)	2012	2013
Gas final energy [kWh/m²a]	22.13	45.29	48.66
18Electrical final energy for equipment technology [kWh/m²a]	1.11	2.01	1.86
Gas primary energy [kWh/m²a]	21.95	44.92	48.27
Electrical primary energy for equipment technology [kWh/m²a]	2.9	5.22	4.85
Total primary energy [kWh/m²a]	24.85	50.15	53.12

D.6.17. Renovation Costs: Total and per m²

D.6.17.1. Total

5.4 Mio € (1214 €/m²) for Retrofitting the School Building (Table D.22).

D.6.17.2. Cost for Each Measure

See Table D.23.

Table D.22 Total renovation cost

Measure	Net-costs	
	€	€/m² NGF
Renovation and development	10,177	3
Structural design	3,303,194	744
Equipment technology	1,436,157	385
Outside facilities	52,689	14
Equipment	2355	1
Extra costs	584,494	157
Total	5,389,065	1214

Table D.23 Cost of non-energy-related and energy-related measures (Friedrich-Fröbel-School Olbersdorf, Germany)

Measure	Net-costs		Energy related/non-energy related*
	€	€/m² NGF	
Building structure	387,880	87	Non-energy related
Facade construction work	928,941	209	50% energy related
Roofing membrane and covering	193,924	44	10% energy related
Paintwork	227,552	51	Non-energy related
Locksmith	63,458	14	Non-energy related
Floor covering	435,045	98	Non-energy related
Drywall construction	391,388	88	Non-energy related
Woodwork	323,627	73	Non-energy related
Others	351,378	79	Non-energy related
Sanitary	130,237	29	Non-energy related
Heating	338,893	76	50% energy related
Ventilation	46,469	10	Energy related
Electrical installation	353,306	80	Non-energy related
Lighting	150,395	34	Non-energy related

(continued)

Table D.23 (continued)

Measure	Net-costs		Energy related/non-energy related*
	€	€/m² NGF	
Telecommunication system	89,228	20	Non-energy related
Conveying system	63,271	14	Non-energy related
User-specific equipment	142,078	32	Non-energy related
Building automation	146,292	33	50% energy related
Others	0	0	Non-energy related
Total	4,763,362	1073	

*Differentiation about energy and non-energy related cost was made by lessons learned

D.6.18. Business Models and Funding Sources

D.6.18.1. Description of the Funding Sources Chosen

Federal ministry for environment through the funding program "Energieoptimiertes Bauen, EnEff:Schule."

D.6.19. Cost Effectiveness of Energy Part of the Project (NPV, SIR, Etc.)

None.

D.6.20. User Evaluation

D.6.20.1. Description of User Training Programs Within the Refurbishment

The teachers were informed how to use the building; they subsequently taught the students in physics lessens.

A user acceptance survey was made before and after the refurbishment; the main results are:

- Improved overall impression of the new building.
- Improvement in the room ambient temperature during the summer.
- Improvement in the acoustic of the rooms.
- Compromised lighting (light shading due to automatic jalousie).
- Lack of fresh air in the rooms.

The positive aspects lead to the fact that more users are aware of energy saving.

The jalousie dims the room such that the occupants have to turn on the lights. Some of the occupants find this fatiguing.

D.6.21. Experiences/Lessons Learned

D.6.21.1. Impact on Indoor Air Quality

There have been compliments about the air quality after the installation of the ventilation system. CO_2 is not sufficient to measure the air quality. A VOC sensor was installed in addition.

D.6.22. References

- http://www.eneff-schule.de/
- http://www.enob.info/en/refurbishment/projects/details/comprehensive-refurbishment-of-a-school-listed-as-a-historic-monument/
- Reiß, Erhorn, Geiger, Roser, Gruber, Schakib-Ekbatan, Winkler, Jensch: Energieeffiziente Schulen, EnEff:Schule, Fraunhofer IRB-Verlag, 2013.

D.7. Office Passive House Retrofit

D.7.1. Name of the Project, Location (City, Country)

- Name: Institut Wohnen und Umwelt IWU.
- City: Darmstadt.
- Country: Germany.

D.7.2. Pictures

These pictures show the building in its original and post-retrofit states, illustrating key features of the retrofit.

Office building IWU, Rheinstraße 65 in Darmstadt		
	Original	Post-retrofit
View		
Heating	No picture available!	
Ventilation	No picture available!	

Source IWU

D.7.3. Project Summary

D.7.3.1. Project Objectives

This building at Rheinstraße was constructed in 1962, and is one part of a property of three building parts that were used by the former district administration of Darmstadt-Dieburg. To act as the new seat of the Institut Wohnen und Umwelt, the mostly unrenovated building got a comprehensive refurbishment. The goal was to create a modern working environment that meets the current and the future energetic requirements. The superior target was a modernization with passive house components to nearly passive house standard.

D.7.3.2. Project Energy Goals

- Calculated with PHPP: all values per m^2 heated net floor area.
- Electricity: net energy value: 31.6 kWh/(m^2 a).
- Heating: net heat energy value: 16.2 kWh/(m^2 a).
- Primary energy value: 169 kWh/(m^2 a).
- Other: no active cooling (building and server), high building automation.

D.7.3.3. Short Project Description

- Energy aspects:
  - Energy modernization of building envelope: the exterior walls were insulated with a thermal insulation composite system up to 30 cm thick.
  - The old windows were replaced through modern triple-glazed windows with passive house suitable plastic frame. The very large window area of the existing building has been slightly reduced in size, resulting in a significant cost savings and the reduction of heat input on these large windows in the summer months.
  - The flat roof was insulated with two layers of insulation and a total thickness of 40 cm and re-covered.
  - The ceiling of the basement was insulated with 14 cm thick insulation boards made of mineral wool.
  - In the windows of the south facade an external sun protection in form of blinds with daylight control was installed.
  - At all windows an internal manually operated anti-glare was installed.
  - To provide summer heat protection, automatic night ventilation was achieved by opening the windows in the 1st and 2nd floors.

  - The building lighting was modernized and provided with automate control (occupancy sensors, automatic switch off and light sensor). For cost reasons, no constant light control was implemented.
  - A central HVAC system with heat recovery (heat recovery efficiency > 80%) and a new distribution network of ventilation ducts was installed.
  - The server cooling takes place with outside air. Active cooling is demonstrably not necessary.
- Adapting the building to the requirements of the tenant:
  - Central access to the building on a new internal staircase.
  - Integration of the Institute's own library and print shop in the building.
  - Creation of modern meeting rooms.
  - Optimization of the foyer: loosening, optimized for receiving visitors.
- Economic aspects:
  - Long-term assessment of the modernization costs on the rent.
  - Agreement on a warm rent for the building.

D.7.3.4. Stage of Construction

Operating phase in the third year: The modernization was completed in March 2011 and the building has been in operation since then.

D.7.3.5. Point of Contact (POC) Information

KEA Klimaschutz- und Energieagentur Baden-Württemberg.
info@kea-bw.de

D.7.3.6 Date of the Report

May 2014.

D.7.4. Site

- Coordinates: 49.8731395, 8.6343576, 15.21
- Climate zone: ASHRAE 90.1-2007: 5/warm temperate
- Cooling Degree Days (based on 65 F),
- Heating Degree Days (based on °C): $GT_{20/15:}$ 3688 Kd.

D.7.4.1. Type (Office, Barracks, Etc.)

Office building, heated and ventilated.

D.7.4.2. Typology/Age

1950–1970.

D.7.4.3. General Information

- Year of construction: 1962.
- Year of renovation (as described here): 2010/2011.
- Total floor area (m^2): 1931 (heated gross floor area), 1680 (heated net floor area).
- Area of unconditioned space included above (m^2): 0.
- Other information as appropriate:
  - Building mass (category) c_{eff}: medium-weight: c_{eff} = 90 Wh/(m^2K) (assumption).
  - Thermal bridges: many (0.10 W/(m^2K)) (assumption).
  - Air tightness n50: 2.81 h^{-1} according to the test report (assumption, no measurement).

D.7.5. Architectural and Other Relevant Drawings

Available, can be send on request.

D.7.6. National Energy Use Benchmarks and Goals for Building Type Described in the Case Study

D.7.6.1. Benchmark, According to the Standard National Average, Min, and Max

Support of the Hessian state for refurbishment with passive house components, to achieve 25 kWh (m^2 heated net floor area *a) heating demand, which is equivalent to the current EnerPHit Standard.

D.7.6.2. National Energy Target for This Type of Building (If Any)

The national objective, meeting the EnEV (energy saving regulations) requirements, could be far exceeded with the refurbishment with passive house components. The building with its excellent energy standard is an example of how an energy efficient and at the same time economic retrofit of existing buildings (similar requirements like EPBD EU directive) can be achieved.

D.7.7. Awards or Recognition

D.7.7.1. Awards

Winner of Green Building Award Frankfurt Rhein/Main 2013. http://www.greenbuilding-award.de/index.php?id=115.

D.7.7.2. Publications

Schaede, Margrit; Hörner, Michael: Sanierung eines Verwaltungsgebäudes mit Passivhauskomponenten: Betriebserfahrungen. In: Gebäude energetisch optimieren, Tagungsband der 16. Eckernförder Fachtagung, Eckernförde 2013, 48–58.

Hörner, Michael; Schaede, Margrit: Das IWU-Haus: Modernisierung mit Passivhauskomponenten – Konzept, Kosten, Betriebserfahrungen. In: 17. Internationale Passivhaustagung, Tagungsband zur Konferenz in Frankfurt 2013, 503–508.

Schaede, Margrit, Hörner, Michael: Hocheffiziente Bürogebäude – Konzept, Kosten und Betriebserfahrungen einer Modernisierung. In: Effizienztagung 2013, Tagungs-CD zur Konferenz in Hannover 2013.

D.7.8. Site Energy Cost Information (Electricity, \$/kWh; Gas, \$/m³, Etc.)

IWU is paying an all-inclusive price for their energy consumption.

D.7.8.1. Electricity

719.88 €.

D.7.8.2. Natural Gas

805.58 €.

D.7.8.3. Other (E.g., Biomass)

Operation costs: 2022.53 €.

D.7.9. Pre-renovation Building Details

D.7.9.1. Envelope Details: Walls, Roof, Windows, Insulation Levels

Solid construction, strip-structured facade, flat roof.

D.7.9.2. Heating, Ventilation, Cooling, and Lighting Systems

Condensing gas boiler, ventilation system with heating function, fluorescent lamp bar-shaped with conventional ballast.

D.7.10. Description of the Problem: Reason for Renovation

The institute IWU is, among others, doing research on energy efficiency in buildings. So because the building standard was obsolete, the disused office building was completely retrofitted with passive house components. A special challenge, thereby was to achieve the targeted energy efficiency level despite the structural restrictions of an existing Building, while still complying with the given budget.

D.7.11. Renovation SOW (Non-energy and Energy Related Reasons)

- Insulation of outer walls: thermal insulation composite system, 30 cm.
- Insulation of flat roof: two layers of insulation, in total 40 cm.
- Insulation of basement, below: 14 cm.
- Windows: triple glazing with passive house compatible plastic frames; sound insulation windows at southern facade.
- New ventilation system with efficient heat recovery.

- Lighting system with electronic ballast, motion and light sensor, automatic switch-off.
- Summer heat protection: external sun protection on the south side and sun protection glazing on the north side plus automatic night ventilation (window ventilation with sensors for temperature, wind and rain) for cooling down the building.
- Server cooling: passive concept with outdoor air.

D.7.12. Energy Saving/Process Improvement Concept and Technologies Used

D.7.12.1. Building Envelope Improvement

See Table D.24.

Table D.24 Construction building envelope characteristic (Office Passive House Retrofit, Darmstadt, Germany)

Building element	Component	Pre insulation measures	Post insulation measures	Insulation thickness (cm, λ-value)
Windows south	U-value, U_w [W/m^{2}K]	3.30	0.85	/
Windows north	U-value, U_w [W/m^{2}K]	3.30	1.10	/
Windows north (congress hall 2. Floor)	U-value, U_w [W/m^{2}K]	1.90	1.90	/
Outer wall against outdoor air	U_{EW} [W/m^{2}K]	1.36	0.11	30 cm, WLS 035
Outer wall against outdoor air (opaque panels congress hall 2. Floor)	U_{EW} [W/m^{2}K]	0.60	0.60	/
Outer wall against unheated	U_{EW} [W/m^{2}K]	2.67	2.67	/
Ceiling	U_R [W/m^{2}K]	0.70	0.09	40 cm, WLS 035
Basement	U_{GND} [W/m^{2}K]	1.09	0.26	14 cm, WLS 035

D.7.12.2. New Generation/Distribution System

Due to the limited budget and the relative new condensing boiler, no changes.

D.7.12.3. New HVAC System or Retrofits to Existing

See Table D.25.

Table D.25 HVAC characteristics

Ventilation system 3	Description	Pre retrofit	Post-retrofit
Which zone is ventilated?	New installation of a modern ventilation system for supplying the zones inside the thermal envelope	/	1–6, 8–9
Function of the ventilation Air change heating cooling others		/	Cover the minimum incoming air temperature of 19 °C by damper register
Type:		/	Incoming and outgoing air
Year of construction		/	2011
Nominal volume flow rate (outgoing/incoming air [m^3/h]		/	5000/5000
Nominal electric capacity [kw]		/	1.18/1.09
Recirculated air [%]		/	0%
Volume flow regulation • constant • variable • demand driven		/	Variable (time regulated and in addition demand driven)
Damper register kw: cooling register kw:		/	Damper register
Heat recovery [%]		/	81% (from manufacturer)
Fire protection must be fulfilled?		/	Yes
Sound insulation must be fulfilled?		/	Yes
Hygiene must be fulfilled?		/	Yes

Operation time Regulation	Percentage from nominal volume flow Variable	Electrical power [kW] /	Hours per day /	Days per week /
concept: constant Demand driven				
Step 1	100%	2.27	1	1
Step 2	39%	0.43	10	5
Step 3	16%	0.15	13	5

	Description	Pre retrofit	Post-retrofit
Which zone is ventilated?	New installation of a ventilation system for the server room	/	10
Function of the ventilation Air change Heating Cooling Others		/	None
Type		/	Outgoing air
Year of construction		/	2011
Nominal volume flow rate (outgoing/incoming air [m^3/h]		/	200
Nominal electric capacity [kW]		/	1.1
Recirculated air [%]		/	0%
Volume flow regulation • constant • variable • demand driven		/	Variable (cooling load)
Damper register kW: cooling register kW:		/	Without
Heat recovery [%]		/	No
Fire protection needs fulfilled?		/	Yes
Sound insulation fulfilled?		/	Yes
Hygiene fulfilled?		/	Yes

Operation time	Percentage from nominal volume flow	Electrical power [kW]	Hours per day	Days per week
Regulation concept: constant	100	0.1	8	7

D.7.12.4. New Lighting System

Luminescent screen tubes with electronic ballasts that use direct and indirect illumination techniques, occupancy sensor and dimming switch off.

D.7.12.5. Daylighting Strategies

Southern facade: external blinds that react automatically of solar radiation. There is a light steering in the upper third to drive daylight into the offices.

D.7.13. Energy Consumption

D.7.13.1. Pre-renovation Energy Use (Per m^2 Heated Gross Floor Area)

- Electricity: 111,578 kWh/a → 57.78 kWh/m^2*a.
- Heat: 363,914 kWh/a → 188.46 kWh/m^2*a.

D.7.13.2. Predicted Energy Savings (Site, Source, GHG), Total and per m^2/year (Net Floor Area)

See Table D.26.

Table D.26 Predicted energy savings (Darmstadt, Germany)

	Energy consumption pre retrofit, calculated (final energy) with TEK-Tool (kWh/m^{2}EBZa)		Energy consumption post-retrofit, calculated (final energy) with TEK-Tool (kWh/m^{2}EBZa)		Energy consumption measured (year 2013)	
	Total MWh/a	kWh/m^{2}bgfa	Total MWh/a	kWh/m^{2}bgfa	Total MWh/a	kWh/m^{2}bgfa
Heating	405.0	241.1	60.5	35.8	79.9	47.5
Hot water	19.2	11.4	1.7	1.0	Included in others	Included in others
Cooling	9.1	5.4	0.0	0.0	0.0	0.0
Ventilation	3.9	2.3	3.4	2.0	5.0	2.98
Lighting	26.9	16.0	14.2	8.4	6.2	3.64
Others	34.8	20.7	22.0	13.0	22.1	13.1

D.7.13.3. Measured Energy Savings (Thermal, Electrical), Total and per m²/year (Heated Net Floor Area)

See Table D.26.

D.7.13.4. Annual Energy Use Reduction

See Table D.27.

D.7.14. Energy Cost Reduction

D.7.14.1. Split in All Energy Forms—Electricity, Oil, District Heating

- Energy costs today:
  - Gas: 6.5 Cent/kWh
  - Electricity: 15 Cent/kWh.
- Cost reduction electricity = 11,736 Euro.
- Cost reduction gas/heating = 18,462 Euro.

Table D.27 Energy reduction

	a − 1	a − 1	A	A	a + 1	a + 1	a + 2	a + 2
	Total kWh/a €/a	kWh/m²/ a €/m²/a	Total kWh/a €/a	kWh/m²/ a €/m²/a	Total kWh/a €/a	kWh/m²/ a €/m²/a	Total kWh/a €/a	kWh/m²/ a €/m²/a
Electricity								
Consumption	111,578	66.42	None	Data	34,249	20.39	33,338	19.84
Heating including distribution loss								
Consumption	363,914	216.62	None	Data	None	Data	79,870	47.54
Electricity: energy use reduction a − 1 to a + 2	78,240	46.58	–	–	–	–	–	–
Heating: energy use reduction a − 1 to 1 + 2	284,044	169.08						

D.7.15. Non-energy Related Benefits Realized by the Project

- Improvement of the connection to the building with the central stairway.
- Integration of its own library and printer.
- Realization of a prestigious reception.
- Barrier free connection to the ground and the first floor with coke wharfs and elevator.
- Very good sound insulation.
- Comfortable room temperatures, especially in summer.

D.7.16. Renovation Costs: Total and per m²

D.7.16.1. Total

1,811,328.20 Euro.

D.7.16.2. Non-energy Related

906,684.10 Euro.

D.7.16.3. Energy Related

904,644.10 Euro.

D.7.16.4. Cost for Each Measure (Without VAT)

See Table D.28.

Table D.28 Renovation cost of each measure

Passive house standard	Investment (Euro)
Windows	168,601.00
Entrance system	7965.00
Sun protection	75,074.00
Scaffoldings	23,625.50
External wall cladding	172,490.00
Roof layers	115,014.00
Insulation basement ceiling	24,440.00
Lighting	105,893.60
Ventilation	169,740.00
Supports/binding joists	13,686.00
Summer night ventilation	29,135.00

D.7.17. Business Models and Funding Sources

The retrofit was financed by the building owner.

D.7.18. Cost Effectiveness of Energy Part of the Project (NPV, SIR, Etc.)

- Interest rate: 4%.
- Life cycle periods: measures building envelope = 30 years; measures HVAC, technology = 15 years.
- Observation period = 30 years.
- Energy cost increase: heat = 5.5%/a; electricity = 3.5%/a.
- Retrofit with passive house components was more economical than a usual retrofit to meet the national energy saving requirements.

D.7.19. User Evaluation

D.7.19.1. Description of User Training Programs Within the Refurbishment

- Guided tour for the employees to show the main technical components of the building, especially of the offices.
- Short handout with a summary of the most important functions.
- A contact person for questions and problems was named.

D.7.19.2. Integration of Users Demands in the Planning Process

The IWU institute was intensively involved in the planning process and set the requirements for their working conditions.

D.7.20. Experiences/Lessons Learned

D.7.20.1. Energy Use

Heating consumption is higher than calculated, possible reasons can be:

- Flooding of the heating cables.
- Heater has to be switched on relatively early and could be switched off late.
- (Unknown) high the losses over the basement ceiling. (A relevant error in measurement is supposed.)

D.7.20.2. Impact on Indoor Air Quality

CO_2 and VOC measurements show a very good air quality; only air humidity not optimal.

D.7.20.3. Practical Experiences of Interest to a Broader Audience

There had been some guided tours for interested visitors and an opening day for the broad public to dedicate the building. A tour through the building can be arranged on request.

D.7.20.4. Resulting Design Guidance

- More potential for optimization was found in planning for heating system, ventilation, sun protection, and lighting.
- A comprehensive planning including all subsections is necessary.
- Basis for the planning has to be a clear, detailed and reproducible functional description.
- Also important for planners and operating engineers ensure the described functions.

D.7.20.5. Space Utilization Changes

Central stairway for a good connection with library, meeting rooms, printing room, etc. was integrated in the planning process from the beginning.

D.7.20.6. Follow Up on the Renovation

A follow-up should include an evaluation of how the users actually operate the system. It might also include an review of seismic upgrades and consequences. In this case, some buildings features could not be implemented because seismic upgrades requirements.

IWU started its own project for the monitoring and optimization of the operations. Internal user surveys showed general satisfaction with the building operation and environmental conditions.

D.8. Town Hall—Bavaria, Germany

D.8.1. Name of the Project, Location (City, Country)

Town Hall in North of Bavaria, Germany.

D.8.2. Pictures

None available.

D.8.3. Project Summary

D.8.3.1. Project Objectives

Refurbishment and modernization of the building construction and building services, that guarantees an efficient maintenance and energy management for a long time.

Retrofitting the fire protection. Considerable deficits caused a 2-year vacancy before the refurbishment.

D.8.3.2. Project Energy Goals

- Electricity:31.04 kWh/m^2 a.
- Heat:44.06 kWh/m^2 a.
- Cooling:6.44 kWh/m^2 a.

D.8.3.3. Short Project Description

- Comprehensive refurbishment, renewal of building services and fire protection concept, retrofitting noise protection, energetic refurbishment.
- No main tasks remained unsettled.

D.8.3.4. Stage of Construction

- Construction complete.
- Operation phase: third year.

D.8.3.5. Point of Contact (POC) Information

KEA Klimaschutz- und Energieagentur Baden-Württemberg GmbH.
martina.riel@kea-bw.de, Germany.

D.8.4. Site

- Coordinates: 49.4362322, 10.9610411.
- (ASHRAE 90.1-2007: 5/warm temperate).

D.8.5. Building Description/Typology

D.8.5.1. Typology/Age

Atrium house.

D.8.5.2. Type (Office, Barracks, Etc.)

Town hall, office building.

D.8.5.3. Typology/Age

1970.

D.8.5.4. General Information

- Year of construction: 1975.
- Year of renovation (as described here): 2010/2011.
- Total floor area (m^2): 9.941 m^2 heated gross floor area.
- 8.959 heated net floor area.

D.8.6. Architectural and Other Relevant Drawings

Not available.

D.8.7. National Energy Use Benchmarks and Goals for Building Type Described in the Case Study

D.8.7.1. Benchmark, According to the Standard National Average, Min, and Max

EnEV 2007 (energy saving regulations) with EnEV 2009 and further strengthened requirements in mind.

D.8.7.2. National Energy Target for This Type of Building (If Any)

If an existing building will be renovated, the requirements on the renovated building are 40% less strength than on a new building of this type.

D.8.8. Awards or Recognition

D.8.9. Site Energy Cost Information (Electricity, \$/kWh; Gas, \$/m^3, Etc.)

D.8.9.1. Electricity

0.07 €/kWh.

D.8.9.2. District Heating

0.05 €/kWh.

D.8.9.3. Cooling

0.07 €/kWh.

D.8.10. Pre-renovation Building Details

D.8.10.1. Envelope Details: Walls, Roof, Windows, Insulation Levels

- Wall Solid construction, strip structure facade, U-value 1.58 W/(m^{2}K), area: 2159 m^2.
- Wall (adjacent to soil/ unheated rooms): U-value 1.58 W/(m^{2}K), 658 m^2.
- Windows U_w-value 4.3 W/(m^{2}K), area: 1369 m^2.
- Flat roof and top story ceiling: U-value 0.6 W/(m^{2}K), area: 2075 m^2.
- Floor (adjacent to soil/ unheated rooms): U-value 0.77 W/(m^{2}K), area: 1683 m^2.

D.8.10.2. Heating, Ventilation, Cooling, and Lighting Systems

- Ventilation system with no more information available.
- Connection to district heating.

D.8.11. Description of the Problem: Reason for Renovation

- Large deficits in fire protection made it impossible to use the building; it was closed for 2 years.
- Poor noise insulation.
- Poor heat insulation.
- High energy consumption.

D.8.12. Renovation SOW (Non-energy and Energy Related Reasons)

Refurbishment and modernization of the building construction and building services to today's requirements to be able to use the building again. It was vacant for 2 years because of a lack of fire protection. The goal was to realize a refurbishment that guarantees efficient maintenance and energy management in the long term.

D.8.13. Energy Saving/Process Improvement Concept and Technologies Used

D.8.13.1. Building Envelope Improvement

- Windows: Uw = 0.7 W/m^2*K (in average).
- Rooflights dome: 1.06 W/m^2*K.
- Panel field: 0.3 W/m^2*K.
- Facade: insulation thickness 200 mm, λ = 0.04 W/m K, Uw = 0,185 W/m^2*K.
- Facade to soil: insulation thickness 120 mm, λ = 0.04 W/m K, Uw = 0.31 W/m^2*K.
- Ceiling: insulation thickness 170 mm, λ = 0.04 W/m K, Uw = 0.22 W/m^2*K.
- Roof: insulation thickness 200 mm, λ = 0.04 W/m K, Uw = 0.19 W/m^2*K.
- Floor (to air): Uw = 0.3 W/m^2*K.
- Floor (to soil): insulation thickness 160 mm, λ = 0.04 W/m K, Uw = 0.23.

D.8.13.2. New HVAC System or Retrofits to Existing

- Air systems with heat recovery, demand driven.
- Construction year: 2010.

- Volume flow: (office and small conference rooms 24,000; conference room 10,100; foyer 4000; service area, ground floor 2000) m^3/h.
- Fire protection requirements: yes.
- Sound insulation: yes.
- Hygiene: yes.

D.8.13.3. New Lighting System

Luminescent screen tubes with electronic ballasts, illumination technique direct and occupancy sensor, in the foyer halogen lamps in addition.

D.8.13.4. New Generation/Distribution System

Continued utilization of the district heating, complete new distribution system with new energy efficient pumps.

D.8.14. Energy Consumption

D.8.14.1. Pre-renovation Energy Use (Total and per m^2/year)

- Electricity: no information available.
- District heating: no information available.
- Cooling: no information available.

D.8.14.2. Predicted Energy Savings (Site, Source, GHG), Total and per m^2/year

See Table D.29.

Table D.29 Predicted energy savings (Town Hall, Bavaria, Germany)

	Energy consumption pre retrofit, calculated (final energy) (kWh/m²EBZa)		Energy consumption post-retrofit, calculated (final energy) (kWh/m²EBZa)		Energy consumption measured (average of the years 2012 + 2013)	
	Total MWh/a	kWh/m²bgfa	Total MWh/a	kWh/m²bgfa	Total MWh/a	kWh/m²bgfa
Heating	No data available		304.2	38.97	200.1	25.64
Hot water			30.4	3.91	Included in others	Included in others
Cooling			119.1	15.26	104.6	13.4
Ventilation			34.7	4.45	Included in others	Included in others
Lighting			149.4	16.67	238.8	26.65
Others			83.7	10.72	132.2	16.94

D.8.14.3. Measured Energy Savings (Thermal, Electrical), Total and per m²/year

Please see table above.

D.8.14.4. Annual Energy Use Reduction

See Table D.30.

D.8.15. Energy Cost Reduction

Energy consumption before refurbishment is not known.

Table D.30 Annual energy use reduction (Town Hall, Bavaria, Germany)

	a − 1	a − 1	A	A	a + 1	a + 1	a + 2	a + 2
	Total kWh/a €/a	kWh/m²/a €/m²/a	Total kWh/a €/a	kWh/m²/a €/m²/a	Total kWh/a €/a	kWh/m²/a €/m²/a	Total kWh/a €/a	kWh/m²/a €/m²/a
Electricity								
Consumption total			None	Data	465,496	59.62	485,644	62.21
Consumption only cooling					103,699	13.28	105,494	13.51
Heating including distribution loss								
Consumption	373,411*	47.83	None	Data	196,248	25.14	204,010	26.14

*Assumption, as no data available

D.8.16. Non-energy Related Benefits Realized by the Project, E.g., Improved Productivity, Increased Rent/Lease, Increased Useful Space, Reduced Maintenance, Etc.

- Increase in comfort and improvement in the use of area.
- Previously much less quality of utilization (less natural lightning, no acclimatization, bad utilization concept).
- Retrofitted noise protection.
- More usable area.

D.8.17. Renovation Costs: Total and per m²

D.8.17.1. Total (Gross): Investment Costs

16.3 m €; Running Costs: 10.7 m €.

D.8.17.2. Non-energy Related (Without Tax)

Needed renovation: 11,356,023, other 1,346,540 €.

D.8.17.3. Energy Related (Without Tax)

838,502 €.

D.8.17.4. Cost for Each Measure

See Table D.31.

Table D.31 Cost for energy-related and non-energy-related measures (Town Hall, Bavaria, Germany)

	Non energy related costs (without tax)		Energy related costs (without tax)	
	Total [€]	Per m² [€/m²]	Total [€]	Per m² [€/m²]
Roof	452,087	218	15,563	7.50
External wall	1,508,020	698	24,829	11.50
Windows	446,950	327	191,550	139.93
Insulation basement ceiling	7,780	3.7	70,020	33.74
Other	5,819,525			

(continued)

Table D.31 (continued)

	Non energy related costs (without tax)		Energy related costs (without tax)	
	Total [€]	Per m² [€/m²]	Total [€]	Per m² [€/m²]
Heating delivery system	521,460	61	85,690	9.98
Ventilation	885,350	103	265,450	30.91
Other technique	1,714,850	200	185,400	21.59
Total costs	11,356 023	1,322	838,502	
Additional costs				
Architecture	1,250,690			
Building control system	95,850			

D.8.18. Business Models and Funding Sources

D.8.18.1. Decision Making Process Criteria for Funding and Business Models

- Forfeiting model with a communal waver of objection.
- Predetermined by the initiator, no participation in the decision process.

D.8.18.2. Description of the Funding Sources Chosen

Public-private partnership.

D.8.18.3. Description of the Business Model Chosen (Option)

Public-private partnership.

D.8.18.4. Risk Allocation in the Business Model

Private Partner of the PPP.

D.8.18.5. Planning Process in the Business Model

- Financer: a cooperative bank.
- Technical/economical counsel.
- Judicial counsel.

D.8.18.6. Funding Sources of the Business Model

- Technical counsel.
- Economical counsel.

D.8.18.7. Construction Phase in the Business Model

Private Partner of the PPP.

D.8.18.8. Operation Phase

- It is a PPP-Project (Public-Private-Partnership).
- Maintenance, operating, controlling and optimization of the technical systems is done by the "private partner."

D.8.18.8. Energy Management and Controlling in the Business Model

Private Partner of the PPP.

D.8.19. Cost Effectiveness of Energy Part of the Project (NPV, SIR, Etc.)

Economic comparison: 25 years; 5.25% fixed interest rate, cash value, simple payback period 25 years.

D.8.20. User Evaluation

D.8.20.1. Description of User Training Programs Within the Refurbishment

Previous none, during utilization phase continuously.

D.8.20.2. Integration of Users Demands in the Planning Process

Room and function matrix according to users requirements implemented.
Participation of users in the planning process.

D.8.21. Experiences/Lessons Learned

D.8.21.1. Energy Use

Less energy use than predicted because times of utilization are lower than planned.

D.8.21.2. Space Utilization Changes

Lower flexibility in using rooms caused by high requirements for noise protection.

D.8.21.3. Follow Up on the Renovation

A follow-up should include an evaluation of how the users actually operate the system. It might also include an review of seismic upgrades and consequences. In this case, some buildings features could not be implemented because seismic upgrades requirements.

Users should be involved early in the planning process to achieve higher user acceptance and satisfaction.

D.9. High School Passive House, Nordrhein-Westfalen.GE

D.9.1. Name of the Project, Location (City, Country)

High-School in Nordrhein-Westfalen, Germany.

D.9.2. Pictures

These pictures show the building in its original and post-retrofit states, illustrating key features of the retrofit (Fig. D.17).

Fig. D.17 High-School in Nordrhein-Westfalen, Germany. *Source* KEA Klimaschutz- und Energieagentur Baden-Württemberg, 2014

D.9.3. Project Summary

D.9.3.1. Project Objectives

- Refurbishment of the High-School, to a certified passive house.
- Energetic and functional optimization of the school with a high quality in design.

D.9.3.2. Project Energy Goals

- Primary Energy demand max. 120 kWh/m^2 heated net floor area.
- Net energy demand for heating15 kWh/m^2 heated net floor area.

D.9.3.3. Short Project Description

Energy retrofit of the building envelope and refurbishment of the technical facilities.

D.9.3.4. Stage of Construction

End of the construction of tracts of the school. First year of the operating phase of all tracts.

D.9.3.5. Point of Contact (POC) Information

KEA Klimaschutz- und Energieagentur Baden-Württemberg GmbH.
martina.riel@kea-bw.de

D.9.3.6. Date of the Report

May 2014.

D.9.4. Site

- Climate zone: ASHRAE 90.1-2007: 5/warm temperate, $GT_{20/15:}$ 3415 Kd.
- Cooling Degree Days (based on 65 F), Heating Degree Days (based on 65 F).

D.9.5. Building Description/Typology

D.9.5.1. Type (Office, Barracks, Etc.)

School.

D.9.5.2. Typology/Age

1970.

D.9.5.3. General Information

- Year of construction: From 1968.
- Year of renovation (as described here): 2013.
- Total floor area. For tract 2:Heated net floor area after refurbishment: 1016 m^2.
- Area of unconditioned space included above (m^2): 0 m^2.
- Other information as appropriate: A/V-ratio before and after: 0.43.

D.9.6. Architectural and Other Relevant Drawings

See Figs. D.18 and D.19.

D.9.7. National Energy Use Benchmarks and Goals for Building Type Described in the Case Study

D.9.7.1. Benchmark, According to the Standard National Average, Min, and Max

New building passive house standard.

D.9.8. Awards or Recognition

One of three winner of the German Competition "Kommunaler Klimaschutz 2010."

D.9.9. Site Energy Cost Information (Electricity, \$/kWh; Gas, \$/m^3, Etc.)

None.

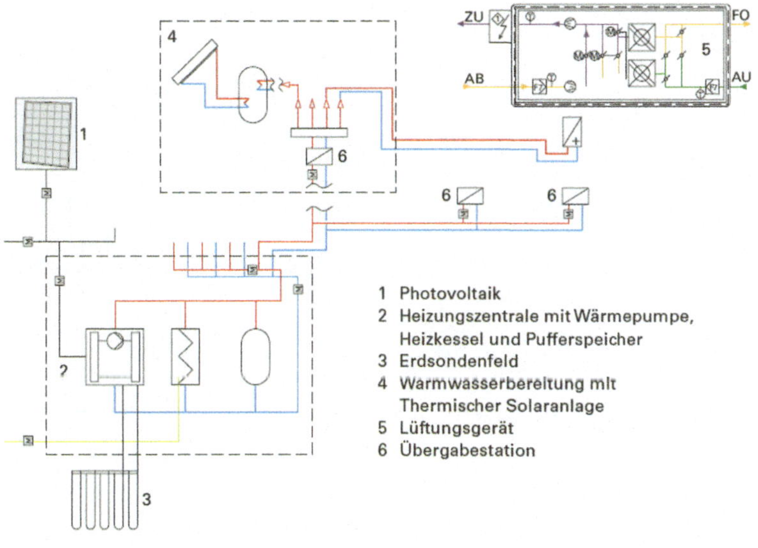

Fig. D.18 Energy concept. *Source* DBZ ES 11_2011

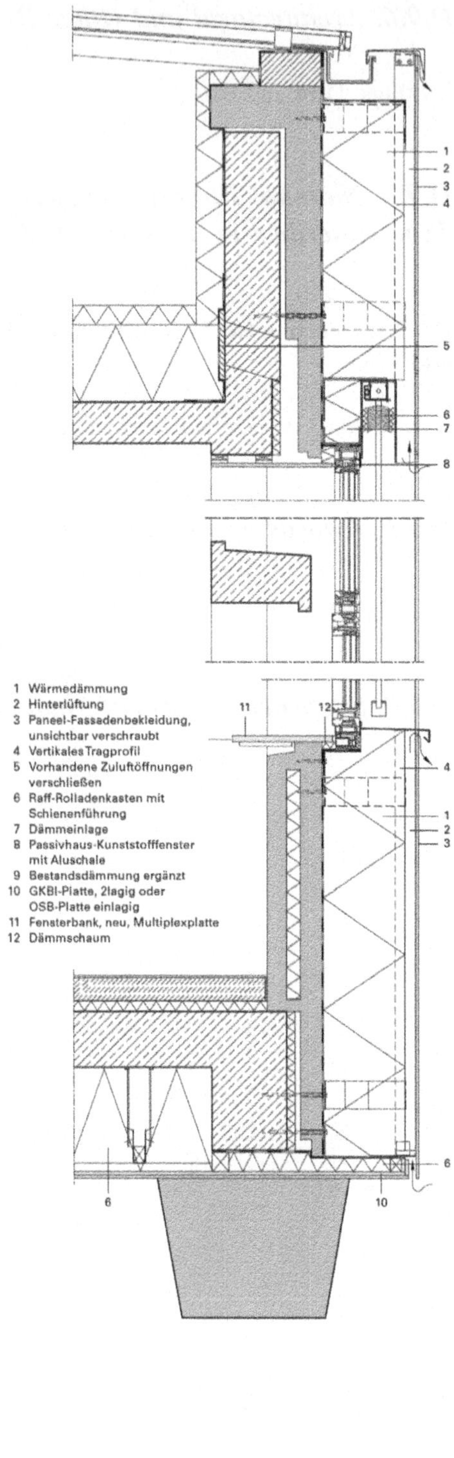

Fig. D.19 Detail of the facade. *Source* DBZ ES 11_2011

D.9.10. Pre-renovation Building Details

D.9.10.1. Envelope Details: Walls, Roof, Windows, Insulation Levels

- (Area, only for track 1).
- Solid construction, strip-structured facade, U-value 1.4 W/(m^2K).
- Area: 769 m^2.
- Windows U-value 3.4 W/(m^2K), area: 312 m^2.
- Roof: U-value 0.65 W/(m^2K), area: 572 m^2.
- Ground plate to earth: U-value 1.0 W/(m^2K), area: 164 m^2.
- Floor to outdoor air: U-value 1.0 W/(m^2K), area: 406 m^2.

D.9.10.2. Heating, Ventilation, Cooling, and Lighting Systems

- Heating: 127 kW, gas boiler.
- Ventilation: Exhaust air system in classrooms, rest rooms.

D.9.11. Description of the Problem: Reason for Renovation

High energy consumption, renovation of the building was necessary anyway

D.9.12. Renovation SOW (Non-energy and Energy Related Reasons)

Energetic and functional optimization of the school with a high quality in design, ensure accessibility by integration of elevators.

D.9.13. Energy Saving/Process Improvement Concept and Technologies Used

D.9.13.1. Building Envelope Improvement

- Solid construction, strip-structured facade, U-value 0.13 W/(m^2K), 30 cm insulation.
- Windows U-value 0.77 W/(m^2K).

- Roof: U-value 0.07 W/(m^2K), 50 cm insulation.
- Ground plate: U-value 0.14 W/(m^2K), 18 cm insulation.
- Floor to outdoor air: U-value 0.9 W/(m^2K), 36 cm insulation.

D.9.13.2. New HVAC System or Retrofits to Existing

- Heating: Heat pump with geothermal earth probes, 30.15 kW and a condensing gas boiler (for the whole school complex).
- Ventilation: Air system with heat recovery 75% in classrooms and rest rooms.
- Natural cooling with automatically opening windows in the nights for and geothermal cooled supply air in summer.

D.9.13.3. New Lighting System

Performance of the system for each Light: 9.8 W/m^2 with daylight controller and motion detector.

D.9.13.4. Renewable Energy

PV on the roof.

D.9.13.5. Daylighting Strategies

Sun protection outside with a sun protection control.

D.9.14. Energy Consumption

D.9.14.1. Pre-renovation Energy Use (Total and per m^2/year)

220 kWh/m^2a for heating.

D.9.14.2. Predicted Energy Savings (Site, Source, GHG), Total and per m^2/year

More than 90%, heating demand after renovation: 20.3 kWh/m^2a.

D.9.15. Energy Cost Reduction

None available.

D.9.16. Non-energy Related Benefits Realized by the Project, E.g., Improved Productivity, Increased Rent/Lease, Increased Useful Space, Reduced Maintenance, Etc.

None available.

D.9.17. Renovation Costs: Total and per m^2

None available.

D.9.18. Business Models and Funding Sources

D.9.18.1. Description of the Funding Sources Chosen

Subsidy of the DBU (Deutsche Bundesstiftung Umwelt, German Environment Foundation).

D.9.18.2. Description of the Business Model Chosen (Option)

Conventional by the city council with architect.

D.9.18.3. Funding Sources of the Business Model

Conventional.

D.9.18.4. Construction Phase in the Business Model

- By Architect office, engineering office.
- In each construction phase the craftsmen had to attend special trainings to learn the important details of passive house construction.

D.9.18.5. Energy Management and Controlling in the Business Model

Monitoring is controlled by the architect office and the city energy manager.

D.9.19. Cost Effectiveness of Energy Part of the Project (NPV, SIR, Etc.)

None available.

D.9.20. User Evaluation

D.9.20.1. Integration of Users Demands in the Planning Process

Yes.

D.9.21. Experiences/Lessons Learned

D.9.21.1. Energy Use

None available.

D.9.21.2. Impact on Indoor Air Quality

The measured CO_2 concentration is rarely higher than 800 ppm.

D.9.21.3 Space Utilization Changes

In T1: Administration (change of ground plan), in T3 + 4: new refectory integrated.

D.9.22. References

DBZ, Energie spezial 11 2011.

Appendix E: Case Study—Ireland

E.1. EnerPHit

E.1.1. Name of the Project, Location (City, Country)

EnerPHit Social Housing in Dun Laoghaire Rathdown, Ireland.

E.1.2. Pictures

Figures E.1, E.2, E.3 and E.4 show the building in its original and post-retrofit states, illustrating key features of the retrofit.

E.1.3. Project Summary

See Figs. E.3 and E.4.

E.1.3.1 Project Objectives

To complete a deep retrofit of an assisted living facility for elderly people in south Dublin.

R. Lohse et al. (eds.), *Deep Energy Retrofit—Case Studies*,
https://doi.org/10.1007/978-3-031-17517-6

Fig. E.1 Building before retrofit

Fig. E.2 Key figure during renovation process

E.1.3.2. Project Energy Goals

To reduce the heating energy demand to that of the EnerPHit standard, namely 25 kWh/m^2 year.

Fig. E.3 Key figure during renovation process

Fig. E.4 Building after retrofit

E.1.3.3. Short Project Description

The proposed development comprises the addition of another whole floor on top of the existing building as well as minor extensions to accommodate vertical circulation. Due to increased accommodation size the resulting number of apartments will remain at 34. The retrofitting of the existing structure involves external wall insulation and Passive house certified windows and doors, as well as the new top level comprising walls of aerated concrete block structure with external insulation and lightweight metal deck roof with the same windows and doors. A MVHR unit will be installed in each apartment as well as for each community areas and circulation areas.

E.1.3.4. Stage of Construction

As of September 2015, all major construction work is complete and the first blower door tests are being carried out.

E.1.3.5. POC

Mariana Moreira, Conservation Architect.
MosArt Architects.
Wicklow, Ireland.
E-mail mariana@mosart.ie.
Telephone +353 (0)404 25777.

E.1.3.6. Date of the Report

September 3rd 2015.

E.1.3.7. Acknowledgement

- Dun Laoghaire Rathdown County Council, owners of the project.
- This data is being provided as part of the www.EuroPHit.eu research project co-funded by the Intelligent Energy Europe Programme of the European Union.

E.1.4. Site

- Location: Sallynoggin Road Upper, Dun Laoghaire, County Dublin, Ireland.
- Grid coordinates:53.274304, − 6.145218
- Elevation: 25 m.
- Climate zone: Temperate.
- Heating degree hours (metric) = 67 kKh/year (Table E.1).

Table E.1 Design temperature

*Cooling design temperature—0.4% occurrence**	
Dry bulb temp C (F)	Mean coincident wet bulb temp C (F)
25 °C	
*Heating design temperature—99.6% occurrence***	
Dry Bulb Temp C (F)	
20 °C	

E.1.5. Building Description/Typology

E.1.5.1. Typology/Age

Two stories block of apartments built in the 1970s.

E.1.5.2. Type (Office, Barracks, Etc.)

Small apartments for the elderly.

E.1.5.3. Typology/Age

Early 1970s.

E.1.5.4. General Information

- Year of construction: early 1970s.
- Year of previous major retrofit—if known. None.
- Year of renovation (as described here): 2014–2015.
- Total floor area (m^2): 1613 m^2.
- Area of unconditioned space included above (m^2): zero.
- Other information as appropriate: one additional floor (new build) being added to the existing 1970s two story building.

E.1.6. Architectural and Other Relevant Drawings

Not available.

E.1.7. National Energy Use Benchmarks and Goals for Building Type Described in the Case Study

E.1.7.1. Benchmark, According to the Standard National Average, Min, and Max

Current modeled heating energy demand for the project is 354 kWh/m^2 year according to the Passive House Planning Package (PHPP). Average oil bill per month in 2012 was €1350.

E.1.7.2. National Energy Target for This Type of Building (If Any)

Such buildings in Ireland have no energy targets.

E.1.8. Site Energy Cost Information (Electricity, \$/kWh; Gas, \$/m^3, Etc.)

E.1.8.1. Electricity

Data not available.

E.1.8.2. Natural Gas

Data not available.

E.1.3.3. Distillate Oil (Kerosene, Heating Oil, Etc.)

Average oil bill per month in 2012 was €1350.

E.1.9. Pre-renovation Building Details

E.1.9.1. Envelope Details: Walls, Roof, Windows, Insulation Levels

Existing U-values: Floor slab = 3.85 W/m^{2}K, External walls = 3.75 W/m^{2}K, Windows = 1.71 W/m^{2}K, Roof 4.36 W/m^{2}K (Figs. E.5, E.6, E.7 and E.8).

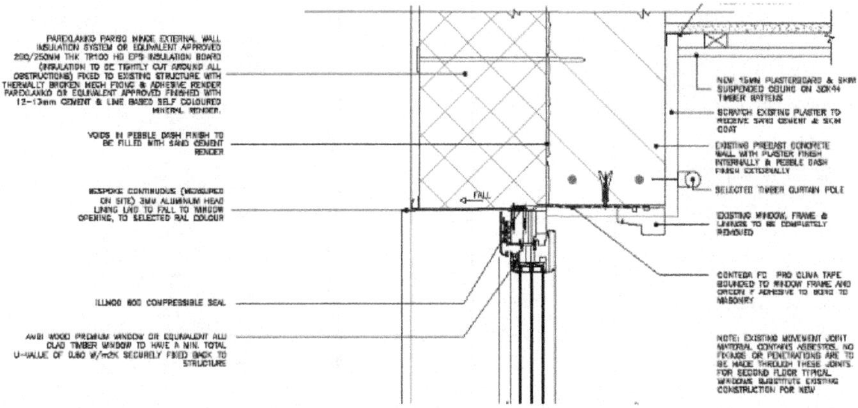

Fig. E.5 Head window detail on the existing building (not to scale) [DRL Co.Co.2014]

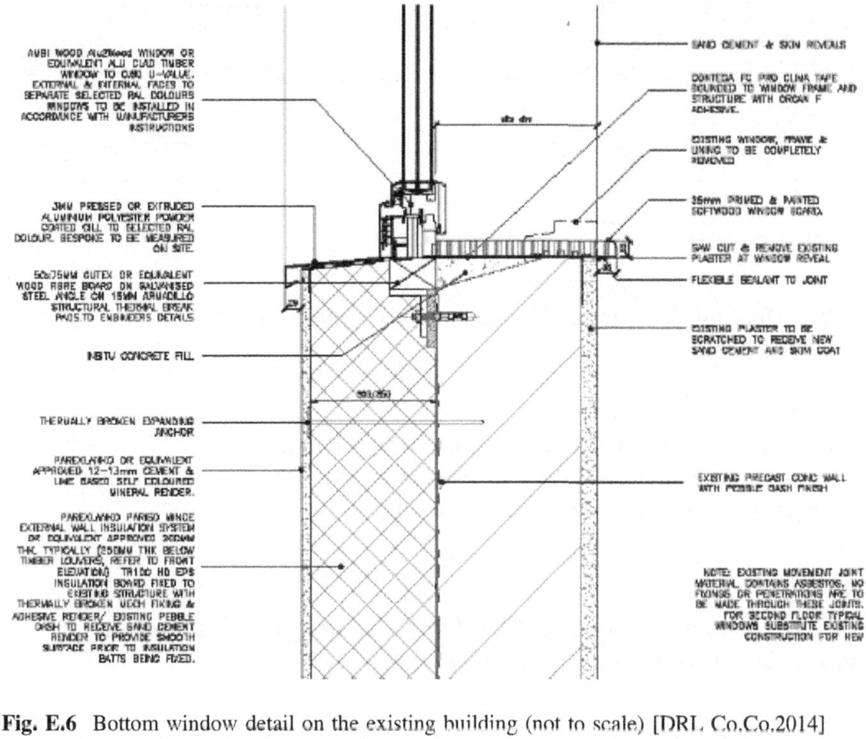

Fig. E.6 Bottom window detail on the existing building (not to scale) [DRL Co.Co.2014]

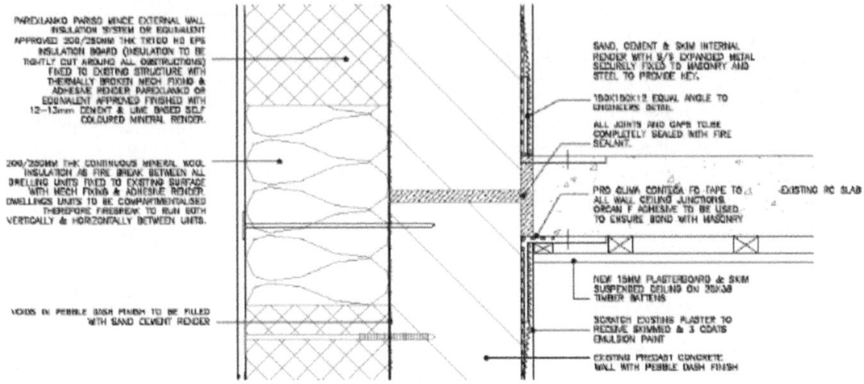

Fig. E.7 Wall floor slab junction (not to scale) [DRL Co.Co.2014]

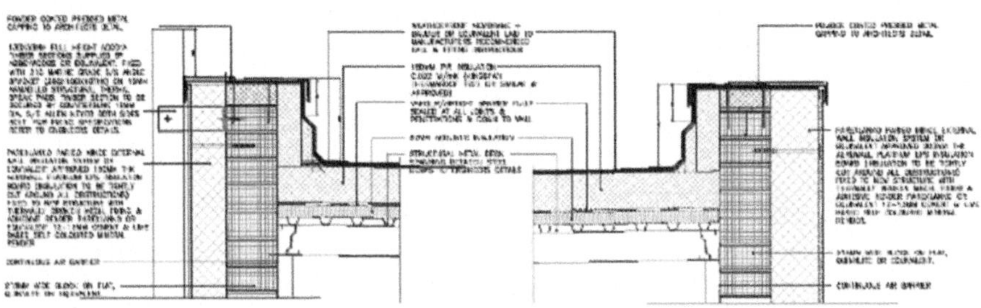

Fig. E.8 Parapet details (not to scale) [DLR Co.Co.2014]

E.1.9.2. Heating, Ventilation, Cooling, and Lighting Systems

- Cooling is not required in this location.
- There is no mechanical ventilation in existing building.
- Heating and hot water are provided using an old oil boiler and additional input for DHW using direct resistance electricity.

E.1.10. Description of the Problem: Reason for Renovation (Non-energy and Energy Related Reasons)

Extremely poor condition in terms of comfort and energy demands; also the apartments were too small by modern standards.

E.1.11. Renovation SOW (Non-energy and Energy Related Reasons)

Not available data.

E.1.12. Energy Saving/Process Improvement Concept and Technologies Used

E.1.12.1. Building Envelope Improvement

- Floor slab existing = 3.85 W/m^2K, proposed = 0.18 W/m^2K.
- External walls existing = 3.75 W/m^2K, proposed = 0.13 W/m^2K.
- Windows existing = 1.71 W/m^2K, proposed = 1.00 W/m^2K.
- Roof existing = 4.36 W/m^2K, proposed = 0.13 W/m^2K.

E.1.12.2. New HVAC System or Retrofits to Existing

- Mechanical ventilation with heat recovery will be provided to all units.
- ConfoAir 160 HRV—Zehnder to be installed in each of the 34 apartments and five more units to be installed in the circulation and communal areas. Heat recovery efficiency will be 89%.

E.1.12.3. New Generation/Distribution System

Micro CHP with gas condenser boiler.

E.1.12.4. Renewable Energy

It is planned in the longer term to install the following renewable energy systems:

- In 2020, to install 136 m^2 solar thermal collector with a 2000 L stratified solar storage tank.
- In 2035, to install solar PV on the roof to generate electricity.

E.1.13. Energy Consumption

E.1.13.1. Pre-renovation Energy Use (Total and per m^2/year)

- Primary energy of entire building (including plug loads) prior to renovation = 688 kWh/m^2 year (using a primary energy factor of 2.6), post renovation = 109 kWh/m^2 year.
- Space heating demand prior to renovation = 354 kWh/m^2 year, post renovation = 24 kWh/m^2 year.

E.1.13.2. Predicted Energy Savings (Site, Source, GHG), Total and per m^2/year

84% reduction in Primary Energy demand and 94% reduction in space heating demand.

E.1.13.3. Measured Energy Savings (Thermal, Electrical), Total and per m^2/year

Data not available—project is still in construction.

E.1.13.4. Annual Energy Use Reduction

See answers under Pre-renovation energy use (total and per m^2/year) and under Predicted energy savings (site, source, GHG), total and per m^2/year.

E.1.14. Energy Cost Reduction

E.1.14.1. Split in All Energy Forms—Electricity, Oil, District Heating

- Current heating costs approximately €1350 per month = €16,200 per year. Expected cost saving = 90% = €14,580 per year.
- Costs for electricity unknown.

E.1.15. Non-energy Related Benefits Realized by the Project, E.g., Improved Productivity, Increased Rent/Lease, Increased Useful Space, Reduced Maintenance, Etc.

Anticipated improvement of overall quality of life for inhabitants due to extended apartment sizes, greater comfort and enhanced indoor air quality.

E.1.16. Renovation Costs: Total and per m²

Note that all costs below are in Euro (€) and exclude Value Added Tax (VAT).

E.1.16.1. Total

€1986 million/€1260 per m^2.

E.1.16.2. Non-energy Related

€1669 million/€1065 per m^2.

E.1.16.3. Energy Related

€0.318 million/€203 per m^2.

E.1.16.4. Cost for Each Measure

See Table E.2.

Table E.2 An itemized breakdown for costs is provided here

Element details	Total project costs (ex. VAT) €	New build costs (ex. VAT)	Refurbishment costs (ex. VAT)	E.O costs EuroPHit
(06) Preliminaries	294,852.00	121,712.47	173,139.53	28,756.20
(07) Insurances				
(10) Demolitions and alterations	136,943.28	17,812.00	119,131.28	

(continued)

Table E.2 (continued)

Element details	Total project costs (ex. VAT) €	New build costs (ex. VAT)	Refurbishment costs (ex. VAT)	E.O costs EuroPHit
(19) Substructures	21,722.09	15,953.79	5,768.30	2674.00
(21) External walls and finishes	422,900.92	235,177.99	187,722.93	39,000.00
(22) Internal walls	81,137.27	40,051.76	41,085.51	
(23) Floor structure	54,275.18	54,275.18	0.00	5100.00
(24) Stairs, balustrades and handrails	39,555.00	25,044.00	14,511.00	
(27) Roof structure	22,776.96	22,776.96	0.00	
(28) Frame	136,327.89	116,543.71	19,784.18	
(31) External wall completions	166,022.75	55,258.13	110,764.62	57,500.00
(32) Internal wall completions	164,117.39	54,623.96	109,493.43	
(37) Roof completions	8430.00	8430.00	0.00	
(42) Wall finishes internally	135,214.01	45,003.91	90,210.10	9700.00
(43) Floor finishes	140,448.53	46,746.14	93,702.39	7500.00
(45) Ceiling finishes	60,558.14	20,155.85	40,402.29	5500.00
(47) Roof finishes	93,904.50	31,254.67	62,649.83	13,000.00
(74) Sanitary fittings	63,238.66	21,048.02	42,190.64	
(79) Building fittings and furniture	125,596.00	41,802.70	83,793.30	
(52) Drainage and refuse disposal	2770.00	921.95	1848.05	
(10) Site preparation	13,418.50	4466.14	8952.36	
(20) Site structures	27,949.32	9302.50	18,646.82	
(30) Site enclosures	800.00	266.27	533.73	
(40) Roads paths pavings	90,394.40	30,086.39	60,308.01	
(60) Site services—(piped and ducted)	45,311.73	15,081.32	30,230.41	
(60) Site services—(mainly electrical)	14,353.02	4777.18	9575.84	
(70) Site fittings	3000.00	998.50	2001.50	
(80) Landscaping	36,993.05	12,312.57	24,680.48	
(59) Mechanical installation******	677,700.00	225,562.06	452,137.94	124,283.00
(66) Lift installation	21,379.00	7115.67	14,263.33	
(69) Electrical installation	253,065.00	84,228.81	168,836.19	24,750.00

(continued)

Table E.2 (continued)

Element details	Total project costs (ex. VAT) €	New build costs (ex. VAT)	Refurbishment costs (ex. VAT)	E.O costs EuroPHit
Total construction cost excl. VAT	3,355,154.59	1,368,790.63	1,986,363.96	317,763.20
Floor areas m^2	2347	781.16	1566	1,565.84
Construction cost per m^2 (excl. VAT)	1429.55	1752.25	1268.56	202.94
% uplift on refurbishment works				16.00%
% uplift on total project				9.47%

Source 'E.O' costs EuroPHit Report. Walsh associates

E.1.17. Business Models and Funding Sources

This is a publically funded project supported by local and central government.

E.1.18. Cost Effectiveness of Energy Part of the Project (NPV, SIR, Etc.)

Unknown at this time.

E.1.19. User Evaluation

This project is underway and therefore there have been no user training programs developed as yet.

E.1.20. Experiences/Lessons Learned

This information is not available as the project is under construction. Expected completion date is end of 2015.

E.1.21. References

For project description see: http://www.europhit.eu/cs01-rochestown-home-elderly

Appendix F: Case Study—Latvia

F.1.1. Name of the Project, Location (City, Country)

Energy efficiency improvement of multifamily building on Mastu street 8k-1, Riga. Location: Riga, Latvia.

F.1.2. Pictures

Figure F.1 shows the building in its original and post-retrofit states, illustrating key features of the retrofit

F.1.3. Project Summary

F.1.3.1. Project Objectives

The modernization and deep retrofit of poor quality Soviet-era multi-family building through energy conservation based on Energy Performance Contracting (EPC).

The purpose of the renovation is to conserve the building and protect it from weatherization thereby securing the existence and usage of people's home for another one or two generations.

F.1.3.2. Project Energy Goals

The ex-post energy consumption of this building has been significantly reduced by RENESCO.

Calculated energy consumption from energy audit (normalized year—203 heating days, average outdoor temperature 0 °C): total heat consumption 88

R. Lohse et al. (eds.), *Deep Energy Retrofit—Case Studies*,
https://doi.org/10.1007/978-3-031-17517-6

Fig. F.1 Key figures of the energy retrofit project

Fig. F.1 (continued)

Fig. F.1 (continued)

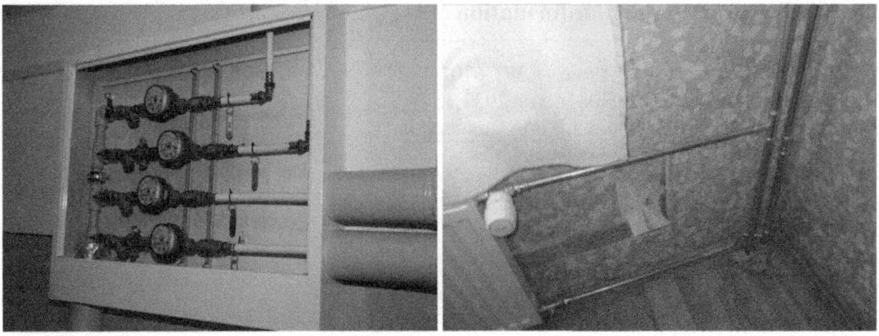

Fig. F.1 (continued)

kWh/m^2 year: for domestic hot water 31 kWh/m^2 year, for space heating 56.7 kWh/m^2 year.

Complete data and detailed analyses will conduct on Spring in 2015.

F.1.3.4. Short Project Description

This project was undertaken to quite literally, "Save your house by saving energy!" by financing and realizing the modernization of mass housing stock through energy conservation.

F.1.3.5. Stage of Construction

Project had been finished on July, 2014.

F.1.3.6. Point of Contact Information

RENESCO Ltd 31 Baznicas Street (4th floor) Riga, LV-1010 Latvia	Phone: + 371 6702 7427 Fax: + 371 2937 1545 e-mail: info@renesco.lv www.renesco.lv
Dzintars Jaunzems Project Manager	Phone: + 371 67,506,010 Mobile: + 371 26,108,494 e-mail: dzintars@renesco.lv
Claudio Rochas Member of the board	Phone: + 371 6702 7427 e-mail: claudio@ekodoma.lv

F.1.3.7. Date of the Report

January, 2015.

F.1.3.8. Acknowledgement

Project partly financed from the European Regional Development Fund.

F.1.4. Site

- Riga, Latvia, Europe
- 56° 58′ 13.6`` N 24° 06′ 01.0' E
- 56.970431, 24.100276
- https://www.google.lv/maps/place/56%C2%B058%2713.6%22N+24%C2%B006%2701.0%22E/@56.970431,24.100276,19z/data=!3m1!4b1!4m2!3m1!1s0x0:0x0

F.1.5. Climate Zone: Temperate/Cold Climate Zone

F.1.6. Normalized Space Heating Season

- 203 days, 0 °C.
- 3721 Heating degree days (based on 18.33 °C [65 F])

F.1.7. Building Description/Typology

F.1.7.1. Typology/Age

Specific/individual project, 21 years.

F.1.7.2. Type (Office, Barracks, Etc.)

Multifamily building (39 flats, heated area 1536.5 m^2).

F.1.7.3. Typology/Age

Multifamily building.

F.1.7.4. General Information

- Year of construction: 1993.
- Year of renovation (as described here): 2014.
- Total floor area (m^2): 2756.4.
- Area of unconditioned space included above (m^2): 1220.1 (staircase, basement, and attic).

F.1.8. Architectural and Other Relevant Drawings

See Figs. F.2, F.3 and F.4.

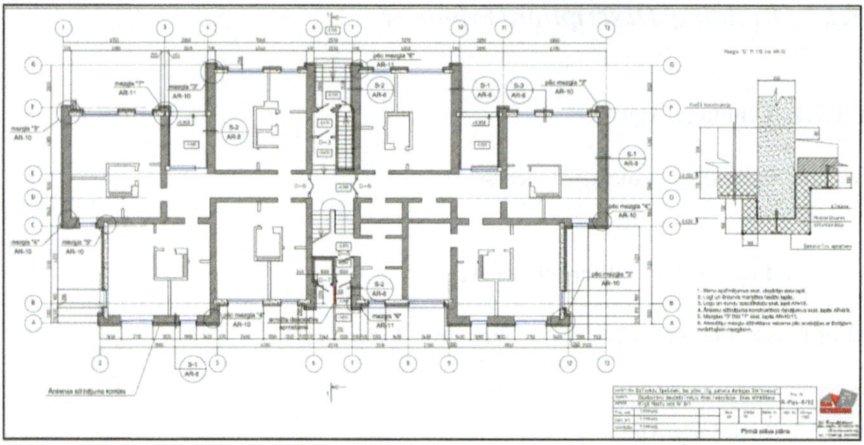

Fig. F.2 Scheme of 1st floor

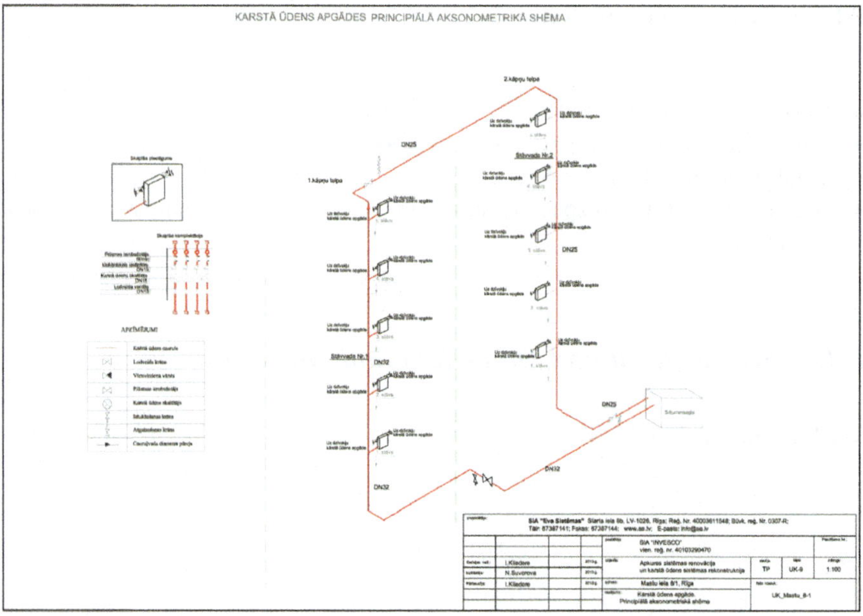

Fig. F.3 Axonometric scheme of new domestic hot water system

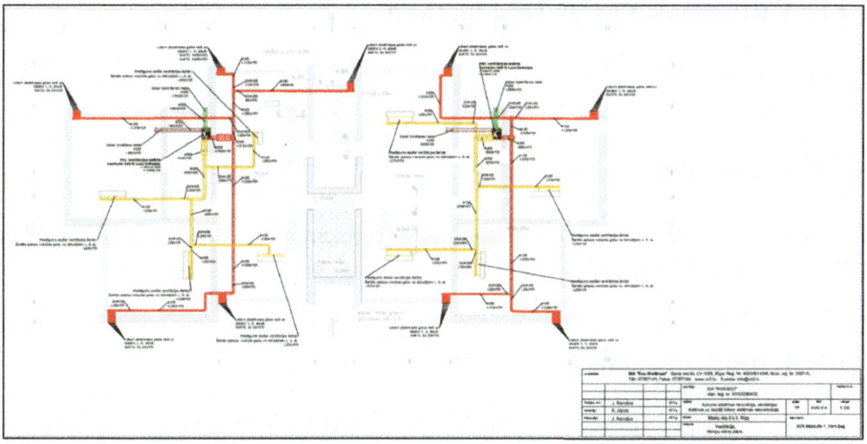

Fig. F.4 Channels of mechanical ventilation system with heat recovery in the attic

F.1.9. National Energy Use Benchmarks and Goals for Building Type Described in the Case Study

F.1.9.1. Benchmark, According to the Standard National Average, Min, and Max

- Class D—heat energy consumption for space heating less than 60 kWh/m^2 a.
- European Regional Development Fund: at least 25% heat savings comparing to pre-renovation heat consumption level.

F.1.10. Awards or Recognition

None.

F.1.11. Site Energy Cost Information (Electricity, \$/kWh; Gas, \$/m^3, Etc.)

F.1.11.1. Electricity

Electricity cost varied based on several factors (individual or legal person, type of tariff (e.g., one zone or three zone) etc.): 0.14–0.2 \$/kWh, without taxes.

F.1.11.2. Renewable Energy Feed-In Tariff

Under revision.

F.1.11.3. Natural Gas

Depending on users type and Natural gas consumption: 363.54 $/1000m^3$ for consumers > 25,000 m^3/a.

F.1.11.4. District Heating

The cost of District Heating was 50.75–96.88 $/MWh without taxes, in this particular case—71.38 $/MWh (57.40 EUR/MWh) without taxes (on December, 2014).

F.1.12. Pre-renovation Building Details

F.1.12.1. Envelope Details: Walls, Roof, Windows, Insulation Levels

Overall condition of building envelope was unsatisfactory (crumbling bricks, cracks etc.). Roof was with poor hydro-isolation and spoilt. Almost half of old wooden windows were changed with new two glass plastic windows. There was no insulation in the building envelope.

F.1.12.2. Heating, Ventilation, Cooling, and Lighting Systems

There was one-pipe space heating system in bad technical and operational condition (e.g., existing insulation were damaged and with low thermal resistance, lack of any balancing possibilities etc.). Larger radiators were installed In some apartments, which resulted in unequal heat delivery. Domestic hot water system was inefficient and old-fashioned with very high circulation loses and relatively high electricity consumption for circulation pump.

There was natural, uncontrolled ventilation system in the building.

F.1.13. Description of the Problem: Reason for Renovation (Non-energy and Energy Related Reasons)

Bad aesthetic look of building elements, unsatisfied technical condition of building envelope. High energy consumption, un-evenly distributed heat through building (under heated and overheated apartments, etc.) and unacceptable ventilation system. Worn-out and old domestic hot water system with high circulation loses.

F.1.14. Renovation SOW (Non/Energy and Energy Related Reasons)

None.

F.1.15. Energy Saving/Process Improvement Concept and Technologies Used

F.1.15.1. Building Envelope Improvement

- Strengthening of major walls and foundations by replacing the crumbled bricks, fixing cracks for further damage, and doing other repairs.
- Insulation of outer walls (100 mm rock wool).
- Insulation of basement ceiling (100 mm rock wool/polystyrene).
- Insulation of attic floor (300 mm bulk wool).
- Hydro-isolation of basement walls and insulation of socle (50 mm extruded polystyrene).
- Renovation of roof (hydro-isolation etc.).
- Replacement of old wooden-frame windows in apartments and in staircase.
- Renovation of main entrance and backyard doors.
- Repairing of staircases.

F.1.15.2. New HVAC System or Retrofits to Existing

- New and efficient (30 mm insulation) domestic hot water system with reduced circulation losses.
- Two pipe space heating system with differential pressure controller pair in the bottom of each riser.

- Space heating system designed for low temperature regime (supply temperature ~ 50 °C at − 25 °C outdoor temperature).
- New radiators with thermostatic radiator valves (factory set max. temperature: 21.5 °C).
- Mechanical ventilation system with heat recovery. Two air handling machines Zehnder ComfoAir 550 R Luxe Enthalpie (Certificated Passive house component, Certificate No. 0329vs03, www.passivehouse.com).

F.1.15.3. New Lighting System

Lighting system in staircase with motion sensor.

F.1.16. Energy Consumption

F.1.16.1. Pre-renovation Energy Use (Total and per m^2/year)

- Total: 351.28 MWh/year.
- Specific: 190.2 kWh/m^2/year.

F.1.16.2. Predicted Energy Savings (Site, Source, GHG), Total and per m^2/year

- Total savings: 351.28 – 162.53 MWh = 188.72 MWh/year.
- Total specific savings: 190.2 – 88 = 102 kWh/m^2/year.
- GHG savings: 351.25 – 162.53 = 188.72 MWh/year or 49.82 tCO_2/year*.
- Used emission factor: 0.264 tCO_2/MWh.

F.1.16.3. Measured Energy Savings (Thermal, Electrical), Total and per m^2/year

N/A (deep retrofit finished in July, 2014).

F.1.16.4. Annual Energy Use Reduction

Expected ~ 54% (from energy audit).

F.1.17. Energy Cost Reduction

No data available.

F.1.18. Non-energy Related Benefits Realized by the Project, E.g., Improved Productivity, Increased Rent/Lease, Increased Useful Space, Reduced Maintenance, Etc.

Normal building maintenance costs (e.g., $\$/m^2$ per month) are usually reduced by 5%, and unexpected repair costs are reduced as well.

F.1.19. Renovation Costs: Total and per m^2

F.1.19.1. Total

331,095.88 $ (€265,702.51).

F.1.19.2. Non-energy Related

€34,005.16 (roof, Staircases and other cosmetic work, Energy audit, Technical appraisal, Construction Design, Engineering networks Design, Supervision).

F.1.19.3. Energy Related

€231,697.86.

F.1.19.4. Cost for Each Measure

Construction cost:	€199,631.04
Roof:	€11,736.83
Insulation Facade:	€112,442.77
Insulation Attic:	€9,252.30
Insulation Basement:	€23,107.32
Windows:	€19,173.30
Doors:	€9,748.88
Staircases and other cosmetic work:	€8,950.25
Work on behave of ventilation:	€5,219.39
Networks cost:	€52,753.40
Space heating system (radiators, regulators, accessories):	€22,845.69
DHW system:	€19,281.55
Mechanical Ventilation system with heat recovery:	€10,626.16
Supervision and documentation:	€15,155.01

F.1.20. Business Models and Funding Sources

F.1.20.1. Decision Making Process Criteria for Funding and Business Models

The decision on renovation and the measures to be taken was made by the company based on the existing technical condition of the building, the energy usage track record, and the interest and awareness of apartment owners. Specific criteria for decision making were:

- Size and technical condition of building.
- Baseline heat energy consumption of building.
- Heat energy price.
- Subsidies or grants.
- Loan availability.

F.1.20.2. Description of the Funding Sources Chosen

EU grants (European Reconstruction and Development fund); Loans: Citadele Bank—commercial loan; Dutch International Guaranties for Housing (DIGH)—subordinated loan.

F.1.20.3. Description of the Business Model Chosen (Option)

RENESCO has developed and implemented a business model for housing renovation in post-soviet countries, which has enabled a commercial ESCO to attract bank financing to renovate existing multi-family buildings without any financial input, collateral, or guarantees from the existing flat-owners.

RENESCO's projects are financed solely on the basis of the cash-flow generated by saved energy during a 20 years contract.

F.1.20.4. Risk Allocation in the Business Model

All investments and risks take place on the balance sheet of RENESCO. Putting ALL implementation risks in one professional hand is in our opinion a game-changer.

F.1.20.5. Planning Process in the Business Model

The business model was created based on a combination of world expertise in the field, and project management and consulting experience in post-soviet countries (Ukraine, Latvia, Lithuania, Czech Republic, Slovenia). This was adapted to the local situation 2008-2009.

F.1.20.6. Funding Sources of the Business Model

Business model development was financed by shareholders private capital, as well as by financial assistance of Dutch International Guaranties for Housing (DIGH) organization.

F.1.20.7. Construction Phase in the Business Model

Implementation of business model was started 2008–2009 and was done in two main steps:

- Development of principle of Energy Performance (or ESCO) Contract, including legal opinion.
- Initiation of first pilot projects in Valmiera and Cēsis (rural towns of Latvia). Pilot projects covered renovation of 36 apartment building (Valmiera, Gaujas 13) and installation of local boiler-house for 70 apartment building (Cēsis, Kovārņu 31).

F.1.20.8. Operation Phase

Comprehensive renovation of: (1) five multifamily buildings in Cēsis 2010–2012; (2) eight multifamily buildings in Cēsis, Sigulda and Riga 2013–2014.

Implementation of additional measures, such as ventilation air heat recovery, use of renewable energy sources (e.g., ground source heat pumps, etc.) for heat supply.

F.1.20.9. Energy Management and Controlling in the Business Model

Temperature data and energy consumption in a building are remotely monitored. The projects cash-flows were analysed.

F.1.21. Cost Effectiveness of Energy Part of the Project (NPV, SIR, Etc.)

Cost effectiveness is considered to include that description of framework data such as calculated interest rates, life cycle period, etc. Figure F.5 shows the results of an investment and financial simulation.

F.1.22. User Evaluation

F.1.22.1. Description of User Training Programmes Within the Refurbishment

After the deep retrofit project ends, residents receive written instructions regarding new domestic hot water system, renovated space heating system and new mechanical ventilation, and all necessary consultations for optimal use of mentioned building service and engineering systems.

F.1.22.2. Integration of Users Demands in the Planning Process

Comprehensive retrofit of the building covers all user demands for high quality living conditions.

F.1.23. Experiences/Lessons Learned

F.1.23.1. Energy Use

Energy use was reduced by at least 50%, and the indoor climate (e.g., temperature) was greatly improved.

Building: Mastu 8/1, Riga

Investments	EURO
Audits and design	4 351
Supervision	2 352
Investment	214 054
Sun Energy Development fee	22 076
Investment, Grand Total	**242 833**
Subsidy	**108 138**
Investment - (minus) subsidy	**134 695**
investment per saved MWh/a	722

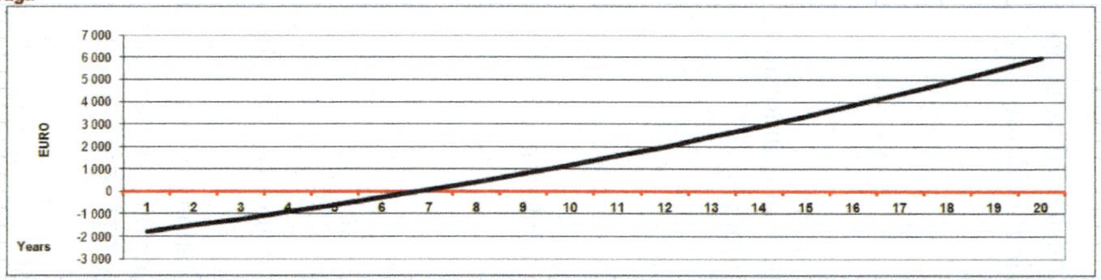

Year:	Total 20	2011	2012	2013	2014	2015	2016	2017	2018	2019	2020
		1	2	3	4	5	6	7	8	9	10
Heat sale											
Heat sale	**667 143**	22 404	23 300	24 232	25 201	26 209	27 258	28 348	29 482	30 661	31 888
Heat Energy Purchase	**311 645**	10 466	10 884	11 320	11 772	12 243	12 733	13 242	13 772	14 323	14 896
Heating Gross result	**355 498**	**11 938**	**12 416**	**12 912**	**13 429**	**13 966**	**14 525**	**15 106**	**15 710**	**16 338**	**16 992**
Annual maintenance/admin costs	**76 721**	2 576	2 679	2 787	2 898	3 014	3 135	3 260	3 390	3 526	3 667
Operational Cashflow	**278 777**	**9 362**	**9 736**	**10 126**	**10 531**	**10 952**	**11 390**	**11 846**	**12 320**	**12 812**	**13 325**
Loan payments											
Annual loan repayment	**234 866**	11 743	11 743	11 743	11 743	11 743	11 743	11 743	11 743	11 743	11 743
Interest payment	***100 171***	*8 082*	*7 862*	*7 629*	*7 382*	*7 121*	*6 843*	*6 549*	*6 238*	*5 907*	*5 557*
Payment on principal	***134 695***	*3 662*	*3 881*	*4 114*	*4 361*	*4 623*	*4 900*	*5 194*	*5 506*	*5 836*	*6 186*
Balance	***134 695***	*131 033*	*127 152*	*123 038*	*118 676*	*114 054*	*109 154*	*103 960*	*98 454*	*92 618*	*86 432*
Net results											
pre tax net return	**32 933**	**-1 786**	**-1 505**	**-1 213**	**-909**	**-593**	**-266**	**77**	**432**	**802**	**1 186**
Net result adjusted for inflation	18 310	-1 733	-1 416	-1 107	-805	-510	-221	62	339	610	875

Year:	2021	2022	2023	2024	2025	2026	2027	2028	2029	2030
	11	12	13	14	15	16	17	18	19	20
Heat sale										
Heat sale	33 163	34 490	35 869	37 304	38 796	40 348	41 962	43 640	45 386	47 201
Heat Energy Purchase	15 492	16 111	16 756	17 426	18 123	18 848	19 602	20 386	21 201	22 049
Heating Gross result	**17 672**	**18 378**	**19 114**	**19 878**	**20 673**	**21 500**	**22 360**	**23 255**	**24 185**	**25 152**
Annual maintenance/admin costs	3 814	3 966	4 125	4 290	4 462	4 640	4 826	5 019	5 219	5 428
Operational Cashflow	**13 858**	**14 412**	**14 989**	**15 588**	**16 212**	**16 860**	**17 534**	**18 236**	**18 965**	**19 724**
Loan payments										
Annual loan repayment	11 743	11 743	11 743	11 743	11 743	11 743	11 743	11 743	11 743	11 743
Interest payment	*5 186*	*4 792*	*4 375*	*3 933*	*3 465*	*2 968*	*2 442*	*1 883*	*1 292*	*665*
Payment on principal	*6 557*	*6 951*	*7 368*	*7 810*	*8 279*	*8 775*	*9 302*	*9 860*	*10 451*	*11 079*
Balance	*79 874*	*72 923*	*65 556*	*57 746*	*49 467*	*40 692*	*31 390*	*21 530*	*11 079*	*0*
Net results										
pre tax net return	**1 586**	**2 002**	**2 434**	**2 884**	**3 351**	**3 838**	**4 343**	**4 869**	**5 417**	**5 985**
Net result adjusted for inflation	1 134	1 389	1 638	1 883	2 122	2 357	2 588	2 814	3 037	3 255

Fig. F.5 Cost effectiveness

F.1.23.2. Impact on Indoor Air Quality

The main task was to improve indoor air quality using constant air-exchange provided by a mechanical ventilation system with heat recovery.

F.1.23.3. Practical Experiences of Interest to a Broader Audience

Deep retrofit and energy efficiency must not considered as goals in themselves, but more as tools to solve problems with existing building stock and all relevant housing issues.

F.1.23.4. Resulting Design Guidance

It is always more challenging to implement new and innovative technologies and solutions in existing buildings than in new buildings.

F.1.23.5. Follow Up on the Renovation

RENESCO provide energy management and maintenance all investments made for all period of Energy Performance Contract.

F.1.24. References

RENESCO's (www.renesco.lv) information materials and most important media coverage:

RENESCO *.PPT presentation with renovation measures taken and financing principle: http://www.publicconsulting.at/uploads/06__presentation_RENESCO.pdf

RENESCO got the European prize, video RU, EN, http://www.youtube.com/watch?v=F1a2mqQX8Do

Ambassador about renovation and RENESCO, video EN, http://www.youtube.com/watch?v=SpxbKzjMofI

Why renovate apartment buildings? (video, with EN subtitles), http://www.youtube.com/watch?v=6f_JwjuJYfY

Post-Renovation Report, Data and Energy Performance Analysis, 13 Gaujas Str., Valmiera (December 2010, pdf, English), http://RENESCO.lv/images/stories/Valmiera/Valm_Dokumenti/gauja_13_post_renov_analysis_v8_en_compr_pic_scr.pdf

Brochure “How To Renovate Your Building? Technical Measures for Housing Renovation” (pdf, English), http://RENESCO.lv/images/stories/Brosuras/2011-09-10_RENESCO_measures_en_m.pdf

Brochure “Why To Renovate Your Building? ESCO Model—A New Way of Financing Renovations” (pdf, English), http://RENESCO.lv/images/stories/Brosuras/2011-09-10_RENESCO_esco_model%20en_euros_fin_m.pdf

“Marktentwicklung für EEDL. Energieeffizienzdienstleistungen erfolgreich gestalten” (“Development of energy efficiency market. Successful development of energy efficiency services”), German magazine ew – das magazin für die energiewirtschaft, 4/2013 (pdf, German), http://RENESCO.lv/images/stories/Publikacijas/2013-04-01_magazine_ew.pdf

“Sovjetflats creatief renoveren” (“Creative renovation of Soviet-time flats”) The Dutch Financial Times, 28 April 2012, Het Financieele Dagblad (pdf, Dutch), http://RENESCO.lv/images/stories/Publikacijas/2012-04-28_dutch_financial_times_RENESCO.pdf

“Save your house by saving energy,” Newsletter by Intelligent Energy Europe “Energy Efficiency Services. Good Practice Business Models and Successful Market Developments” (English, see pages 20–21), http://wupperinst.org/uploads/tx_wupperinst/ChangeBest_brochure_EN.pdf

Brochures in other languages:

http://www.changebest.eu/index.php?option=com_content&view=article&id=56%3Atest-article&catid=1%3Anews&lang=en

“Commercial energy services: a win–win combination,” Newsletter SQ Consult —June 2012, (English), http://www.vm01.net/sq-consult/mailing/52-ddl6o/14el-2ywyj40

“Energy Performance Contracting is an important measure to achieve the EU climate goals,” Berlin Energy Agency, 21 June, 2012, http://www.berliner-e-agentur.de/en/press/energy-performance-contracting-important-measure-achieve-eu-climate-goals.

Appendix G: Case Studies—Montenegro

G.1. Salko Aljkovic Primary School

G.1.1. Name of the Project, Location (City, Country)

Salko Aljkovic Primary School, Plevlja, Montenegro.

G.1.2. Picture

See Figs. G.1 and G.2.

G.1.3. Project Summary

G.1.3.1. Project Objectives

Energy efficient refurbishment retrofit of a primary school.

G.1.3.2. Project Energy Goals (Average of the First 2 years After the Retrofit)

- Quantified energy savings of at least 20%—based on a modeled baseline consumption.
- Minimum temperature of the targeted buildings during winter is 20 °C.

R. Lohse et al. (eds.), *Deep Energy Retrofit—Case Studies*,
https://doi.org/10.1007/978-3-031-17517-6

Fig. G.1 Facade before. *Source* Fichtner

Fig. G.2 Facade after. *Source* Fichtner

G.1.3.3. Short Project Description

The building is located in the mountainous region in the north of the country. It has a ground floor, and a 1st floor. The boiler room is located in the basement. The school has two buildings that are connected and Gymnasium/Sport hall. Overall conditioned area of the building: 3170 m^2 for 1080 students and 81 staff. It was built in 1971. The facade and the technical equipment is mainly > 40 years old. The energy (power and heating) is currently provided by electricity and lignite fired boilers. The roof was reconstructed in 2002, and is now a timber roof structure covered with boarding d = 2.5 cm, plies and ribbed plate for roof tiling. The ceiling is made of pressed compound of sawdust and lime. The space between ceiling and roof is hollow. Doors and windows are in a poor condition and do not close well. The school has two meters. One meter registers electricity consumption within the school and the other at the open playground. The school is also equipped with surveillance system. The lighting is extremely poor from technical aspect. The classrooms are fitted with old and inefficient lights with annealing metal wire, which produce indirect lighting of pupils' desks, and which are unacceptable for this type of buildings.

Investments in the school building in the past period:

- in July–August 2002, around 650m^2 of outdated tiles (in six cabinets) was replaced with parquet and flooring, and the library and school workers' club was renovated.
- In July–August 2003—school roof was replaced, NEO VULKAN III boiler was purchased and central heating system renovated.
- In October 2004—NEO VULKAN III was purchased.

G.1.3.4. Stage of Construction

Completed in 2015.

G.1.3.5. Point of Contact Information

Mr. Dewi Evans and Mr. Udo Becker of Fichtner BauConsulting.

G.1.3.6. Date of the Report

July 2015.

G.1.3.7. Acknowledgement

KfW
Ministry of Economy, Montenegro.

G.1.4. Site

- Location: Meše Selimovića bb-Pljevlja Montenegro.
- Latitude: 44° 52′ 27″ N.
- Longitude: 18° 48′ 39″ E.
- Elevation: 770 m above sea.

G.1.5. Building Description/Typology

G.1.5.1. Typology/Age

Old hollow brick construction, 43 years old (built in 1970).

G.1.5.2. Type (Office, Barracks, Etc.)

Dormitory.

G.1.5.3. General Information

- Year of construction: 1971.
- Year of renovation (as described here): 2015.
- Total floor area (m^2): gross floor area: 3170, net floor area: 3170 (2744 school, 432 m^2 gymnasium).

G.1.6. Architectural and Other Relevant Drawings

See Figs. G.3 and G.4.

Fig. G.3 School in winter (January 2012). *Source* Fichtner

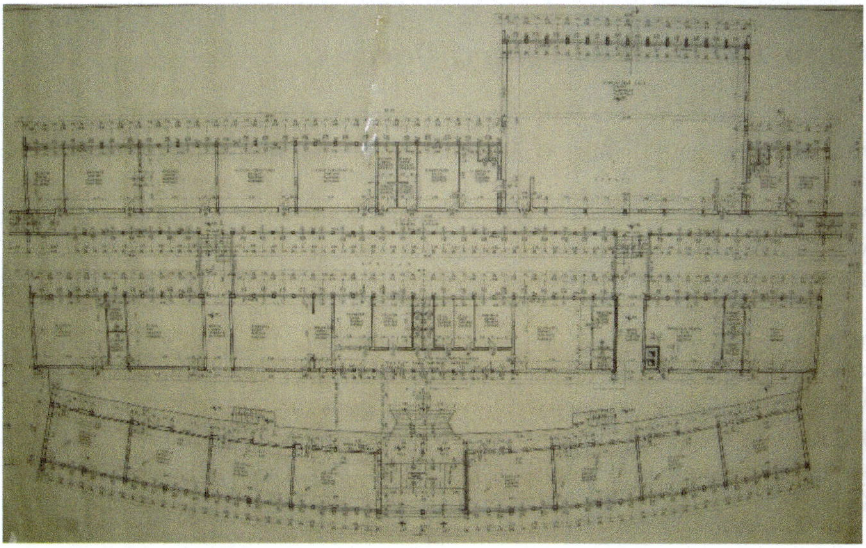

Fig. G.4 Historical layout drawing

G.1.7. National Energy Use Benchmarks and Goals for Building Type Described in the Case Study

G.1.8. Awards or Recognition

None.

G.1.9. Site Energy Cost Information (Electricity, \$/kWh; Gas, \$/m^3, Etc.)

G.1.9.1. Electricity

0.106 €/kWh.

G.1.9.2. District Chilled Water

Other (e.g., biomass): Lignite: 0.0185 €/kWh (does not include approx. 0.021 €/kWh for slag and ash disposal).

G.1.10. Pre-renovation Building Details

G.1.10.1. Envelope Details: Walls, Roof, Windows, Insulation Levels

Walls: Hollow brick walls d = 33 cm, External skin: mortar; internal plaster. Some walls are pebble dashed.
Roof: Type R1: Flat roof: Ribbed reinforced concrete panels with bituminous water proofing, and screed.
Windows: Generally in poor condition. Mixture of wood and PVC (old and newer). Some single glazed, some double-glazed. External roller shutters are mostly broken. Windows on the ground floor were mostly broken.
Floor Slab: In general, the floor was in an acceptable condition. Concrete slab, screed, tiles/parquet. No insulation (Figs. G.5, G.6, G.7, G.8, G.9, G.10, G.11, G.12, G.13, G.14, G.15, G.16, G.17, G.18, G.19, G.20, G.21, G.22, G.23 and G.24).

G.1.10.2. Heating, Ventilation, Cooling, and Lighting Systems

- Heating:
  - Central heating with a two-pipe distribution system.
  - Three Boilers: 1 × 315 kW from 1971; 1 × 315 kW from 2002; 1 × 315 kW from 2004.
  - Fuel: Lignite/wood coal.

Fig. G.5 Eastern facade. *Source* Fichtn

- No automatic control.
- Rooms heated by means of cast iron ribbed radiators and additionally electric heaters as required (158 kW). No thermostatic radiator valves.
- Domestic hot water is heated using electricity. The boilers however are not operational.
- During the heating season of 2011/2012, with temperatures getting as low as − 16 °C, in the morning hours, the boilers could not provide optimal comfort conditions in the rooms (tun = 13–15 °C).
- Ventilation: A total of three fans for air discharge are installed, as well as two heat pumps. Total installed power 4.3 kW.
- Cooling: No cooling.

Fig. G.6 Western facade. *Source* Fichtn

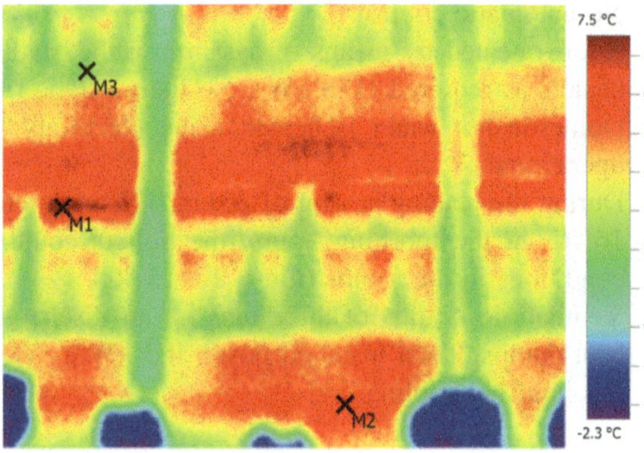

Fig. G.7 Southern facade of the school

Fig. G.8 Southern facade of the school

Fig. G.9 Damages on the facade. *Source* Fichtn

Fig. G.10 Damages on the flashing on roof eaves. *Source* Fichtner

Fig. G.11 Pebble dashing at the side entrance into hallway in front of the gym. *Source* Fichtn

Fig. G.12 Eaves at the side entrance into hallway in front of the gym. *Source* Fichtn

Fig. G.13 Floor in a corridor. *Source* Fichtn

Fig. G.14 Floor in a classroom. *Source* Fichtn

Fig. G.15 Southern side windows. *Source* Fichtner

Fig. G.16 School's front entrance doors. *Source* Fichtner

Fig. G.17 School roof. *Source* Fichtner

Fig. G.18 Inside of the school roof structure. *Source* Fichtn

Fig. G.19 Radiators in a classroom. *Source* Fichtner

Fig. G.20 Radiators in the corridors. *Source* Fichtner

Fig. G.21 Distribution pipes and circulator pumps. *Source* Fichtner

Fig. G.22 Hot-water boilers. *Source* Fichtner

Fig. G.23 Hot-water boiler —temperature and pressure measuring and hollow boiler rib. *Source* Fichtner

Fig. G.24 Water boiler—temperature and pressure measuring and hollow boiler rib. *Source* Fichtner

- Lighting systems: No active daylight-system. Average: 6 W/m^2. The existing lux values were not measured (Table G.1).

  Miscellaneous:
  See Figs. G.25 and G.26; Table G.2

Outdoor consumers:

Outdoor consumers	Total power (kW)	Operating hours		Time control (yes/no)	In operation since (year)	Comments
		h/week	Week/year			
External lighting	1	70	52	Yes	2007	

G.1.11. Description of the Problem: Reason for Renovation (Non-energy and Energy Related Reasons)

Problems included:

- High energy costs.
- Low indoor temperatures in winter.
- High indoor temperatures in summer.
- Bad air-condition in rooms.
- Poor general state of repair of the building and the heating system.

Table G.1 Lighting system

Lights	Bulb/tube power (W)	No. of bulbs per light (kΩ)	Light power (W)	No. of lights (kΩ)	Total power (kW)	Type of control
With annealing metal wire	100	1	100	105	10.5	Manual
With annealing metal wire	150	1	150	65	9.75	Manual
With annealing metal wire	200	1	200	50	10.0	Manual
Fluorescent	36	2	72	21	1.51	Manual
Fluorescent	36	4	144	1	0.14	Manual
Compact flue (energy efficient)	12	1	12	6	0.07	Manual
Total				248	31.97	

Lighting			
Total, average power (W/m^2)	6	Hours in function (h/week)	24
Max simultaneous power (W/m^2)	10	Hours in function (weeks/year)	36

Fig. G.25 Classroom lighting. *Source* Fichtner

Fig. G.26 Classroom lighting. *Source* Fichtner

Table G.2 Equipment

Various equipment — exploitable	Amount (pieces)	Power of one unit (W)	Total power (kW)	Average power (W/m^2)	Hours in function (h/week)	In operation since (year)	Comments
Computers	34	250	8,5	2.68	24	2008	
Copy machines	1	500	0.5	0.16	8	2007	
Other (TV, radio)		3550	3.55	1.12	10	2006	Printers, players
Other (kitchen)		8500	8.5	2.68	17	2005	Boilers, and appliances
Total			21.05	9.00			

Various equipment—exploitable			
Total, average power (W/m^2)	2.5	Hours in function (h/week)	23
Max simultaneous power (W/m^2)	2.5	Hours in function (weeks/year)	36

G.1.12. Renovation SOW (Non-energy and Energy Related Reasons)

None available.

G.1.13. Energy Saving/Process Improvement Concept and Technologies Used

G.1.13.1. Building Envelope Improvement

- Roof: not recommended.
- Outer walls: external thermal insulation system consisting of expanded polystyrene (10 cm) (Figs. G.27, G.28, G.29 and G.30).

Fig. G.27 Facade before. *Source* Fichtner

Fig. G.28 Facade after. *Source* Fichtner

Fig. G.29 Windows before. *Source* Fichtner

Fig. G.30 Windows after. *Source* Fichtner

Table G.3 Building envelop characteristic

Building element	U-value before renovation	U-value after renovation	Thickness of new insulation
Windows	≤ 3.2 – 5.6	1.5	
Outer wall	1.39	0.49	10 cm
Roof	0.95	0.95	–
Cellar ceiling	–	–	
Base plate	–	Not renovated	N/A
Walls against earth	N/A	N/A	
Walls against unheated space		Not renovated	N/A

Windows and doors:

- Windows and glass door replaced by double-glazed U = 1.5 W/m^2K.
- Inner walls (Table G.3).

G.1.13.2. New HVAC System or Retrofits to Existing

- Heating:
  - The base load boiler = 200 kW pellet boiler.
  - 750 l heating water storage.
  - Peak load and back up:
  - 2 × 170 kW Solid Fuel (Coal and Wood) Boilers as Back-Up.
- Ventilation: No mechanical/electrical ventilation. The building will be manually ventilated by means of school staff who will receive appropriate training.

G.1.13.3. New Lighting System

- New lights provide illumination exceeding 300 lx, which is in line with European standards for classrooms.
- All conventional ballasts were replaced by electronic starters.
- Classrooms: bright parabolic reflector and single light source (T 16 2 × 28 W) 350 (no) (source color is neutral white), with electronic control.

G.1.13.4. New Generation/Distribution System

- The existing system was kept, with minor improvements.
- Thermostatic valves are installed on all radiators and the complete system was hydraulically balanced.

G.1.13.5. Renewable Energy

Wood pellet boiler installed (see above section "Heating" under "New HVAC system or retrofits to existing").

G.1.14. Energy Consumption

G.1.14.1. Pre-renovation Energy Use (Total and per m^2/year

See Table G.4.

G.1.14.2. Predicted Energy Savings (Site, Source, GHG), Total and per m^2/year

Table G.5 lists calculated savings in delivered energy and related reductions in CO_2 emissions.

Table G.4 Pre-renovation energy demand

	2009		2010		2011	
	kWh/a	kWh/m^2 a	kWh/a	kWh/m^2 a	kWh/a	kWh/m^2 a
	€/a	€/m^2/a	€/a	€/m^2/a	€/a	€/m^2/a
Electricity	0.12	€/kWh				
Consumption	34,169	10.76	36,790	11.58	34,185	11
Costs*	6983.00	2.2	5124.00	1.6	3635.0	1.1
Heat (charcoal)	0.098	€/kWh				
Consumption	868,638	274	879,107	277	836,228	263
Costs	16,083.00	0.09	16,277.00	0.09	15,483	0.08

Table G.5 Predicted energy use

	Energy carrier		
	Electricity	Coal	Pellets
Current situation (kWh/year)	81,663	1,208,600	
After measures (kWh/year)	32,172		508,605
Savings (kWh/year)	49,491	1,208,600	
Savings (kWh/m^2 a)	15,582	380.54	− 160.14
CO_2 emission coefficient (kg/kWh)	0.6171	0.364	0.037
CO_2 emission—base line (kg/year)	490,316		
CO_2 emission—measures (kg/year)	38,669		
Reduction in CO_2 emission (kg/year)	451,648		

G.1.14.3. Measured Energy Savings (Thermal, Electrical), Total and per m^2/year

The energy savings will be verified at a later date.

G.1.15. Energy Cost Reduction

G.1.15.1. Split in All Energy Forms—Electricity, Oil, District Heating

See Sect. G.1.14.2 (Table G.6).

G.1.16. Non-Energy Related Benefits Realized by the Project, E.g., Improved Productivity, Increased Rent/Lease, Increased Useful Space, Etc.

- Toilets and sanitary facilities are now usable.
- Sanitary facilities for handicapped were included.
- Pleasant working atmosphere for staff and students.
- Constant and controllable conditions.

Table G.6 Energy cost

	Total (€/a)	Per (€/m^2a)
Heat and electricity	12,050	3.79

G.1.17. Renovation Costs: Total and per m²

G.1.17.1. Total

€458,920/€144/m^2.

G.1.17.2. Non-energy Related

€85,540/€26.93/m^2.

G.1.17.3. Energy Related

€374,380/€117.88/m^2.

G.1.17.4. Cost for Each Measure

See Table G.7.

G.1.18. Business Models and Funding Sources

KfW grant and loan to the Client (Ministry of Economy).

Table G.7 Cost for energy and non-energy related measures

EE measures		Investment
		[€]
1	Reconstruction of lighting in school and gym	63,031
2	Energy management	960
3	Thermal insulation of facade walls	79,191
4	Installation of thermostatic and automatic balancing valves	43,730
5	Replacement of doors and windows	123,247
6	Replacement of boilers 2	64,221
7	Non EE	84,540
	Total	458,920

G.1.19. Cost Effectiveness of Energy Part of the Project (NPV, SIR, Etc.)

See Table G.8.

G.1.20. User Evaluation

See Table G.9.

G.1.21. Experiences/Lessons Learned

G.1.21.1. Energy Use

As yet, reduction in electricity and heating cost is definite, but unquantifiable.

G.1.21.2. Practical Experiences of Interest to a Broader Audience

EE measures must be combined with general refurbishment (non-EE) measures. The general level of care of the refurbished building has increased (no graffiti; damage is reported and repaired immediately). The impact on the users is positive (staff, students and parents).

Table G.8 Cost effectiveness

OS “SALKO ALJKOVIĆ,” PLJEVLJA			Conditioned area: (school and gym)		
EE measures		Investment	Net savings		Payback
		[€]	[kWh/year]	[€/year]	[year]
1	Reconstruction of lighting in school and gym	63,031.00	49,482	4,121.00	15.30
2	Energy management	960.00	28,044	687,00	1.40
3	Thermal insulation of facade walls	79,191.00	150,892	3,199.00	24.75
4	Installation of thermostatic and automatic balancing valves	43,730.00	27,441	582,00	75.14
5	Replacement of doors and windows	123,457.00	284,951	6,041.00	20.44
6	Replacement of boilers[2)]	64,221.00	749,486	12,050.00	5.33
Total		374,590	1,290,296	26,680.00	14.04
Renovation measures			Investment [€]		
1	General refurbishment (facade repair etc.)		43,083		
2	Special measures for improved comfort (repairs of stairs, new boiler house etc.)		41,158		
	Total		84,241		

Table G.9 Overview user evaluation

EE measures	Brief description of measure	Measure applied	Savings kWh a
Energy management	Introduction of energy monitoring system	x	28,044
Walls	External thermal insulation system consisting of 8 cm expanded polystyrene with λ_{max} = 0.040 W/mK	x	150,892
Windows and doors	New PVC windows: double-glazed with U = 1.5 W/m^{2}K	x	284,951
	New PVC doors: double-glazed with U = 1.5 W/m^{2}K		
Boilers	Replacement of boilers and equipment in the boiler room and piping installation	x	749,486
Heating system	Installation of thermostatic valves	x	27,441
Lighting system	Replacing incandescent luminaries with fluocompact (energy saving) bulbs	x	49,482
Total			1,290,296

G.1.21.3. Resulting Design Guidance

Deep refurbishment projects often require updating the building according to newer building regulations (e.g., inclusion of sanitary facilities for disabled).

G.1.21.4. Follow Up on the Renovation

Users have received initial training, but still revert to old habits. Regular training of operational staff (janitors) as well as users (teachers and pupils), in particular with regard to ventilation (opening of windows etc.) is paramount.

G.2. "Spasic Masera," Montenegro

G.2.1. Name of the Project, Location (City, Country)

Student Dormitory "Spasic Masera," Kotor, Montenegro.

G.2.2. Pictures

These pictures show the building in its original and post-retrofit states, illustrating key features of the retrofit (Figs. G.31, G.32, G.33, G.34, G.35 and G.36).

Fig. G.31 Dormitory building—southwest facade (main entrance). *Source* Fichtner

Fig. G.32 Entrance area (analogue photo 1)

Fig. G.33 Dormitory building—northwest facade. *Source* Fichtner

Fig. G.34 Dormitory building northwest facade (analogue Photo 2). *Source* Fichtner

Fig. G.35 Wooden windows with wooden parapets. *Source* Fichtner

Fig. G.36 New windows (analogue photo 5). *Source* Fichtner

G.2.3. Project Summary

G.2.3.1. Project Objectives

Energy efficient refurbishment retrofit of a student hostel.

G.2.3.2. Project Energy Goals (Average of the First 2 years After the Retrofit)

- Quantified energy savings of at least 20%—based on a modeled baseline consumption.
- Minimum temperature of the targeted buildings during winter is 20 °C.

G.2.3.3. Short Project Description

The building is located on the Adriatic coast in the south of the country. It has a basement, ground floor, and four additional stories. The Hostel has:

- 78 beds in triple rooms with private bath.
- 50 beds in double rooms with bathrooms.
- 132 beds in double rooms with bathrooms condominium.

- Restaurant with 200 seats (self-service).
- Training ground for sports: basketball, football, volleyball, badminton.
- Common room with PCs and working stations.
- Recreation Room.
- Gymnasium/Sport hall.

The administration offices and kitchen with restaurant are located on the ground floor, while the rooms on the floors are used for accommodation.

Overall conditioned area of the building: 4100 m^2 for 235 residents and 32 staff.

It was built in 1973/74, was partially damaged in an earthquake in 1979, and has undergone minor repairs over the years. The facade and the technical equipment is mainly > 40 years old. Some doors and windows were replaced between 2001 and 2012. Lighting was partially replaced with fluorescent lights.

G.2.3.4. Stage of Construction

Completed in 12/2014.

G.2.3.5. Point of Contact Information

Mr. Dewi Evans and Mr. Udo Becker of Fichtner BauConsulting.

G.2.3.6. Date of the Report

July 2015.

G.2.3.7. Acknowledgement

KfW; Ministry of Economy, Montenegro.

G.2.4. Site

- Location: Dobrota bb Kotor, Montenegro.
- Latitude and longitude: N 42° 26′ 21″–E 18° 45′ 50″.
- Elevation: Approx. 30 m above sea.
- Climate zone (e.g., ASHRAE 90.1-2004 Climate Zone).

G.2.5. Building Description/Typology

G.2.5.1. Age

43 years old concrete skeleton construction.

G.2.5.2. Type (Office, Barracks, Etc.)

Dormitory.

G.2.5.3. Typology/Age

1970.

G.2.5.4. General Information

- Year of construction: 1973/74.
- Year of previous major retrofit—if known.
- Year of renovation (as described here): 2014.
- Total floor area (m^2): gross floor area: 5115; net floor area: 4175.

G.2.6. Architectural and Other Relevant Drawings

Figures G.37, G.38, G.39, G.40, G.41 and G.42 shows the layouts of the floors.

G.2.7. National Energy Use Benchmarks and Goals for Building Type Described in the Case Study

Not available.

G.2.8. Awards or Recognition

Not available.

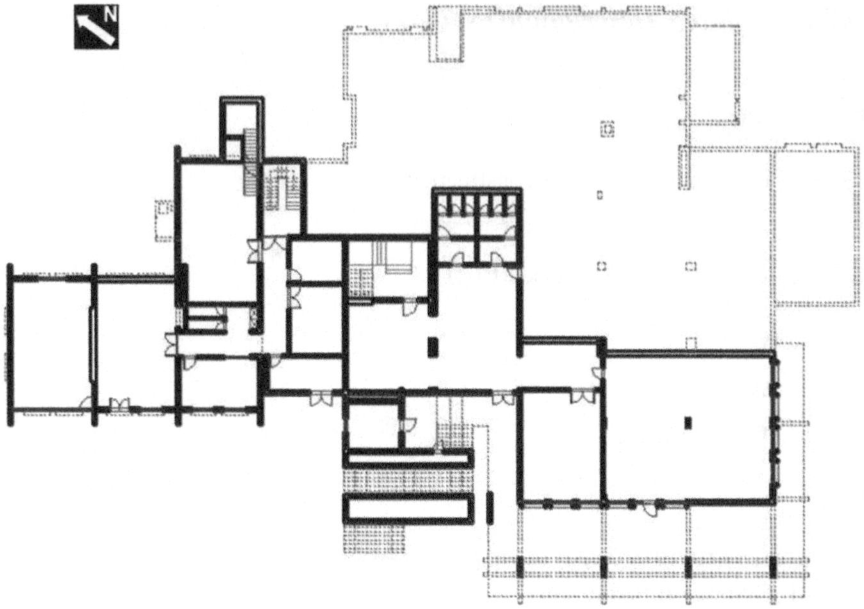

Fig. G.37 Layout of the basement

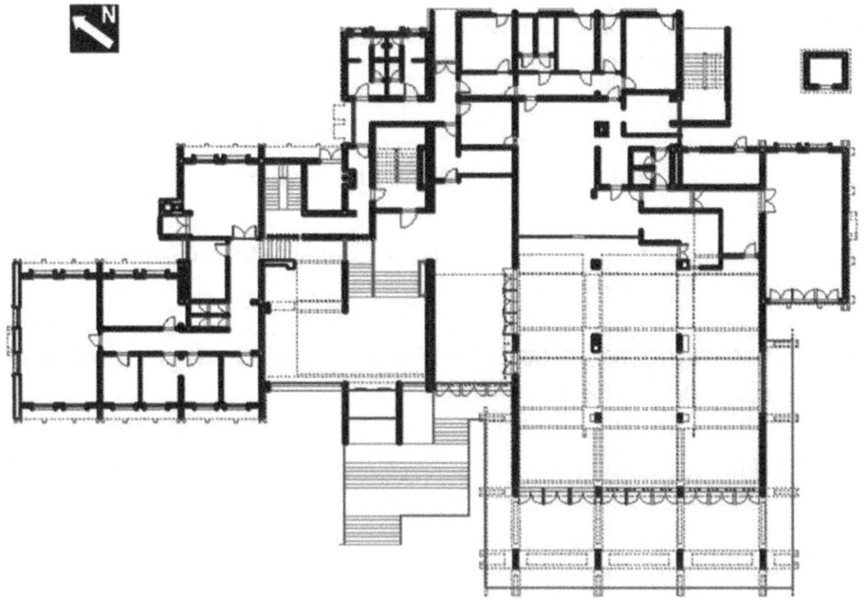

Fig. G.38 Layout of the ground floor

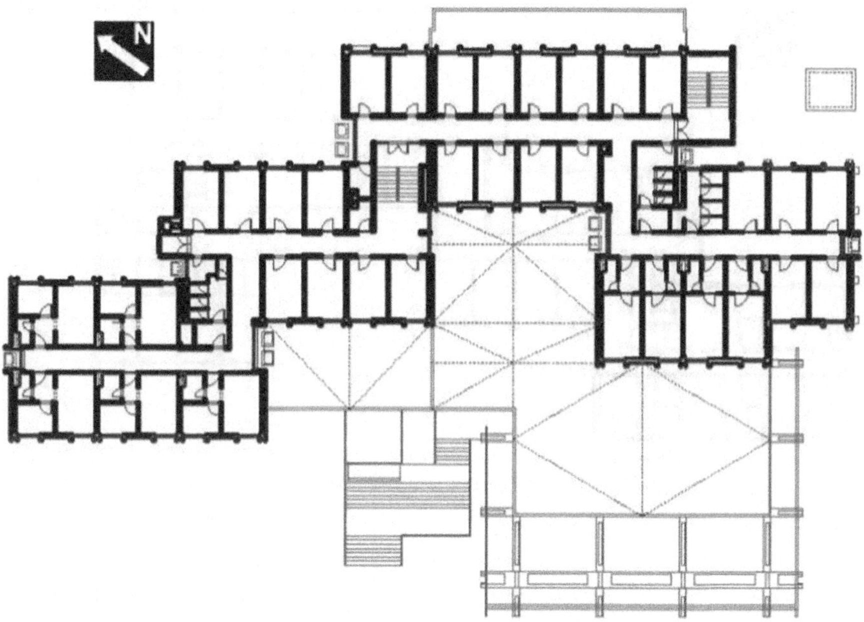

Fig. G.39 Layout of the 1st floor

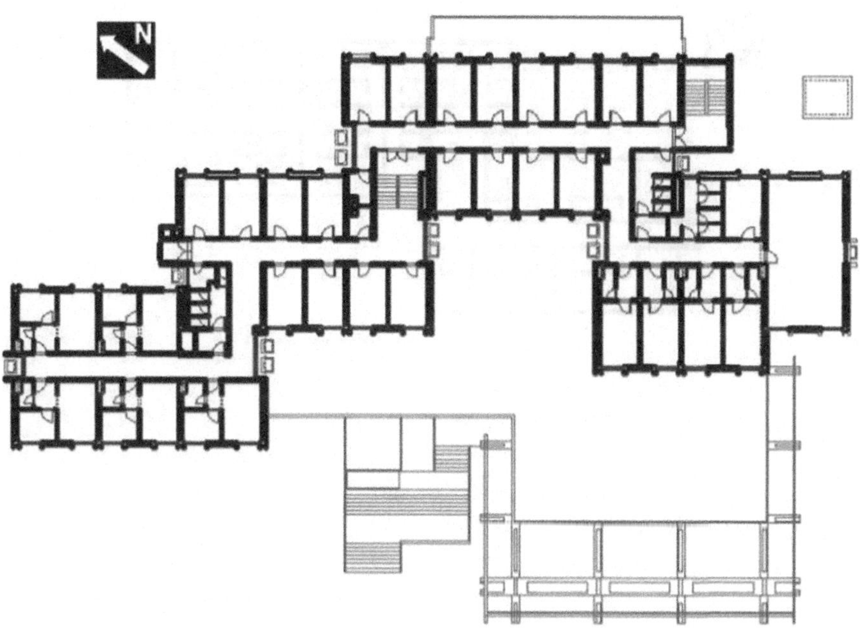

Fig. G.40 Layout of the 2nd floor

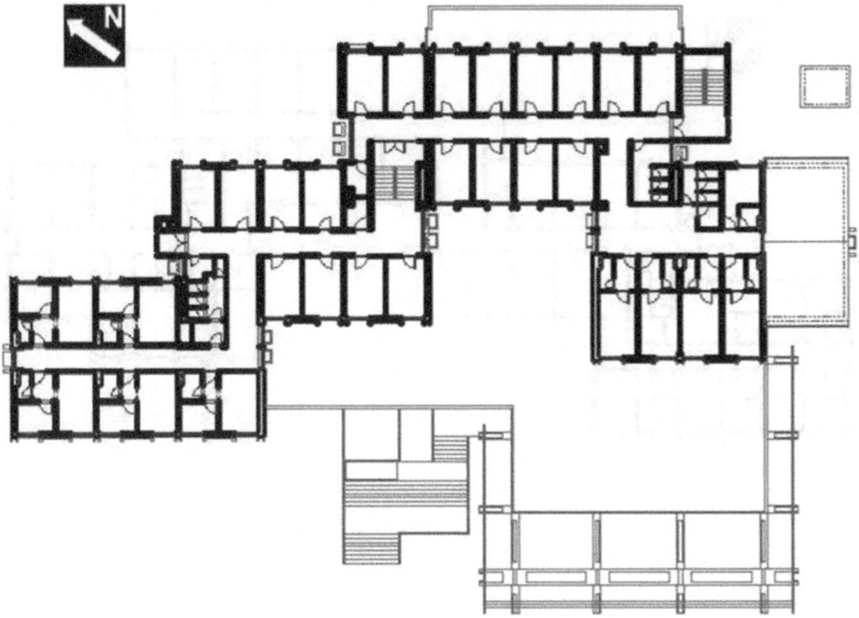

Fig. G.41 Layout of the 3rd floor

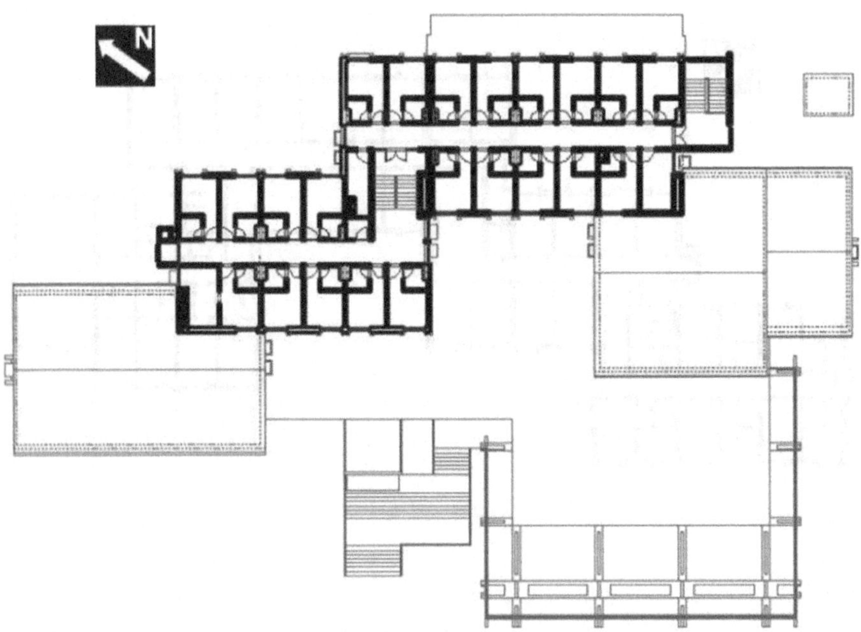

Fig. G.42 Layout of the 4th floor

G.2.9. Site Energy Cost Information (Electricity, \$/kWh; Gas, \$/m^3, Etc.)

G.2.9.1. Electricity

0.120 €/kWh; LFO: 1.20. €/l.

G.2.10. Pre-renovation Building Details

G.2.10.1. Envelope Details: Walls, Roof, Windows, Insulation Levels

- **Walls**:
  - Type 1: Reinforced concrete walls (30%) filled with hollow clay blocks (70%) of the total thickness of 25 cm; External skin: mortar; internal plaster.
  - Type 2: Hollow clay blocks.
  - Type 3: Aerated concrete blocks.
- **Roof**:
  - Type R1: Flat roof: Ribbed reinforced concrete panels with bituminous water proofing, and screed.
  - Type R2: Pitched, with corrugated sheeting and 10 cm mineral wool (Figs. G.43, G.44, G.45 and G.46).

Fig. G.43 Dormitory building—southwest facade (main entrance). *Source* Fichtner

Fig. G.44 Dormitory building—northwest facade. *Source* Fichtner

Fig. G.45 Pitched roof above the floor. *Source* Fichtner

Fig. G.46 Flat roof above restaurant. *Source* Fichtner

- **Windows**: Generally in poor condition. Mixture of wood and PVC (old and newer). Some single glazed, some double-glazed. External roller shutters are mostly broken. Windows on ground floor mostly broken (Figs. G.47, G.48 and G.49).

Fig. G.47 Typical wooden windows on the hostel building southwest side. *Source* Fichtner

Fig. G.48 New PVC windows on the southwest side of the hostel. *Source* Fichtner

Fig. G.49 Wooden windows with wooden parapets. *Source* Fichtner

- Floor Slab: In general in good condition. Concrete slab, screed, tiles/parquet.

G.2.10.2. Heating, Ventilation, Cooling, and Lighting Systems

- **Heating**:
  - Central heating with a two-pipe distribution system. two Boilers: 1 × 300 kW from 1985. 74% efficient 1 × 300 kW from 2001; 80% efficient. Fuel: Light fuel Oil.
  - Minimal automatic control.
  - Rooms heated by means of cast iron ribbed radiators and additionally electric heaters as required (158 kW). No thermostatic radiator valves. Boiler operation set for 6:30–10:30 then 16:30–20:30.
  - 2 × 4000 l water heaters for domestic hot water.
  - The system operates thus: the hot water from boiler, through boiler pipe branch, is distributed to the heating manifold and water heater and heats the sanitary hot water. From the heating manifold, hot water is distributed to the final users (Figs. G.50, G.51, G.52 and G.53).

- **Ventilation**:
  - Ventilation by means of an extraction hood in the kitchen area. Otherwise natural ventilation in the rest of the building.
  - Cooling: Various Air-conditioning units. Total capacity: 26 kW.
  - Lighting systems: Classrooms: 56 × 72 W fluorescent tubes (T36).

Fig. G.50 Boilers "TKT Toplota Zagreb". *Source* Fichtner

Fig. G.51 Burner on old boiler. *Source* Fichtner

Fig. G.52 Typical existing radiator in corridors. *Source* Fichtner

Fig. G.53 Typical "new" radiator in the rooms. *Source* Fichtner

- Hallways and corridors: 560 × 75 W incandescent bulbs. 20% of the bulbs are defective and/or missing.
- External lighting: 6 × 150 W natrium halogen reflectors.
- No active daylight-system.
- Average: 3.75 W/m^2.
- The existing lux values were not measured (Table G.10).

Table G.10 Miscellaneous

Various exploitable	Quantity (pcs)	Unit capacity (kW)	Total capacity (kW)	Avg. power (W/m^2)
Bakery	1	32	32	
Oven furnace	1	30	30	
El. oven (stove)	2	12	24	
Keeper	2	15	30	
Fryer	4	3	12	
Dishwasher	1	6	6	
Electrical boiler	4	2	8	
Aspirator	1	6	6	
Freezer	7	0.5	3.5	
Fridge	2	1	2	
Cooler	2	2	4	
Kitchen equipment	–	10	10	
Computers	20	0.3	6	
AC units	22	–	26.4	
El. heaters-estimate	90	2	158	
Printer, copier	8	0.5	4	
TV	80	0.1	8	
Total			369.9	6.2

Various exploitable			
Total, average power (W/m^2)	6.2	Operation period (h/week)	90
Max, simultaneous power (W/m^2)	114.0	Operation period (weeks/year)	51

Various unexploitable	Quantity (pcs)	Unit capacity (W)	Total capacity (kW)	Avg. power (W/m^2)
Washing machine	5		65	
Drier for laundry	2		6	
Laundry ironing	3		7	
Workshop equipment	–	–	6	
Pumps	5	800	4.0	
Burners	2	850	1.7	
Equip. in boiler room	–	–	1.5	
Total			91.2	16.0

Various unexploitable			
Total, average power (W/m^2)	16.0	Operation period (h/week)	90
Max, simultaneous power (W/m^2)	114.0	Operation period (weeks/year)	51

G.2.11. Description of the Problem: Reason for Renovation

High energy costs, low indoor temperatures in winter, high indoor temperatures in summer, bad air-condition in rooms, poor general state of repair of the building and the heating system.

G.2.12. Renovation SOW (Non-energy and Energy Related Reasons)

None available.

G.2.13. Energy Saving/Process Improvement Concept and Technologies Used

G.2.13.1. Building Envelope Improvement

- **Roof**: thermal insulation of flat roof (10 cm).
- **Outer walls**: external thermal insulation system consisting of expanded polystyrene (8 cm).

- **Window reveal**: external thermal insulation system consisting of expanded polystyrene (3 cm).
- **Windows and doors**: Windows and glass door replaced by double-glazed U_{max}. = 1.5 W/m2K.
- **Entrance**: Glass door replaced by double-glazed Umax. = 2.8 W/m2K.
- **Cellar ceiling**: N.A. (DEMIT envelope includes basement, as all rooms in the basement are heated).
- **Story ceiling**: Min. 75 cm deep, 8 cm thick insulation next to the outer walls to minimize thermal bridges (Figs. G.54, G.55, G.56; Table G.11).

Fig. G.54 Entrance area (analogue photo 1). *Source* Fichtner

Fig. G.55 Dormitory building north west facade (analogue photo 2). *Source* Fichtner

Fig. G.56 New windows (analogue photo 5). *Source* Fichtner

Table G.11 Building envelope characteristic

Building element	U-value before renovation	U-value after renovation	Thickness of new insulation
Windows	≤ 3	1.5	
Outer wall	2.26	0.34	8 cm
Roof	$\gg 0.85$	0.19	20 cm
Cellar ceiling	0.96	N.A	
Base plate	0.9	Not renovated	N/A
Walls against earth	N/A	N/A	
Walls against unheated space		Not renovated	N/A

G.2.13.2. New HVAC System or Retrofits to Existing

- Heating: 3000 l heating water storage.
- Thermostatic valves are installed and the complete system was hydraulically balanced.
- Peak load and back up: The dormitory is still connected to the heating pipeline (backup).
- Ventilation: No mechanical/electrical ventilation. The building will be manually ventilated by means of school staff who will receive appropriate training.

G.2.13.3. New Lighting System

Replacement of the lighting system was subject to a separate project financed by the World Bank.

G.2.13.4. New Generation/Distribution System

- The boilers were replaced by 2 × 235 kW two-stage or modulated light distilled oil burners.
- Radiators were replaced with 239 W 90/70 C ribbed aluminum radiators, including piping (Fig. G.57).

G.2.13.5. Renewable Energy

- For potable hot water, solar panels were installed on the flat roof.
- Total collector area: $A_{kol,uk} \approx 230 * 0.53 \approx 122\ m^2$ (Fig. G.58).

Fig. G.57 Boiler room (analogue photo 8 and 9). *Source* Fichtner

Fig. G.58 Installed solar plant on the roof above the restaurant (analogue photo 7). *Source* Fichtner

G.2.14. Energy Consumption

G.2.14.1. Pre-renovation Energy Use (Total and per m^2/year)

See Table G.12.

G.2.14.2. Predicted Energy Savings (Site, Source, GHG), Total and per m^2/year

Table G.13 lists the calculated savings in delivered energy and related reductions in CO_2 emissions.

G.2.14.3. Measured Energy Savings (Thermal, Electrical), Total and per m^2/year

The energy savings are to be verified at a later date.

Table G.12 Pre-renovation energy use

	2009		2010		2011	
	kWh/a €/a	kWh/m^2/a €/m^2/a	kWh/a €/a	kWh/m^2/a €/m^2/a	kWh/a €/a	kWh/m^2/a €/m^2/a
Electricity (including heat pump)	0.120 €/kwh					
Consumption	450,822	111	481,089	118	475,806	117
Costs**	42,846.00	10.30	42,486	10.20	41,181	10.20
Heat (light fuel oil)	0.098 €/kWh					
Consumption	411,290	101	289,674	71	315,790	78
Costs**	27,656.00	0.02	20,264	0.02	26,119	0.02

**The given costs are the actual costs. These differ significantly from the calculated costs based on consumption and area

Table G.13 Predicted energy saving

	Energy carrier	
	Light fuel oil	Electricity
Present situation (kWh/year)	552,207	468,779
After EE and renovation measures (kWh/year)	96,966	440,223
Savings (kWh/year)	455,241	28,556
Savings (kWh/m^2 a)	112	7.02
Savings (€/year)	44,614	3427
Savings (€/m^2.a)	11	0.84
CO_2 emission reductions (kg/year)	120,184	17,619
CO_2 emission reductions (t/year)	137.80	

G.2.14.4. Annual Energy Cost Reduction

See Sect. G.1.14.2 (Table G.14).

G.2.15. Energy Cost Reduction

None available.

Table G.14 Annual energy cost reduction

	Total (€/a)	Per (€/m^{2}a)
Electricity	3427	0.84
Heat	44,614	11.00

G.2.16. Non-energy Related Benefits Realized by the Project, E.g., Improved Productivity, Increased Rent/lease, Increased Useful Space, Etc.

The building now offers a pleasant working atmosphere for staff and students, and constant, controllable conditions.

G.2.17. Renovation Costs: Total and per m^2

G.2.17.1. Total

€597,000.

G.2.17.2. Non-energy Related

€1000.

G.2.17.3. Energy Related

€596,000.

G.2.17.4. Cost for Each Measure

See Table G.15.

G.2.18. Business Models and Funding Sources

KfW grant and loan to the Client (Ministry of Economy).

G.2.19. Cost Effectiveness of Energy Part of the Project

See Table G.15.

Table G.15 Cost effectiveness

EE measures—energy audit			
Student hostel "Spasic-Masera"—Kotor		Conditioned area: 4065	m^2
EE measures	Brief description of measure	Measure applied	Investment [€]
Energy management	Introduction of energy monitoring system	x	12,000
Walls	External thermal insulation system consisting of 8 cm expanded polystyrene with λ_{max} = 0.040 W/mK	x	125,012
Windows and doors	New PVC windows: double-glazed with U = 1.5 W/m^2K New PVC doors: double-glazed with U = 1.5 W/m^2K	x	89,000
Roof	Thermal insulation of flat roof with material with heat transfer coefficient λ_{max} = 0,035–0.04 W/m^2K	x	26,100
Boilers	Replacement of boilers and equipment in the boiler room and piping installation	x	85,000

(continued)

Table G.15 (continued)

EE measures—energy audit			
Student hostel "Spasic-Masera"—Kotor		Conditioned area: 4065	m^2
EE measures	Brief description of measure	Measure applied	Investment [€]
Heating system	Installation of thermostatic valves	x	7175
Solar water heater	Installation of solar panels for preparing DHW	x	57,000
Lighting system	Replacing incandescent luminaries with fluocompact (energy saving) bulbs	x	4960
Total			406,247

Table G.16 Overview user evaluation

EE measures	Brief description of measure	Measure applied	Savings (kWh a)
Energy management	Introduction of energy monitoring system	x	15,113
Walls	External thermal insulation system consisting of 8 cm expanded polystyrene with λ_{max} = 0.040 W/mK	x	162,987
Windows and doors	New PVC windows: double-glazed with U = 1.5 W/m^2K	x	68,477
	New PVC doors: double-glazed with U = 1.5 W/m^2K		
Roof	Thermal insulation of flat roof with material with heat transfer coefficient λ_{max} = 0,035–0.04 W/m^2K	x	22,958
Boilers	Replacement of boilers and equipment in the boiler room and piping installation	x	88,012
Heating system	Installation of thermostatic valves	x	17,570
Solar water heater	Installation of solar panels for preparing DHW	x	13,442
Lighting system	Replacing incandescent luminaries with fluo-compact (energy saving) bulbs	x	90,487
Total			232,469

G.2.20. User Evaluation

See Table G.16.
All given values are subject to validation

G.2.21. Experiences/Lessons Learned

G.2.21.1. Energy Use

As yet, reduction in electricity and heating cost is definite, but unquantifiable.

G.2.21.2. Practical Experiences of Interest to a Broader Audience

EE measures must be combined with general refurbishment (non-EE) measures. The general level of care of the refurbished building has increased (no graffiti; damage is reported and repaired immediately).

Feedback from the users (staff, students and parents) is positive.

G.2.21.3. Resulting Design Guidance

Deep refurbishment projects often require updating the building according to newer building regulations (e.g., inclusion of sanitary facilities for disabled).

G.2.21.4. Follow Up on the Renovation

Regular training of operational staff (janitors) as well as users (teachers and pupils), in particular with regard to ventilation (opening of windows etc.) is paramount.

Appendix H: Case Study—The Netherlands

H.1. Leeuwarden, The Netherlands

H.1.1. Name of the Project, Location (City, Country)

Shelter home "Veilige Veste," Leeuwarden, the Netherlands.

H.1.2. Pictures

Figures H.1, H.2, H.3, H.4, H.5, H.6 and H.7 show the building in its original and post-retrofit states, illustrating key features of the retrofit.

H.1.3. Project Summary

H.1.3.1. Project Objectives

The 'Veilige Veste' is a sanctuary for women fleeing from maltreatment, estranged suitors, forced prostitution and honor-related violence. In the 'Veilige Veste', care organization Fier Fryslân wants to create a place where these young women can feel safe. As a consequence of all the modern media, these women are often literally hunted. Before, the victims were hidden away anonymously; now they are in a fortress where nobody can enter that does not belong there. A building that says: 'here we are!."

R. Lohse et al. (eds.), *Deep Energy Retrofit—Case Studies*,
https://doi.org/10.1007/978-3-031-17517-6

Fig. H.1 Facade before retrofit

Fig. H.2 Entrance, renovated facade

Fig. H.3 Detail of paneling

H.1.3.2. Project Energy Goals

A radical renovation transformed the previous police station in the Dutch town of Leeuwarden to a shelter home for women. According to passivhaus standards in the Netherlands, it is the first repurposing of an office in this scale. Together with the client and the end user we decided that, if we were going to make the building energy efficient, we were going all the way to passivhaus level. Passivhauses score much better on an energetic evaluation, and have a more pleasant indoor climate. And when taken into account from the start, it does not necessarily mean higher costs.

H.1.3.3. Short Project Description

In the Netherlands, design, energy reduction and the fight against human trafficking come together in one revolutionary project: the 'Veilige Veste'.

Literally translated this means 'safe fortress', and that is exactly what it is. You cannot miss the bright white building, cut like a diamond, subtly gleaming in the sun. It is the new home for girls from all around the world that have been victims of human trafficking. Bold, brave, and safe is the concept for the new shelter for girls that have been victims of human trafficking. Not tucked away in anonymous houses

Fig. H.4 Old cell complex is restaurant after renovation

Fig. H.5 Renovated restaurant

Fig. H.6 Roof terrace

Fig. H.7 Old cell doors are kept and painted

in back alleys anymore, which is the way these girls are normally treated. No, these girls do not have to fear their perpetrators any more in their new home that is standing fierce in the midst of Frieslands' capital Leeuwarden. In their safe fortress, they send out a clear message: we are no longer on the run, game over, giving their perpetrators the finger. 'Veilige Veste' provides security and protection, so the girls can build up their lives again.

What is revolutionary about the 'Veilige Veste', is that this is the first large office block in the Netherlands to be renovated according to the Passive house standard. 'Passive house' is a standard for energy efficiency in a building, reducing its ecological footprint. It results in ultra-low energy buildings that require little energy for space heating or cooling. In this case, the fact that the former police stations' substructure was placed outside the building, meant an enormous energy abuser to be dealt with. The substructure created a thermal bridge that works exactly like a tunnel sucking in the cold outside air. By wrapping the building with the diamond-cut square panels, the substructure is now within the building and the whole building is covered by a thick layer of insulation. At some points, the facade is over 3 ft thicker now. Thanks to optimal insulation, draft proofing and the use of very little, highly energy-efficient equipment, the 'Veilige Veste' consumes exceptionally little power.

H.1.3.4. Stage of Construction

Finished.

H.1.3.5. Point of Contact (POC) Information

Linda Terpstra, CEO, 0031 58 215 70 84.
Holstmeerweg 1, 8936 AS Leeuwarden, The Netherlands.

H.1.3.6. Date of the Report

4-6-2015.

H.1.4. Site

- Location, latitude, longitude. 53.191610, 5.824527.
- Elevation.
- Climate zone (e.g., ASHRAE 90.1-2004 Climate Zone), moderate climate.
- Heating Degree Days (based on 16 °C): 2480.

H.1.5. Building Description/Typology

H.1.5.1. Typology/Age

1975.

H.1.5.2. Type (Office, Barracks, Etc.)

Office block.

H.1.5.3. Typology/Age

1970.

H.1.5.4. General Information

- Year of construction: 1975.
- Year of previous major retrofit—if known. No retrofit.
- Year of renovation (as described here): 2012.
- Total floor area (m^2): 5340 m^2.
- Area of unconditioned space included above (m^2): 0.

H.1.6. Architectural and Other Relevant Drawings

See Figs. H.8, H.9, H.10, H.11 and H.12.

H.1.7. National Energy Use Benchmarks and Goals for Building Type Described in the Case Study

None available.

H.1.8. Awards or Recognition

- Winner duurzaamheidsaward 2012 (sustainability award).
- Nominee PassiefBouwen Awards 2012 (passive building award).

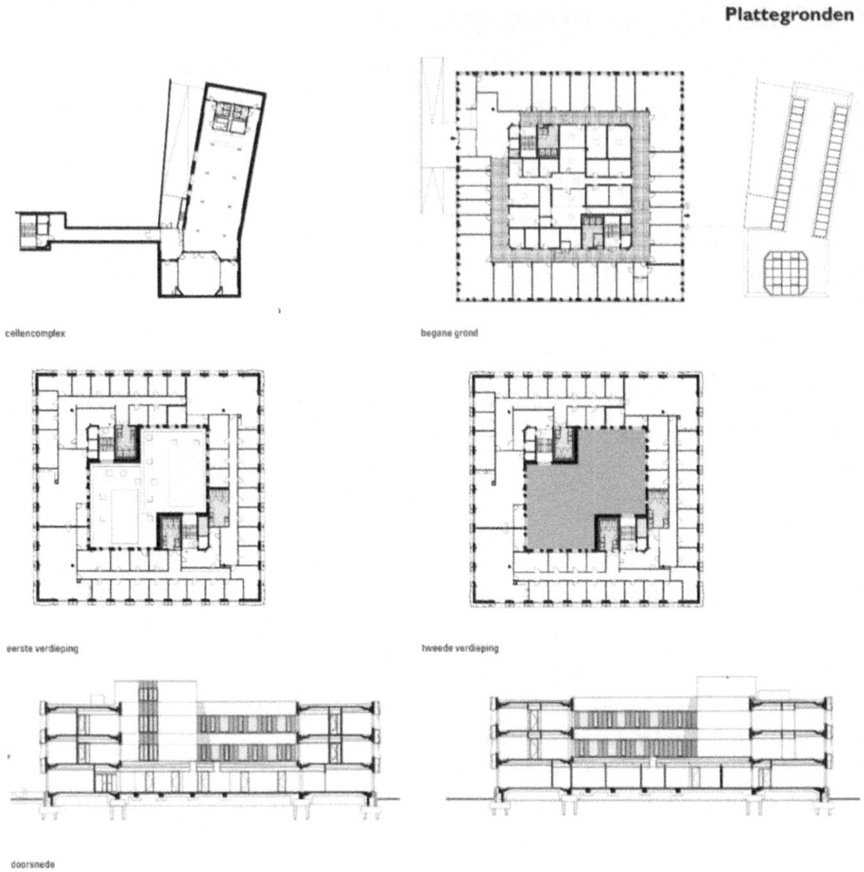

Fig. H.8 Plants of each floor and main sections

- Winner Hedy d'Anconaprijs 2014 (healthcare 7 architecture award).
- Nominee Vredeman de Vriesprijs 2014 (Frisian architecture award).

H.1.9. Site Energy Cost Information (Electricity, \$/kWh; Gas, \$/m³, Etc.)

H.1.9.1. Electricity

€0.12/kWh.

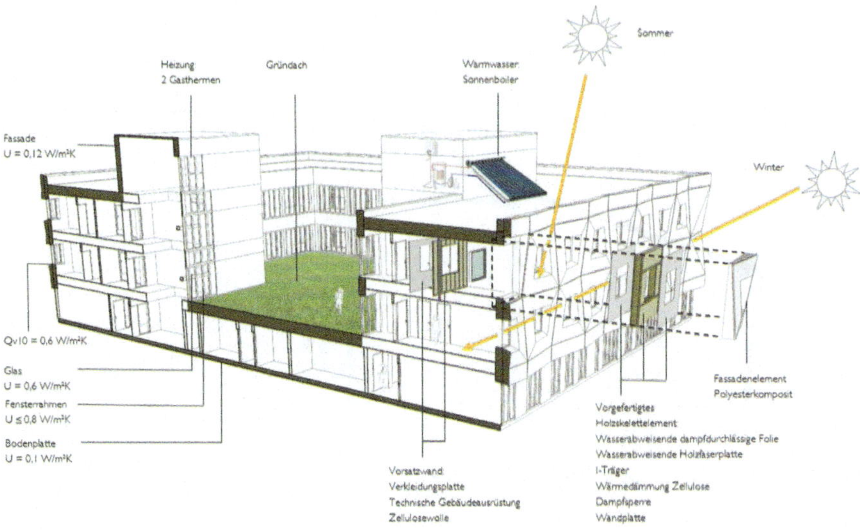

Fig. H.9 Section of the building and description of the components

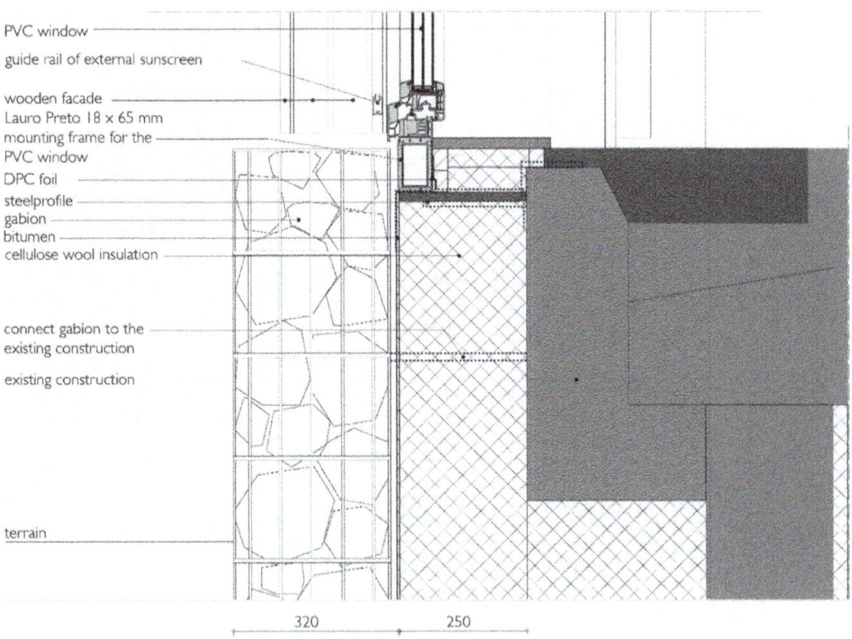

Fig. H.10 Key junction of window with terrane in ground floor

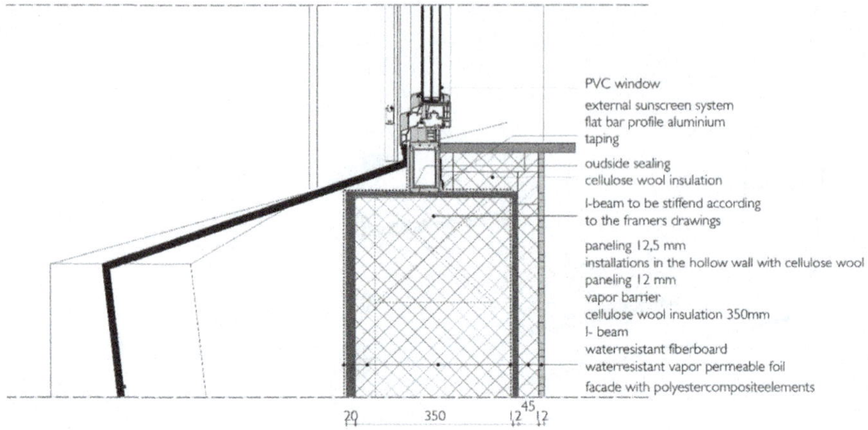

Fig. H.11 Key junction of window with facade

Fig. H.12 Detail of panel module

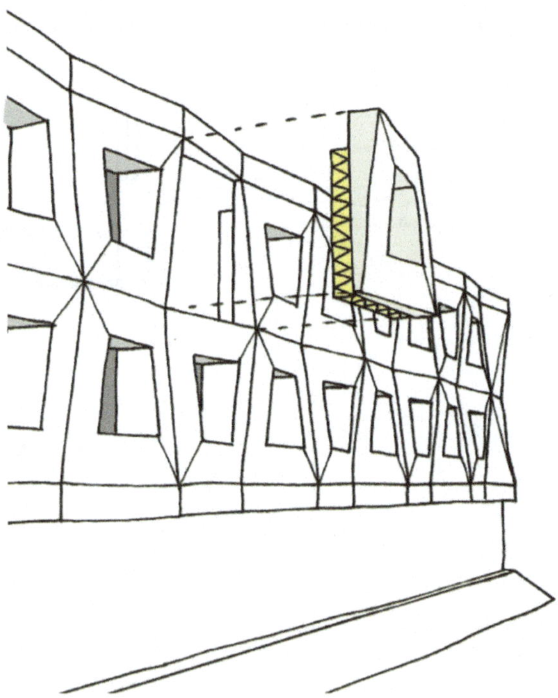

H.1.9.2. Natural Gas

€0.55/m3.

H.1.10. Pre-renovation Building Details

H.1.10.1. Envelope Details: Walls, Roof, Windows, Insulation Levels

- Height (m): 11.
- Net volume (m^3): 8900.
- Building type: concrete with prefab facade.
- Roof construction: concrete roof.

H.1.11. Description of the Problem: Reason for Renovation

The old police station was in its essence a solid concrete building, with the support structure in the facade. That and the oversized floor height offered a multitude of possibilities to remove everything inside and build the interior up completely from scratch. The problem was in the concrete structure, which was one big thermal bridge. The substructure created a thermal bridge that works exactly like a tunnel sucking in the cold. The only solution in such a case is to wrap the building completely.

H.1.12. Renovation SOW (Non-energy and Energy Related Reasons)

None available.

H.1.13. Energy Saving/Process Improvement Concept and Technologies Used

H.1.13.1. Building Envelope Improvement

To solve the problem of the thermal bridge, the structure was wrapped to Rc values of 10.0—three times the standard for new buildings. The elements are carefully masked on the concrete structure. Subsequently a retention wall for timber frames and concrete are placed along. This prevents the leaking of air when driving in screws and withdraws the taped up seams from sight. German passivhaus window frames, with triple draft proof and triple glazing, are placed in the timber frames. The floor has over 10 in. of insulation. The building is air-tight, according to a blower test. Thermal pictures prove that the facade is very well insulated. The actual use will be checked after the building has been in use for a year, but according to calculations there is a heat demand of 15 W/m^2 per year, far under the passivhaus norm of 25 W/m^2. The inner roof on the first floor is partly finished with sedum, giving a pretty view from the upper floors. At the same time, it functions as a buffer for heat and water.

H.1.13.2. New HVAC System or Retrofits to Existing

All new heating, ventilation and air-conditioning system. Old systems were 35 years old and nearly "dead."

H.1.13.3. New Lighting System

With LED light with "occupancy sensors."

H.1.3.4. New Generation/Distribution System

The building is fitted with three small central heating boilers; one of them being responsible for the major part of the heating demand throughout the year. The other two boilers are used when demand is higher in colder periods. The ventilation is organized with heat recovery and summer night ventilation system. The existing cooling system is used for the offices ventilation. Finally, the building is equipped with energy efficient lighting with a occupancy sensor.

H.1.13.5. Renewable Energy

The building is equipped with solar boilers that heat the tap water and a heat recovery system for air ventilation.

H.1.13.6. Daylighting Strategies

All exterior frames on the facades that receive sun are fitted with automated solar protection.

H.1.14. Energy Consumption

H.1.14.1. Pre-renovation Energy Use (Total and per m^2/year)

210,000 m^3 natural gas.

H.1.14.2. Predicted Energy Savings (Site, Source, GHG), Total and per m^2/year

25,000 m^3 natural gas.

H.1.14.3. Measured Energy Savings (Thermal, Electrical), Total and per m^2/year

21,000 m^3 natural gas.

H.1.14.4. Annual Energy Use Reduction

90%.

H.1.15. Energy Cost Reduction

H.1.15.1. Split in All Energy Forms—Electricity, Oil, District Heating

Cost of natural gas were reduced 80–90%. These savings fluctuate a little every year, depending on climate changes, e.g., 80% in a cold winter, 90% in a warm winter.

H.1.16. Non-energy Related Benefits Realized by the Project, E.g., Improved Productivity, Increased Rent/Lease, Increased Useful Space, Reduced Maintenance, Etc.

Increased useful space approximately > 400 m^2, by placing the new facade outside the supporting structure of concrete columns unlike the old facade that was inside the supporting structure.

H.1.17. Renovation Costs: Total and per m²

H.1.17.1. Total

€4,100,000, excl. VAT.

H.1.17.2. Non-energy Related

€2,450,000.

H.1.17.3. Energy Related

€710,000 insulations triple glazing and other Passive house technics.
€940,000 installations including both all electrical as mechanical installations.

H.1.18. Business Models and Funding Sources

H.1.18.1. Decision Making Process Criteria for Funding and Business Models

We calculated the extra initial costs of retrofit it to Passive house standard as compared to classical retrofit. The extra costs were €700,000. The payback time by energy savings was less than 10 years. The choice was then easily made.

H.1.18.2. Description of the Funding Sources Chosen

Both the building's owner and tenant saw the advantages, a more durable building with higher comfort level and lower energy costs and a green image. Both paid 50% of the extra investment.

H.1.18.3. Energy Management and Controlling in the Business Model

Energy use is being monitored.

H.1.19. Cost Effectiveness of Energy Part of the Project (NPV, SIR, Etc.)

The facade has a lifetime cycle that is almost endless. So one should always start reducing the energy needs by building a facade with optimal insulation.

H.1.20. User Evaluation

H.1.20.1. Description of User Training Programmes Within the Refurbishment

The people who work and live in the building are being coached on energy-reduction.

H.1.20.2. Integration of Users Demands in the Planning Process

The whole design is based on user demands and wishes.

H.1.21. Experiences/Lessons Learned

H.1.21.1. Energy Use

An enormous 90% heating energy can be saved, relative to buildings that have little insulation.

H.1.21.2. Impact on Indoor Air Quality

Air quality improved markedly. The air was cleaner and humidity more regulated. Last but not least, the airtight building prevents all drafts.

H.1.21.3. Practical Experiences of Interest to a Broader Audience

H.1.21.4. Resulting Design Guidance

Orientation, glass facades, building details. All important in relation to energy need and climate conditions.

H.1.21.5. Space Utilization Changes

A thicker facade, but most of the time the building can be expanded on the outside.

Appendix I: Case Study—United Kingdom

I.1. The Mildmay Centre, London, UK

I.1.1. The Mildmay Centre, London, UK

I.1.2. Pictures

Figures I.1, I.2, I.3, I.4 and I.5 show the building in its original and post-retrofit states, illustrating key features of the retrofit (Figs. I.6, I.7 and I.8).

I.1.3. Project Summary

I.1.3.1. Project Objectives

Through the retrofit of the Mayville Community Centre, the architects wanted to find out if a deep retrofit can reduce a building's energy demand sufficiently to make an all-electric grid-connected building a viable alternative (in spite of electricity grid transmission losses), in terms of operational costs, to one that burns fossil fuel on site to provide heating. The underlying reason for this interest was the wider question: if a large enough overall energy saving could be achieved by replication of the hoped-for results, would a low carbon electricity grid using largely renewable forms of energy be a viable alternative to one that relies on burning fossil fuel?.

I.1.3.2. Project Energy Goals

The aim was to reduce overall energy consumption by at least 80%.

R. Lohse et al. (eds.), *Deep Energy Retrofit—Case Studies*,
https://doi.org/10.1007/978-3-031-17517-6

Fig. I.1 After retrofit: south elevation

Fig. I.2 After retrofit: interior of main hall with offices to left

Fig. I.3 After retrofit: dining/community recreation space

Fig. I.4 Before retrofit: east elevation

Fig. I.5 Before retrofit: boiler room

Fig. I.6 200 XPS below ground mineral wool

Fig. I.7 290 mm EPS above ground

Fig. I.8 w roof, 400 mm

I.1.3.3. Short Project Description

The proposal was to carry out a Passive house retrofit. This approach was virtually unknown in the UK at the time. It was proposed to open up the basement to gain windows and light and thereby get additional office space without needing to build a new extension. The new windows on the south elevation in the basement would flood what were cold and dark spaces with daylight, and along with the new south facing windows at the ground floor and those already at first floor, harvest sufficient warmth from the sun (and occupants) to provide most of the heat required to maintain comfortable temperatures throughout the year.

I.1.3.4. Stage of Construction

Completed 2012.

I.1.3.5. Point of Contact (POC) Information

The Practice Manager.
Bere: Architects.
73 Poets Road.
London N5 2SH.

I.1.3.6. Date of the Report

October 2015.

I.1.3.7. Acknowledgement

Innovate; formerly the UK Technology Strategy Board.

I.1.4. Site

- The Mildmay Centre, Woodville Road, London N16 8NA.
- Latitude: 51.550825 N, longitude: 0.082346.
- Elevation: 27 m.

I.1.5. Building Description/Typology

- Type: Community center containing offices and recreation spaces.
- Typology/Age: Pre 1910, solid brick, 475–595 mm thick, built in the 1890s as a generating station for London's tram network.

I.1.5.1. General Information

- Year of construction: 1890.
- Year of previous major retrofit—if known: 1973.
- Year of renovation (as described here): 2010–2012.
- Total floor area (m^2): 665.

I.1.6. Architectural and Other Relevant Drawings

See Figs. I.9, I.10, I.11, I.12 and I.13.

I.1.7. National Energy Use Benchmarks and Goals for Building Type Described in the Case Study

I.1.7.1. Benchmark

- TM46 benchmark is almost 200kWh/m^2 year
- The Hybrid Econ 19 user-specified benchmark tailored to match the mixed use of the building is 160kWh/m^2year.

I.1.8. Awards or Recognition

- Winner: UK Passivhaus Awards 2012, Retrofit category (UK Passivhaus Trust).
- Winner: 3R Awards (Refurb, Rethink, Retrofit) 2011, Best Public Building (The Architects Journal, Construction News and Civil Engineer).
- Winner: Green Build Awards 2012, Leisure category. "Recognizing excellence in sustainable buildings, with a particular focus on those projects that can illustrate an understanding of the importance of building performance in use."

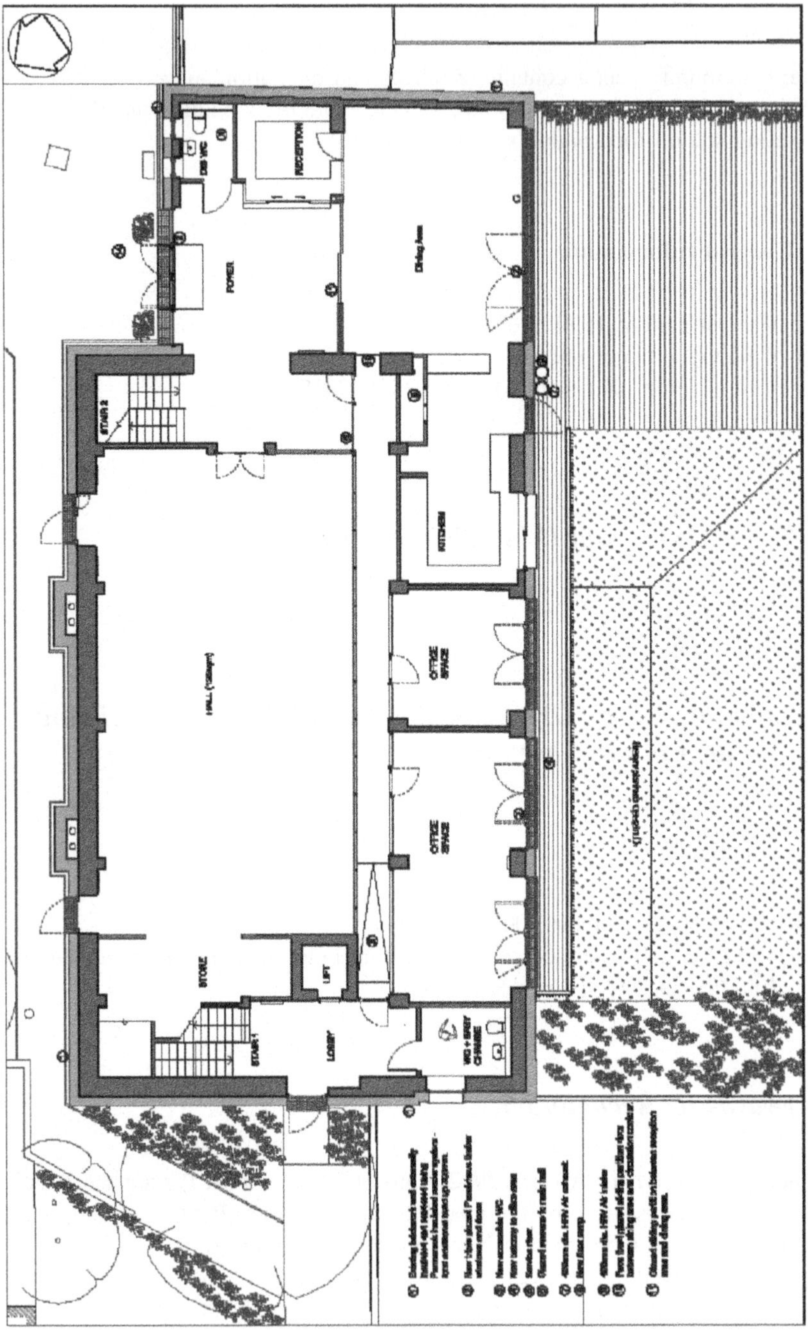

Fig. I.9 Situation plan

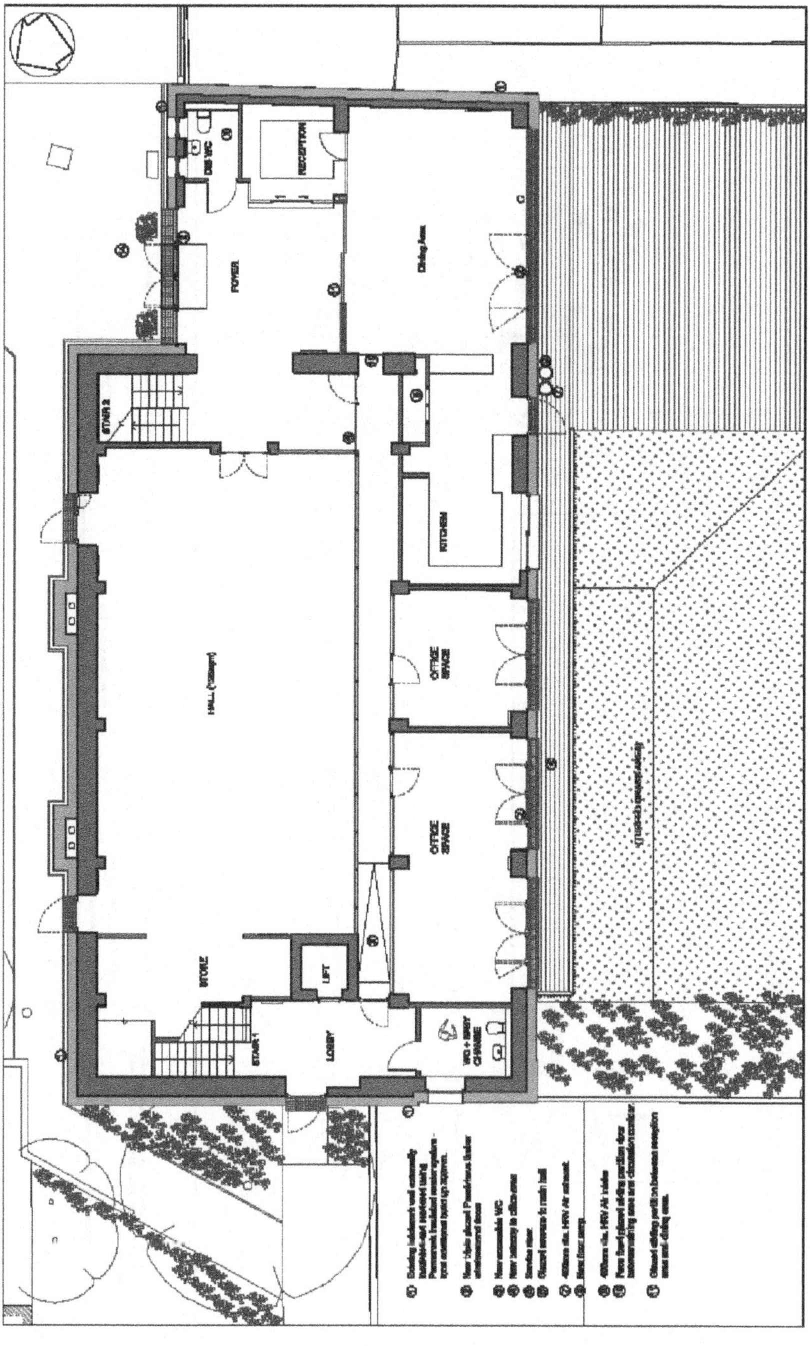

Fig. I.10 Ground floor

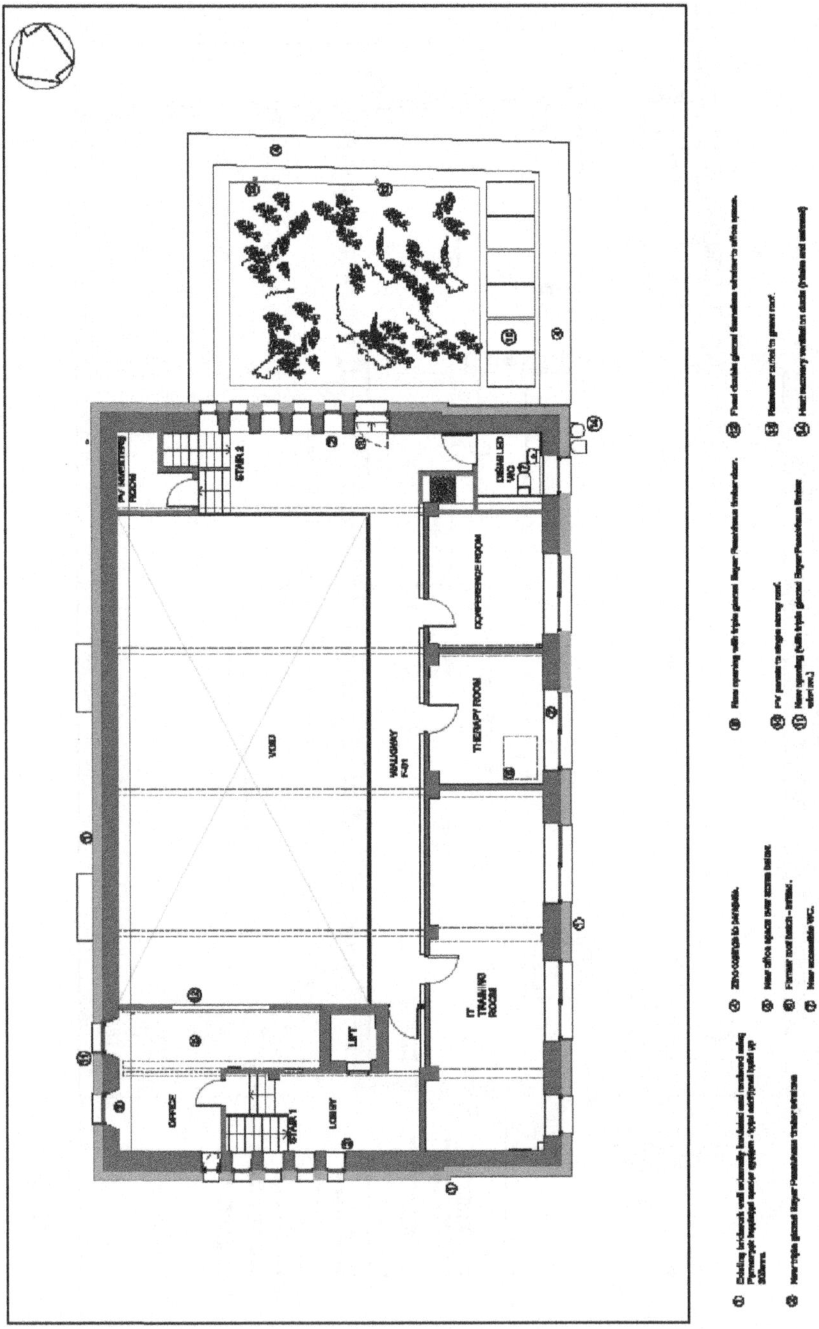

Fig. I.11 Second floor

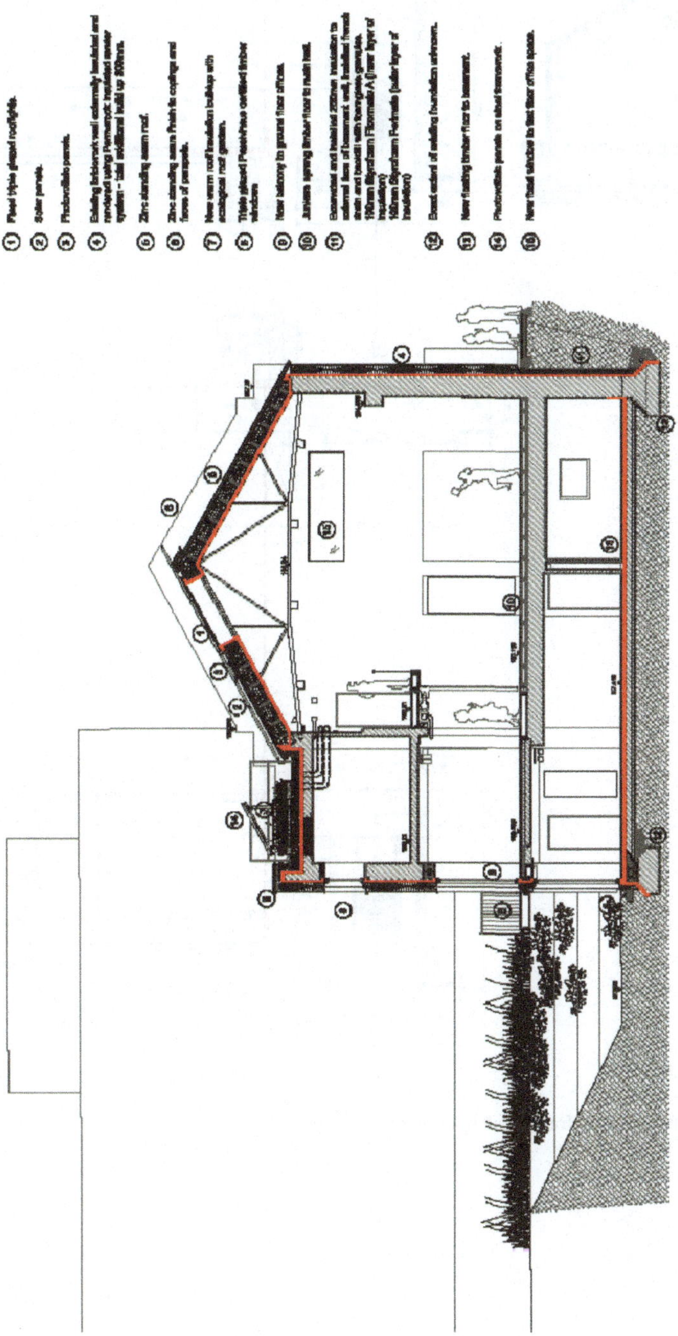

Fig. I.12 Section plan

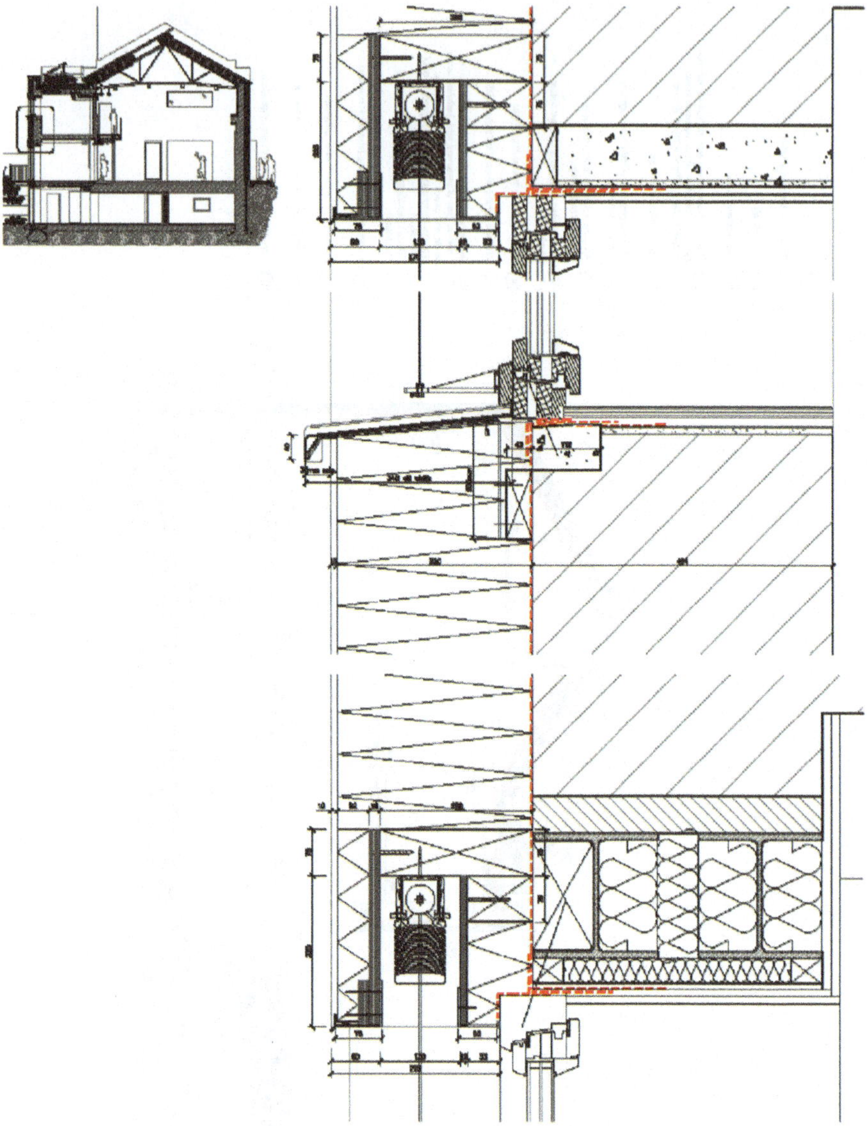

Fig. I.13 Key junction-Horizontal floor with external facade

- Winner: Constructing Excellence Awards, Building Performance in London and SE England 2012.
- Certified Passivhaus: Passivhaus Institut, Darmstadt, Germany.

I.1.9. Site Energy Cost Information

Not available.

I.1.10. Pre-renovation Building Details

I.1.9.1. Envelope Details

Walls: solid brick 475–595 mm thick.
Roof: corrugated asbestos sheeting, twin layer with 20 mm unidentified fiber between layers.
Windows: steel framed, single glazed Crittal windows.
Insulation levels: no insulation.

I.1.10.2. Heating, Ventilation, Cooling, and Lighting Systems

- Heating: large gas boiler supplying hot water radiators.
- Ventilation: opening windows.
- Cooling: none.
- Lighting systems: Fluorescent and incandescent.

I.1.11. Description of the Problem: Reason for Renovation (Non-energy and Energy Related Reasons)

- Users found the old building cold, draughty, dark and uninviting.
- Energy bills of the old building amounted to £10,000 a year, which formed a large proportion of the Centre's annual turnover of £60,000.

I.1.12. Energy Saving/Process Improvement Concept and Technologies Used

I.1.12.1. Building Envelope Improvement

- External wall insulation above ground generally: 290 mm expanded polystyrene insulation (external).
- External wall insulation above ground, single story extension: 320 mm expanded polystyrene insulation (external).
- External wall insulation below ground: 200 mm extruded polystyrene insulation (external).
- Roof insulation (sloping): 400 mm mineral wool insulation.
- Roof insulation (flat): 300 mm Foamglas insulation.
- Basement slab: 75 mm polyurethane foam insulation (internal).
- Ground slab to single story addition: 300 mm Foamglas insulation (beneath slab).
- New triple glazed Passive house windows with insulated timber frames. Tilt-and-turn opening mechanism to facilitate secure summer night-purge ventilation.
- Three new opening triple glazed Velux Passive house roof lights over main hall, electrically operated for summer night-purge ventilation. Together with two large new triple glazed fixed roof lights over main recreation space.
- Secure manual air intake panel for secure summer night-purge ventilation (Fig. I.14).

Fig. I.14 Manual air intake panel for secure night-purge ventilation

- Comprehensive air-tightness measures, which resulted in final airtightness test result of 0.5 h^{-1} (average of compression and decompression under 50_{pa}).
- External retractable and adjustable louvre blinds for summer shading.

I.1.12.2. New HVAC System or Retrofits to Existing

- New Paul Maxi heat recovery ventilation system installed. Constant pressure system, with cascade air supply to the two variable-occupancy recreation areas and CO_2 sensing in the two recreation areas to open valves for direct ventilation of these two areas if CO_2 levels rise above 1100 ppm.
- New Viessmann ground source heat pump installed (8 kWp).

I.1.12.3. New Lighting System

Compact fluorescent lighting throughout.

I.1.12.4. Renewable Energy

- 1 no. Viessmann 3 m^2 vacuum tube solar thermal panel.
- 77 no. Sharp NU-E235E1 photovoltaic panels, power output 18 kWp.

I.1.12.5. Daylighting Strategies

- Strategy was manual switch on, manual switch off, with dimmable daylight-sensing automatic override. However dimmable daylight-sensing automatic override largely omitted to save money.
- Comprehensive sub-metering to monitor energy use (Fig. I.15).

I.1.13. Energy Consumption

I.1.13.1. Pre-renovation Energy Use (Total and per m^2/year)

More than 270 kWh/m^2/year and the building was still freezing in Winter!

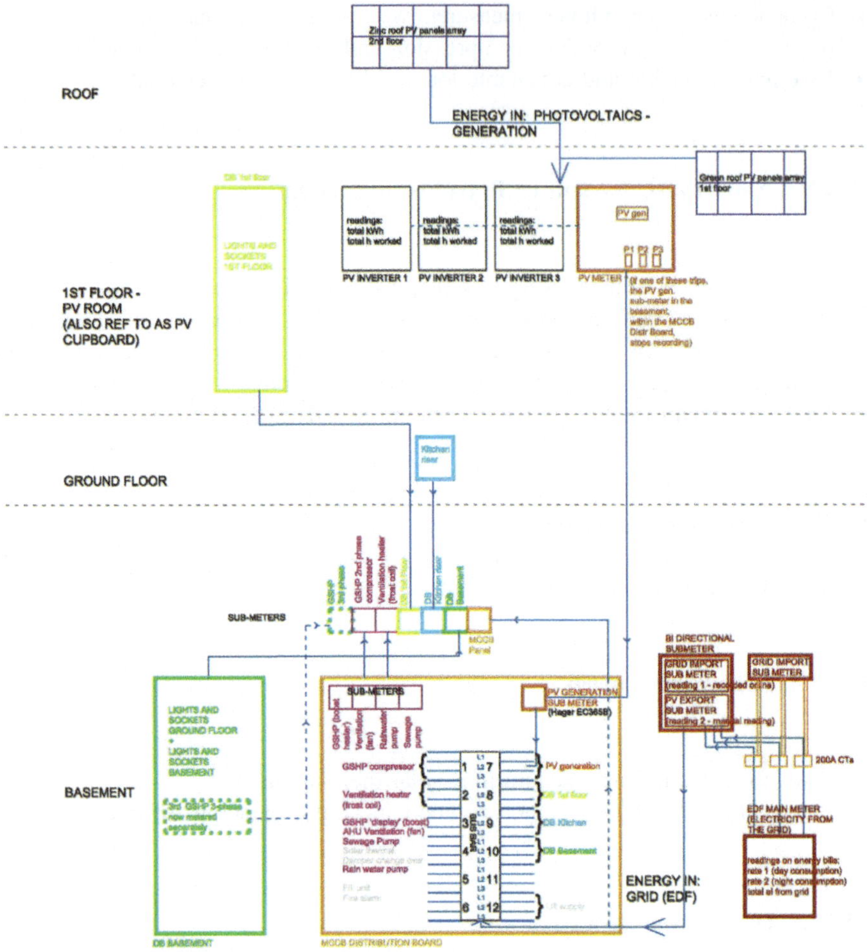

Fig. I.15 Schema of energy services

I.1.13.1. Predicted Energy Savings (Site, Source, GHG), Total and per m^2/year

The renovation provides energy savings of 80%, while at the same time ensuring warm and comfortable winter temperatures.

I.1.13.2. Graph Comparing Predicted and Actual Energy Savings

See Fig. I.16.

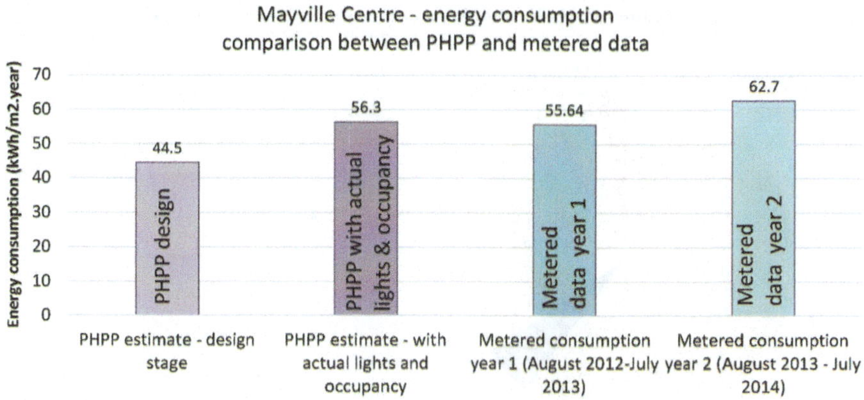

Fig. I.16 Comparison of predicted and monitored energy saving

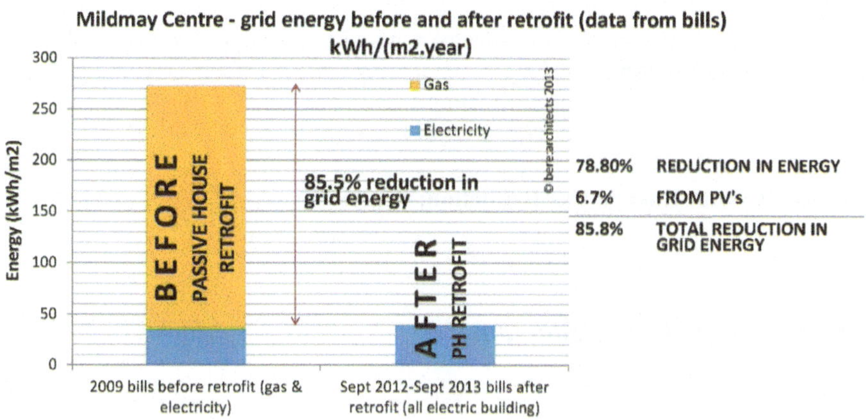

Fig. I.17 Electrical energy before and after retrofit

I.1.13.3. Measured Energy Savings (Thermal, Electrical), Total and per m^2/year

The evaluation team compared the energy savings and the cost of bills, before (in 2009) and after the retrofit (Sept 2012–Sept 2013). The energy used from the grid after the retrofit (electricity only, all-electric building) is 85.5% lower than before the retrofit (gas and electricity), in spite of the large increase in the occupancy level of the post-retrofit building. Figure I.17 shows a graph of the energy savings.

The bills show a 58.5% reduction before and after retrofit, due to reduced energy consumption. An estimation of the energy used before the retrofit (kWh) in current energy costs (if the retrofit had not happened) shows that the current bills represent

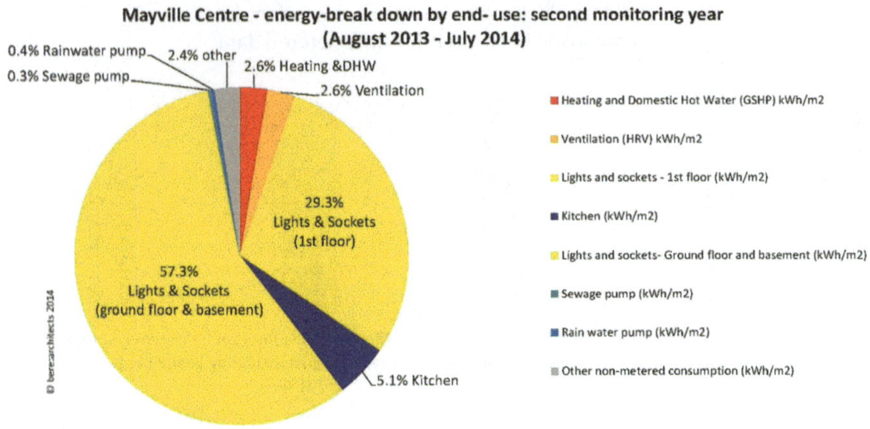

Fig. I.18 Energy use break down

savings of 72.7%. Again this is in spite of a large increase in the occupancy level of the building post-retrofit.

Graph of energy use break-down (Fig. I.18):

I.1.13.4. Annual Energy Use Reduction: 85.5%

I.1.14. Energy Cost Reduction

I.1.14.1. Split in All Energy Forms

- Before retrofit: electricity, gas.
- After retrofit: all electric building (designed to fit the vision of a low consumption consumer environment, served by a low carbon electricity grid, supplied with 100% renewable energy.).
- Cost savings (Fig. I.19).

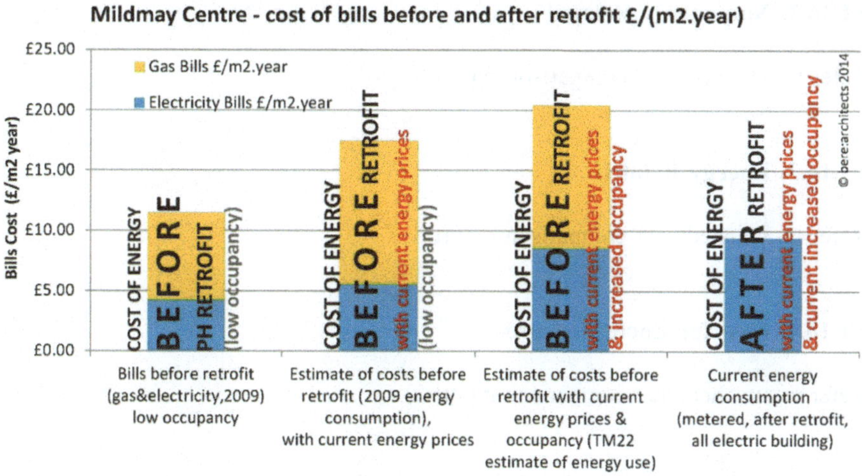

Fig. I.19 Cost of energy before and after retrofit

I.1.15. Non-energy Related Benefits Realized by the Project, (E.g., Improved Productivity, Increased Rent/Lease, Increased Useful Space, Reduced Maintenance, Etc.

- Increased rent.
- 35% additional lettable office space.
- Increased quality of physical environment.
- Increased comfort.
- Improved appearance.
- Reduced maintenance.
- Increased asset value.

I.1.16. Renovation Costs: Total and per m^2

I.1.16.1. Total

Total costs amount to £1,600,000, of which only 7.84% relates to additional measures beyond the requirements of the UK Building Regulations for a general refurbishment.

I.1.16.2. Non-energy Related

Refer to attached cost comparison report.

I.1.16.3. Energy Related

Refer to attached cost comparison report.

I.1.16.4. Cost for Each Measure

Refer to attached cost comparison report.

I.1.17. Business Models and Funding Sources

None available.

I.1.18. Cost Effectiveness of Energy Part of the Project (NPV, SIR, Etc.) (Including Description of Framework Date Such as Calculated Interest Rates, Life Cycle Period)

None available.

I.1.19. User Evaluation

I.1.19.1. Description of User Training Programs Within the Refurbishment

- User walk-around with design team.
- Presentation and discussion forum of initial performance results after 1st year of monitoring.
- User Guide (Fig. I.20).

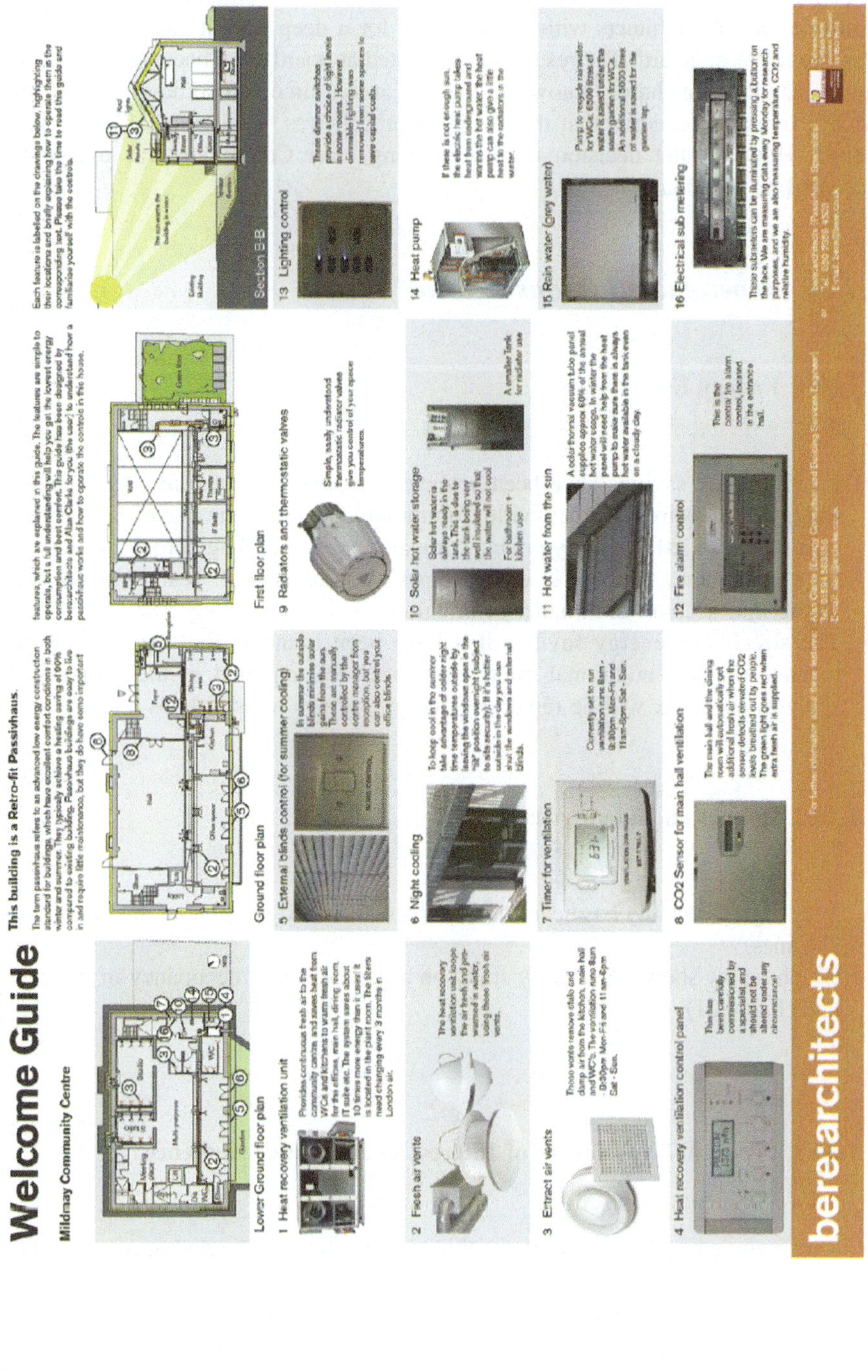

Fig. I.20 User guide for training programs

I.1.19.2. Integration of Users Demands in the Planning Process

A panel of building users interviewed the three architects competing for the project and selected the architects with the proposal for a deep retrofit carried out to the Passive house standard. The result was contested at board level because the Passive house standard was hardly known in the UK at that time and some trustees of the MCP expressed concerns that the design approach was "experimental" and "very risky." However the decision of the Building User Committee was eventually accepted by the Board.

I.1.20. Experiences/Lessons Learned

I.1.20.1. Energy Use

- 85.5% energy savings have been achieved, in spite of increased occupancy.
- Energy savings from the Mayville Community Centre, at the rate of £20,000 per year (based on 2014 energy prices) compared to the building before retrofit, will save more than £1million in energy costs over the next 50 years without even taking into account the impact of inflation.
- The size of the energy savings that have been found in this research project indicate that the additional build costs of 3–8% reported in the capital cost analysis appendix will be repaid by energy savings in as little as 3 years.

I.1.20.2. Impact on Indoor Air Quality

- Very high air quality has been found to be maintained throughout the building at all times.
- Graph below shows air quality study on a typical day of occupancy in the main hall (Fig. I.21).

I.1.20.3. Practical Experiences of Interest to a Broader Audience

- 85.5% energy savings have been achieved, in spite of increased occupancy. The results endorse the substantial benefits that can be achieved by a rigorously applied deep fabric-first retrofit strategy.

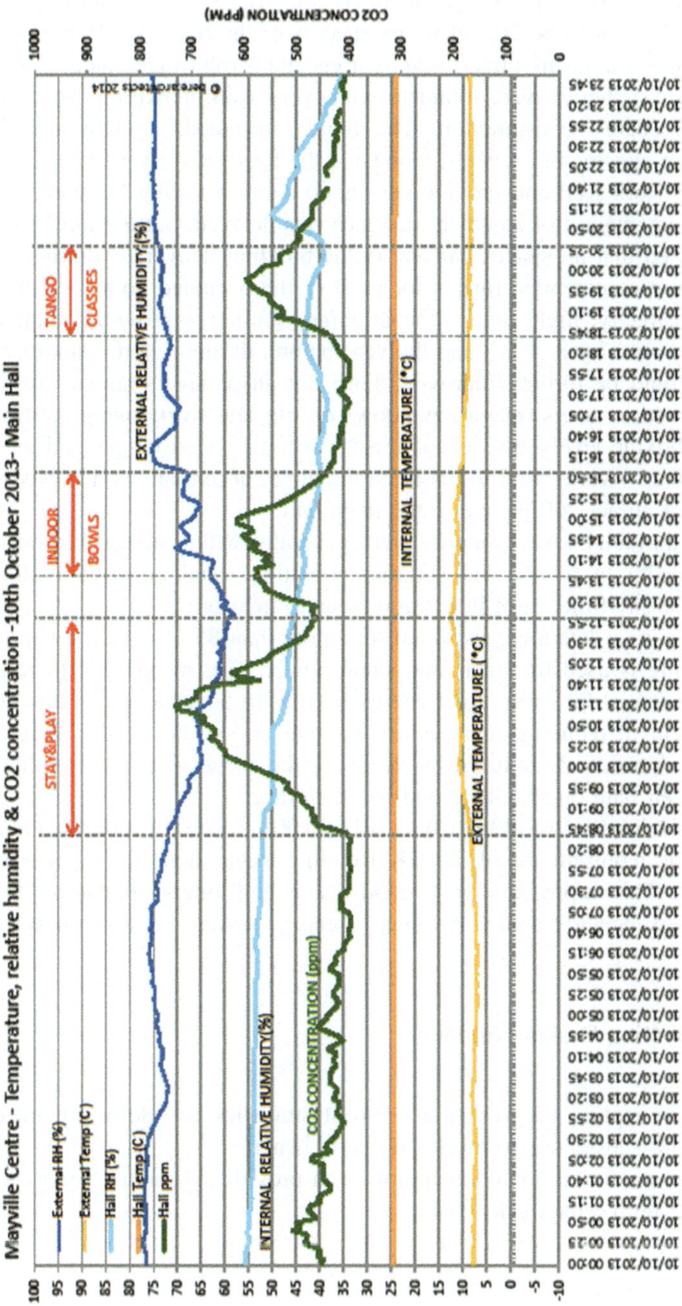

Fig. I.21 Monitors air quality

- The simple controls strategy was found to be successful. The design decision to avoid having a Building Management System (BMS) was found to be a good decision, thought to be critical to the success of the building. The simple controls strategy was easily understood by users and is robust and maintenance-free.
- The winter strategies were found to work very well, delivering good comfort with low energy consumption. The highly insulated, draught and thermal bridge-free construction details, alongside the heat recovery ventilation system proved to work well and provide very good comfort during the winter.
- The implementation of summer strategies was delayed, but the third summer's data showed that all spaces bar one in the building delivered summer comfort within the recommended ranges for most of the occupied hours. When implemented correctly, night-time summer purge ventilation provided temperatures generally lower by 3–4 °C than the year before, during a hotter summer. Even so, this could be further improved, since not all passive strategies were fully implemented by users (leaving windows on tilt, effectively using the blinds).
- Carbon dioxide levels are well within the recommended ranges and the ventilation strategy works well to dissipate the high concentrations achieved during events that attract 200 people to the main hall.
- Relative humidity levels indoors were generally within acceptable ranges for human health in winter and summer conditions, and no condensation was reported at any time, even in bathrooms and kitchens.
- The monitoring data analysis found that more than 85% of the energy used in the building is used for lights and small power. This means that there is substantial potential for even larger energy savings to be achieved with better user engagement in reducing unnecessary lighting and socket loads. Socket loads could be significantly reduced by removing the six servers in the building and putting their data on remote, cloud-based storage.
- Only 7.84% of the cost of the works (£1,600,000) relates to additional measures beyond the requirements of the UK Building Regulations for a general refurbishment. This represents very good value and offers short-term financial pay-back of the additional cost when the energy savings are taken into account.

I.1.20.4. Resulting Design Guidance

- Rigorously applied Passive house retrofit strategies can deliver large energy savings and substantial comfort and health benefits.
- Relatively small additional expenditure can provide substantial energy savings with short-term repayment of costs.

I.1.20.5. Space Utilization Changes

The retrofit was part of a general refurbishment of the building, and efficiencies achieved in space-planning have created an additional 35% of lettable office space.

I.1.20.6. Follow Up on the Renovation

- The simple controls strategy was found to be successful. The design decision to avoid having a Building Management System (BMS) was found to be a good decision, thought to be critical to the success of the building. The simple controls strategy was easily understood by users and is robust and maintenance-free.
- Once users began to understand how to use the external blinds and the summer night purge ventilation strategy, a significant reduction in summer daytime temperatures was achieved, with further potential for improvement.

I.1.21. References

Report commissioned in 2012 and published in January 2015 by the UK Technology Strategy Board, Building Performance Evaluation programme; Non-domestic Buildings Phase 2—Buildings in Operation: Mayville Community Centre, Final Report.

17 no. Appendices attached to the above report.

Appendix J: Case Studies: United States

J.1. Grand Junction, CO

J.1.1. Name of the Project, Location (City, Country)

Wayne N. Aspinall Federal Building and U.S. Courthouse Partial Modernization.

J.1.2. Pictures

J.1.2.1. Historic Photos

See Figs. J.1, J.2, J.3, J.4, J.5 and J.6.

J.1.2.2. Preconstruction Photos

See Figs. J.7, J.8 and J.9.
Post Construction:
See Figs. J.10, J.11, J.12, J.13, J.14, J.15 and J.16.

J.1.3. Project Summary

The General Services Administration (GSA) proposes to protect this historic asset by improving energy efficiency through mechanical components and modernizing systems. This will be accomplished by pursing and contracting with a CM firm who

R. Lohse et al. (eds.), *Deep Energy Retrofit—Case Studies*,
https://doi.org/10.1007/978-3-031-17517-6

Fig. J.1 1918

Fig. J.2 1938 addition, in construction

Fig. J.3 1938 addition post construction

Fig. J.4 1918 interior construction

Fig. J.5 1938 lobby post construction

Fig. J.6 1938 postal workroom

Fig. J.7 2010 1st floor office

Fig. J.8 2010 remaining

will also assist GSA in the procurement of a Design Build firm to address all energy upgrade measures. In addition, GSA will contract with a commissioning contractor for systems evaluations. Award of the DB and Cx is anticipated to take place in June 2010 along with the DB firm.

Fig. J.9 2010 corridors

Fig. J.10 2013 lobby east view

The modernization will be accomplished while the building remains partially occupied. The project will modernize the entire square footage of the building and will be completed in phases that will allow tenants to remain in the building.

Fig. J.11 2013 lobby west view

Fig. J.12 2013 lobby

Fig. J.13 2013 3rd floor corridor

Fig. J.14 2013 south elevation

Fig. J.15 2013 west elevation

Fig. J.16 2013 north elevation

J.1.3.1. Project Objectives

- Provide a pleasant, secure, and safe environment for the FB staff and visitors consistent with guidelines for Federal facilities and specifically the GSA PBS P-100.
- Satisfy current and projected special-use needs, and allow for future expansion of these requirements.
- Design facilities compatible with the tenants' existing character and local context, and present a positive image of U.S. Government facilities (welcoming, but formal).
- Use the Leadership in Energy and Environmental Design (LEED) Green Building Rating System, which is the nationally accepted benchmark for the design, construction, and operation of high-performance green buildings. LEED gives building owners and operators the tools they need to have an immediate and measurable impact on their buildings' performance. LEED promotes a whole-building approach to sustainability by recognizing performance in five key areas of human and environmental health: sustainable site development, water savings, energy efficiency, materials selection, and indoor environmental quality. GSA requires all new construction, renovation, and modernization projects to be certified through the LEED program, with design teams required to achieve Silver ratings.
- Owing to the historic significance of the building, all work will follow the Secretary of the Interior's Standards and Illustrated Guidelines for Rehabilitating Historic Buildings, Revised 1992 (36 CFR 67).
- Achieve the status of a high-performance green building meeting federal energy and water conservation goals and incorporating exceptional integration of architectural form and optimizing building systems.
- Be responsive to the local/regional climatology and explore the use of available on-site renewable energy sources, such as solar, geothermal, and water sources, to optimize building orientation, envelope, and fenestration strategy.
- Use durable, sustainable materials.

- Provide healthy, functional space for the tenants and exceptional comfort in thermal perspectives.
- Facilitate cost effective sustainable operations and maintenance.
- Conform to ISC and all security criteria established for this project.
- Use our nation's most talented architects, engineers, landscape architects, interior designers, and contractors.
- Meet all Federal energy and water conservation goals and security requirements specifically the Guiding Principles for Sustainable New Construction and Major Renovations of Executive Order 13423 and the Energy Policy Act of 2005, as well as the fossil fuel reductions, renewable energy and water conservation goals of the Energy Independence and Security Act of 2007.
- Upgrade by replacing the heating, ventilation, and air-conditioning (HVAC) systems..
- Upgrade by replacing all mechanical piping, heating, and hot water systems.
- Comply with the Minimum Performance Criteria of the American Recovery and Reinvestment Act of 2009 (ARRA).
- Upgrade the electrical power system as needed to accommodate other renovations, provide new energy efficient lighting throughout.
- Modernize the elevator. SHPO and accessibility requirements must be fulfilled.
- Remove all asbestos containing materials disturbed as part of this project (ACMs).
- Implement GSA's workplace procedures and achieve measurable effectiveness results.
- Identify design strategies that work to optimize the project's access to energy from seasonal solar, natural lighting, and wind to reduce undesirable loads and maximize desirable loads to meet desired energy targets for the facility (currently targeting an improvement of 30% over American Society of Heating, Refrigerating and Air-Conditioning Engineers (ASHRAE 90.1-2007).
- Upgrade all public and private restrooms, plumbing systems, and fixtures. Provide new hot, cold, and waste water piping.
- Provide lobby security screening station and infrastructure for security screening equipment.
- Upgrade the Federal Building's main lobby to achieve a positive image of U.S. Government Facilities aesthetically in line with the Federal Building's historical context.
- Upgrade interior finishes within public and common space.
- Reconfigure tenant space on the first and second floors to meet tenant space requirements.
- Repair and upgrade the building exterior and envelope as defined in the Program of Requirements.
- Replace the upper and lower roofing systems.
- Upgrade tenant space core and shell components disturbed by the mechanical work.

J.1.3.2. Project Energy Goals

- Pursue Net Zero Energy.
- Pursue LEED Platinum.
- Improve indoor environmental quality and thermal comfort.
- Reduce water use.
- Use sustainable construction practices.
- Effectively use the technology available today not only for the purposes of deep energy conservation, but also as a tool for historic preservation.

J.1.3.3. Short Project Description

Condensed Version—Design Build contractor to design and install highly efficient mechanical systems that offer deep energy savings in addition to supporting the building's overall historic preservation effort. Design and install new electrical systems, lighting, occupancy sensors, and controls in-line with the building's original design, and supporting the overall building's energy efficiency. Design and install new plumbing systems and fixtures that offer water savings and replace the buildings aging piping systems. Using the existing cab, install a new elevator system with regenerative motor. New cab interior to reflect the building's historic interior. The building's overall preservation will focus on public spaces restoring original finishes. Preservation effort will restore the building's original volumes and re-incorporate the historic Postal lobby within the new layout. Tenant spaces will preserve the historic character of defining features such as windows, ceiling heights, etc., but will provide a modern working environment for agencies, supporting their mission. The design build contractor will propose innovative solutions to achieve net zero energy and LEED Platinum certification.

J.1.3.4. Stage of Construction

Substantial Completion/Contract Close-Out.

J.1.3.5. Point of Contact Information

Project Manager.
1 Denver Federal Center, Bldg. 41, Rm. 240.
Denver, CO 80225.
jason.sielcken@gsa.gov.
303,236,2972.

J.1.3.6. Date of the Report

January 18, 2014.

J.1.3.7. Acknowledgement

Demetra Chavez.
Project Sponsor/Business Center Manager.
1 Denver Federal Center, Bldg. 41, Rm. 240.
Denver, CO 80225.
demi.chavez@gsa.gov.
303,236,3678.

J.1.4. Site

- Location:

400 Rood Avenue.
Grand Junction, CO 81501.

- Coordinates. Latitude: 39.068585 N; Longitude: − 108.565682 W.
- Elevation: 4839 ft. above sea level.
- Climate: ASHRAE Standard 90.1-2007 Zone 5B (Table J.1).

J.1.5. Building Description/Typology

J.1.5.1. Type

- Primary Usage Type: Office Building/Courthouse.
- % Area Assigned to Primary Use: 100%.

Table J.1 Design temperature

Cooling design temperature		
Dry bulb temp C (°F)	Mean coincident wet bulb temp C (°F)	Enthalpy at xF/xF
97.4	61.9°F	27.5 Btu/lb
Heating design temperature		
Dry bulb temp C (°F)	Mean coincident dry bulb (°F)	Enthalpy at xF/xF
65.5	85.4	30.2 Btu/lb

- Most of the building is classified as Business Occupancy by NFPA 101 criteria, and as Use Group B as defined by IBC. The 3rd floor courtroom is designated as a Class C Assembly Occupancy by NFPA 101 criteria, and as Use Group A-3 as defined by IBC.
- The Federal Building is classified as a Business Group B occupancy because of the office and courtroom use. A Business Group B occupancy of Type IIIB construction.

J.1.5.2. Age

- Year of Original Construction: 1918.
- Year of Previous Major Retrofit: 1938; 1961.
- Year of Current Retrofit Complete: 2013.

J.1.5.3. General Information

- Gross Area (m^2): 41,564 GSF; 3861.422 m^2.
- Other information as appropriate: Listed on National Register of Historic Places, 1980.

J.1.6. Architectural and Other Relevant Drawings

See Pictures.

J.1.7. National Energy Use Benchmarks and Goals for Building Type Described in the Case Study

J.1.7.1. Benchmark

ASHRAE 90.1-2007; *Target Energy Performance Rating: 100; Median Energy Performance Rating: 50; Median Site Energy Use Intensity: 64 kBtu/sf/year; Median Source Energy Use Intensity: 214 kBtu/sf/year.*

J.1.7.2. National Energy Target for This Type of Building

Same as ENERGY STAR Target: 100.

J.1.8. Awards or Recognition

- LEED Platinum Certification, September 2013.
- AIA Colorado—2013 Honor Award for Built Architecture.
- AIA Denver—2013 Citation Award for Interior Architecture.
- AGC Colorado—2013 Award for Construction Excellence (ACE Award).
- ENR Mountain States—2013 Best Renovation/Restoration Project Colorado.
- DBIA RMR—2013 Best Project, Rehabilitation/Renovation/Restoration.

J.1.9. Site Energy Cost Information (Electricity, \$/kWh; Gas, \$/m^3, Etc.)

\$0.0934 per kWh, regional average.

J.1.10. Pre-renovation Building Details

J.1.10.1. Envelope Details: Walls, Roof, Windows, Insulation Levels

- Description of Construction Type: The Wayne N. Aspinall Federal Building and U.S. Courthouse is a three-story Second Renaissance Revival style building situated one block north of Main Street at the corner of Rood and 4th Street in Grand Junction, CO. The three-story limestone faced steel structure is rectangular in plan with a second and third floor light court in the rear. Slab construction at the basement, 1st and 2nd floors is concrete. The third floor and attic/roof consist of wood decking.
- Wall Materials: Historic walls consist of plaster, lath, over terra-cotta block, air gap, then limestone facade.
- Wall Insulation: None.
- Wall U Value (W/m^2-K): 0.183 Btu/h-sf-F.
- Attic Above Conditioned Space: Yes. Third Floor.
- Roof Materials: EPDM two ply roofing.
- Roof Insulation: R-15 rigid roof insulation.
- Roof U-value (W/m^2-K): 0.06 Btu/h-sf-F.

- Insulation Location (roof or attic floor): Roof.
- Thermal Bridge Mitigation (Y or N) No.
- Air tightness (ACH @ indicated delta P): Not Available.

J.1.10.2. Heating, Ventilation, Cooling, and Lighting Systems

- Lighting:
  - Primary Lighting Type: Fluorescent.
  - Primary Lighting LPD (W/m^2): 2.0 W/sf.
  - Primary Lighting% of Floor Area: 100%.
  - Secondary Lighting Type.
  - Secondary Lighting LPD (W/m^2).
  - Secondary Lighting% of Floor Area.
  - Occupancy Sensors: No.
  - Type: NA.
  - % of Area Covered by Occ. Sensors: 0%.
  - % of Light Power Controlled by Occ: 0%.
- HVAC System Type: Constant Volume Air Handling Units, roof mounted DX condensing units, hydronic heating system including atmospheric boilers and pumps, unit heaters, convectors and radiators.
- HVAC System Description: The pre-construction HVAC system consisted of three constant volume air handling units (AHU) with two located above the ceiling on the first floor supporting the first and second floors of the building. A third AHU was located in the attic space supporting the third floor. Ducting was run through a plenum between the dropped acoustical ceiling and deck on the first and second floors. Ducting on the third floor was a combination of attic run duct and duct in the plenum between an acoustic dropped ceiling and the deck in the public corridor. Three roof mounted DX refrigeration type condensing units fed the AHU's. Hydronic heating systems supporting the entire building perimeter heating, were managed with atmospheric boilers and pumps located in the boiler room. Air intake and exhaust consisted of louvers located in removed windows on the north side of the building.
- Heating Supply Type (boiler or pump): Boiler.
- Heating Supply Efficiency or COP: Unknown.
- Cooling Supply Type (Chiller or Pump): Chiller.
- Cooling Supply COP: Unknown.
- Chiller Type (Air or Water): Air.
- Heat Pump Type (Air to Air, etc.) Water-to-Water.
- Heat Recovery System (Y or N): No.
- Heat Recovery Type: NA.
- Heat Recovery Effectiveness: NA.

- Economizer Controls (Y or N): No.
- Economizer Type (temp or enthalpy): NA.
- Demand Controlled Ventilation (Y or N): No.
- Max CO_2 limit with DCV: NA.
- Description of the problem: reason for renovation.

No capital expenditure had been made to the building in over 50 years. The mechanical systems, plumbing, electrical, roofing, and elevators had long surpassed their useful life. Funding through the American Recovery and Reinvestment Act of 2009 allowed the GSA to make necessary investments to upgrade the above mentioned systems for energy efficiency and the longevity of the property. Funding also allowed for a substantial historic preservation effort to restore the building following the Secretary of Interior Standards for the Treatment of Historic Properties.

J.1.11. Description of the Problem: Reason for Renovation (Non-energy and Energy Related Reasons)

No capital expenditure had been made to the building in over 50 years. The mechanical systems, plumbing, electrical, roofing, and elevators had long surpassed their useful life. Funding through the American Recovery and Reinvestment Act of 2009 allowed the GSA to make necessary investments to upgrade the above mentioned systems for energy efficiency and the longevity of the property. Funding also allowed for a substantial historic preservation effort to restore the building following the Secretary of Interior Standards for the Treatment of Historic Properties.

J.1.12. Renovation SOW (Non-energy and Energy Related Reasons)

Condensed Version: Design Build contractor to design and install highly efficient mechanical systems that offer deep energy savings in addition to supporting the building's overall historic preservation effort. Design and install new electrical systems, lighting, occupancy sensors, and controls in-line with the building's original design, and supporting the overall building's energy efficiency. Design and install new plumbing systems and fixtures that offer water savings and replace the buildings aging piping systems. Using the existing cab, install a new elevator system with regenerative motor. New cab interior to reflect the building's historic interior. The building's overall preservation will focus on public spaces restoring original finishes. Preservation effort will restore the building's original volumes and re-incorporate the historic Postal lobby within the new layout. Tenant spaces will preserve the historic character of defining features such as windows, ceiling heights, etc., but will provide a modern working environment for agencies, supporting their mission. The design build contractor will propose innovative solutions to achieve net zero energy and LEED Platinum certification.

J.1.13. Energy Saving/Process Improvement Concept and Technologies Used

J.1.13.1. Building Envelope Improvement

- Wall Materials.
  - Historic interior walls consist of plaster, lath, over terra-cotta block. Perimeter walls upgraded through the project (all walls with the exception of the historic courtroom and west stair) consist of lime-stone facade, 2-in. average spray foam insulation, covered by gypsum board. The new perimeter walls are R-12 (average). Interior storm panels were installed inboard of existing single pane windows. A high performance 3 M solar control film is installed on the inner surface of the storm window glazing.
  - Wall Insulation: Perimeter Walls: R-12 average, 2-in. average spray foam insulation (see above).
  - Wall U Value (W/m^2-K): 0.085 Btu/h sf-F.
- Attic Above Conditioned Space (Y or N): Yes.
- Roof Materials: R-30 rigid, tapered roof insulation. Single-ply, white membrane, TPO roofing, SRI value 110.
  - Roof Insulation: R-30 rigid, tapered roof insulation.
  - Roof U Value (W/m^2-K): 0.3 Btu/h-sf-F.
- Thermal Bridge Mitigation: Yes.
- Air Tightness (ACH @ indicated delta P): NA.

J.1.13.2. New HVAC System or Retrofits to Existing

- HVAC System Type

Mitsubishi Variable Refrigerant Flow (VRF) and decoupled Dedicated Outdoor Air (DOAS) unit.

- HVAC System Description

Ventilation and Exhaust—The required ventilation and exhaust for the facility is provided by a roof-mounted, custom, 4500 cfm Dedicated Outdoor Air (DOAS) unit, decoupled from the heating and cooling system. Centrifugal plenum-type supply and exhaust fans with variable frequency drives (VFD). Indirect evaporative cooling coil and heating water coil. Heating water coil is connected the heating water system. Counter flow plate heat exchanger to pre-condition outside air. 2-in. flat pre-filter (MERV-7) and 12-in. cartridge type final filter (MERV-14). Approximately 31 VAV boxes installed and connected to the ventilation ductwork system. The ventilation system is fully ducted, with the building return and exhaust combined and connected to the return connection of the DOAS unit for pre-conditioning the incoming outside air.

Decoupled Heating and Cooling

The heating and cooling system is a Variable Refrigerant Flow (VRF) fan coil system. A variety of fan coil styles are used depending upon the architectural design of the space. The fan coils are a combination of recessed cassette, concealed (ducted) and ductless (horizontal or vertical cabinet style exposed). The VRF fan coil units are connected to 6, 8, and 10-ton water-cooled condensing units installed within the boiler room. A complete refrigerant piping system, connects the VRF fan coil units to the condensing unit. The cooling coil condensate is collected within the basement storage tank and pumped into the indirect evaporative cooling coil of the DOAS.

Heating Plant.

Two high efficient, heating water-to-water heat pumps are installed in the basement boiler room. The heat pumps are sized for the DOAS unit heating coil, and the perimeter heating devices. Perimeter heating devices include first floor finned tube radiation, cabinet unit heaters, and unit heaters.

Ground Loop Heat Exchanger (Geo-Exchange Well Field).

The geo-exchange well field is used to reject/extract heat to/from the facility. The well field consists of 32 vertical wells, spaced 15 ft on center, at 475 ft, 0-in. deep. Geo-exchange pumps circulate water between the heat pumps, water—cooled condensing units and the well field. The geothermal water is distributed by using two alternating, centrifugal vertical inline pumps with VFD.

- Heating Supply Type (boiler or pump): Heat Pumps.
- Heating Supply Efficiency or COP: 5.0.
- Cooling Supply Type (Chiller or Pump): Mitsubishi VRF, Water Cooled.
- Cooling Supply COP: 4.4.
- Chiller Type (Air or Water): Mitsubishi VRF, Water Cooled.
- Heat Pump Type (Air to Air, etc.) Water-to-Water.
- Heat Recovery System (Y or N): No.

- Heat Recovery Type: NA.
- Heat Recovery Effectiveness: NA.
- Economizer Controls (Y or N): No.
- Economizer Type (temp or enthalpy): NA.
- Demand Controlled Ventilation (Y or N) Yes.
- Max CO_2 limit with DCV: 750 ppm.

J.1.13.3. New Lighting System

- Primary Lighting Type: T8 Lamps.
- Primary Lighting LPD (W/m^2): 0.76 W/sf (building average).
- Primary Lighting% of Floor Area: 80%.
- Secondary Lighting Type: LED Lamps.
- Secondary Lighting LPD (W/m^2): 0.76 W/sf (building average).
- Secondary Lighting% of Floor Area: 20%.
- Occupancy Sensors: Magnum Energy Solutions occupancy sensors have been installed throughout the building and are the primary occupancy sensor type.
- Type: Primarily Manual On/Auto Off. In some locations that are public interfacing or in areas with no natural daylight, the sensors are programmed Auto On/Auto Off (public restrooms, interior stairwells, etc.).
- % Of Area Covered by Occ. Sensors: 95%.
- % of Light Power Controlled by Occ: 99%.
- Daylight Dimming Controls: Yes.
- % of area covered by daylight dimming: 50%.
- % of Light Power Controlled by Daylight: 50%.

J.1.13.4. New Generation/Distribution System

All new panelboards, branch wiring, and receptacles were provided. Panelboards interface with the building automation system to provide real-time energy use data at each circuit.

J.1.13.5. Renewable Energy

Renewable energy consists of 385 SunPower E19-320 panels located directly on the roof and above the roof on an elevated canopy. The system has 123 kW DC rating.

J.1.13.6. Daylighting Strategies

Daylight harvesting controls are located on all light fixtures within 15 ft, 0-in. of a window. Since the building was constructed in 1918/1938, daylight strategies are inherent in the original design. This project removed all dropped ceilings covering full height windows to allow for maximum daylighting. A skylight original to the building on the lower roof of the north elevation light well was covered up in the 1960s. This project re-installed a new skylight in the original location to maximize daylight within the first floor north office spaces. A glass wall at the perimeter of office space runs parallel to the south windows to allow for daylight to infiltrate deeper into the buildings first floor office space.

J.1.14. Energy Consumption

J.1.14.1. Pre-renovation Energy Use (Total and EUI per m^2/year)

See Table J.2.

J.1.14.2. Predicted Energy Savings (Site, Source, GHG), Total and per m^2/year

See Table J.3.

J.1.14.3. Measured Energy Savings (Thermal, Electrical), Total and per m^2/year

See Table J.4.

J.1.14.4. Predicted Energy Use Reductions

See Table J.5.

J.1.14.5. Actual Energy Use Reductions

See Table J.6.

J.1.15. Energy Cost Reduction

See Table J.7.

Table J.2 Pre-renovation energy use

Electricity	Total (kWh)—FY2007	EUI (kWh/m^2-year)	Demand (kW)
Electricity	271,520		NA
Natural gas	219,607		NA
Oil	NA		NA
Propane	NA		NA
District cooling	NA		NA
District heating	NA		NA
Biomass	NA		NA
Other	NA		NA

Table J.3 Predicted energy saving

	Total (kWh)—FY2007	EUI (kWh/m^2-year)	Demand (kW)
Electricity	173,800 excluding renewable energy	NA	100
Natural gas	NA	NA	NA
Oil	NA	NA	NA
Propane	NA	NA	NA
District cooling	NA	NA	NA
District heating	NA	NA	NA
Biomass	NA	NA	NA
Other	NA	NA	NA

Table J.4 Measured energy saving

	Total (kWh)—FY2007	EUI (kWh/m^2-year)	Demand (kW)
Electricity	220,000 excluding renewable energy	NA	103
Natural gas	NA	NA	NA
Oil	NA	NA	NA
Propane	NA	NA	NA
District cooling	NA	NA	NA
District heating	NA	NA	NA
Biomass	NA	NA	NA
Other	NA	NA	NA

Table J.5 Predicted energy use reduction

	Total (kWh)—FY2007	EUI (kWh/m^2-year)	Demand (kW)
Electricity	97,720 savings	NA	Unknown
Natural gas	219,607 savings	NA	NA
Oil	NA	NA	NA
Propane	NA	NA	NA
District cooling	NA	NA	NA
District heating	NA	NA	NA
Biomass	NA	NA	NA
Other	NA	NA	NA

Table J.6 Annual energy use reduction

	Total (kWh)—FY2007	EUI (kWh/m^2-year)	Demand (kW)
Electricity	51,520 savings	NA	NA
Natural gas	219,607 savings	NA	NA
Oil	NA	NA	NA
Propane	NA	NA	NA
District cooling	NA	NA	NA
District heating	NA	NA	NA
Biomass	NA	NA	NA
Other	NA	NA	NA

Table J.7 Energy cost reduction

	Total ($)	$/m^2
All energy		
Electricity	Unknown	
Natural gas	6,325	NA
Oil	NA	NA
Propane	NA	NA
District cooling	NA	NA
District heating	NA	NA
Biomass	NA	NA

J.1.16. Non-energy Related Benefits Realized by the Project, E.g., Improved Productivity, Increased Rent/Lease, Increased Useful Space, Reduced Maintenance, Etc.

A daylighting study is currently being conducted at the Aspinall building by Rensler Polytechnic Institute (RMI) to investigate health benefits related to natural daylight in the workplace. This study is ongoing and the results have not been determined. Other benefits have not been tracked and rent/maintenance agreements have not yet been renegotiated by GSA.

J.1.17. Renovation Costs: Total and per m²

The project is design build, and costs will incorporate design costs in addition to construction costs.

J.1.17.1. Total

All Design-Build costs including supplemental agency funds—$14,635,155.03. Construction costs only and no supplemental agency funds—$10,981,075.16.

J.1.7.2. Energy Related

- PV: $1,217,182.00—Includes superstructure costs. PV alone—$764,343.00.

J.1.17.3. Envelope

Construction costs associated with new roofing, R-30 roof insulation, perimeter wall insulation, storm windows with low-e film—$731,789.00.

J.1.17.4. HVAC

Includes all mechanical systems, geo-exchange, ventilation, and functional testing. Demo not included. $3,291,660.00.

J.1.18. Business Models and Funding Sources

J.1.18.1. Description of the Funding Sources Chosen

American Recovery and Reinvestment Act of 2009 provided the majority of funds on this project. Agency provided funds (RWA) were provided to supplement agency specific requirements requested as part of the project Description of the business model chosen (option).

J.1.18.2. Description of the Business Model Chosen

The project was solicited as a firm fixed price design build contract. This procurement method was chosen to allow for a quick delivery method to meet the ARRA funding time-frame for completion.

J.1.18.3. Risk Allocation in the Business Model

All parties assumed risk as a result of this delivery method. The majority of the risk was with the design-build contractor through a firm-fixed price contract with no construction documents or bridging documents at the time of bid.

J.1.18.4. Funding Sources of the Business Model

American Recovery and Reinvestment Act of 2009 provided the majority of funds on this project. Agency provided funds (RWA) were provided to supplement agency specific requirements requested as part of the project.

J.1.18.5. Construction Phase in the Business Model

Construction activities were phased to allow for agencies to remain in the building throughout construction. Phase 1 was demolition of the first floor and construction of temporary swing space for five agencies. One agency remained on floor 2. Phase 2 was the demolition of the 2nd and 3rd floors and construction of the new space for the agency remaining on floor 2. Phase 3 was the demolition of the 2nd floor occupied suite post move in to their final 2nd floor space. Phase 4 was construction of the remainder of floor 2 and floor 3 and move-in from 1st floor swing space. Phase 5 was the demolition and build out of the 1st floor. Exterior and basement work ran concurrently with all the phases above.

J.1.18.6. Operation Phase

Operation phase consisted of an enhanced M&V phase to allow for systems to be monitored for energy consumption and fine-tuning of systems for optimal energy performance. This went beyond what is typically tracked in the commissioning phase to ensure estimated energy use of building systems were being met.

J.1.18.7. Energy Management and Controlling in the Business Model

The GSA has retained the engineer of record for a period of 1 year post construction to assist in the energy management and controlling process. In addition, this model allows for the ongoing education of the property management staff on the operation of the newly installed systems. The engineer will provide bi-weekly updates related to energy consumption of systems and plug loads for property management to address. A monthly meeting will review operational changes made for energy performance and review behavioral changes made by tenants within the building for energy reductions at the plug level.

J.1.19. Cost Effectiveness of Energy Part of the Project (NPV, SIR, Etc.)

None available.

J.1.20. User Evaluation

J.1.20.1. Description of User Training Programmes Within the Refurbishment

Training by contract was provided and videotaped by manufacturers and installers for all major equipment. In addition, the engineer of record has been contracted for a year post DB contract to assist property management as questions arise throughout the operational period after the initial contract.

J.1.20.2. Integration of Users Demands in the Planning Process

Review and response periods were provided to all project stakeholders at conceptual design, design development, and construction document development. Demands and questions were tracked by the Construction Manager and responded to by the Design Build team. Many demands were incorporated, however some were denied either due to conflicts with energy and LEED goals, preservation goals, or budgetary constraints.

J.1.21. Experiences/Lessons Learned

J.1.21.1. Energy Use

Plug load energy management is a process that must be started early and continue on long after the project is completed. IT representatives must be willing to pilot new technology to support over all building energy goals.

Projects pursuing net zero energy should consider these types of projects in two to three stages:

Stage 1—occupant engagement for energy use, including IT representatives.

Stage 2—Investment of deep energy retrofit.

Stage 3—After a period of 1 year post construction/occupancy (allowing for energy tracking in each seasonal condition) install renewable resources to offset tracked energy demand. Earlier purchases can result in the installation of too little/ too much renewable energy to offset predicted energy use.

J.1.21.2. Impact on Indoor Air Quality

The building systems provide a high level of temperature controllability, with digital readouts of temperature at each thermostat. In some ways, this provides too much information, with some tenants very focused on an actual temperature, rather than their own thermal sensation. The building is currently well ventilated with carbon dioxide levels rarely hitting 750 ppm. Shut-off VAV boxes for dedicated outdoor air are typically in their minimum position, to maintain appropriate building pressurization.

J.1.21.3. Practical Experiences of Interest to a Broader Audience

Standby energy use is typically well documented for common IT equipment, like laptops and printers. It is not well documented for HVAC equipment. We discovered that controls power for some equipment can be high, but is not published.

J.1.21.4. Resulting Design Guidance

Variable refrigerant flow systems have many positive features, but do not perform as well for a building with very low actual demand. PBS-P100 guidelines require sizing to a higher level of plug load than would exist in a low energy building.

J.1.21.5. Space Utilization Changes

Recommend consolidating high density IT equipment in a single room and using airside free cooling for conditioning.

J.1.22. References

Jason Sielcken, Project Manager, U.S. General Services Administration.
Roger Chang, Principal, Westlake Reed Leskosky.

J.2. Silver Spring and Lanham, MD

J.2.1. Name of the Project, Location (City, Country)

National Deep Energy Retrofit Project, Silver Spring and Lanham, MD, U.S.

J.2.2. Pictures

Figure J.17 shows the building in its original and post-retrofit states, illustrating key features of the retrofit

J.2.3. Project Summary

J.2.3.1. Project Objectives

In 2012, the U.S. General Services Administration (GSA) competitively challenged energy service companies (ESCOs) to improve the energy performance of 30 GSA-owned buildings through ESPCs featuring innovative solutions to achieve maximum energy savings. The GSA selected Ameresco to develop a

Fig. J.17 Silver Spring and Lanham, MD

comprehensive ESPC addressing over 1 million sq ft of office space at the New Carrollton Federal Building (NCFB) in Lanham, MD and the Silver Spring Metro Center 1 building (SSMC 1) in Silver Spring, MD. The deep energy retrofit ESPC supports the GSA in meeting its commitment towards the President's Better Buildings Challenge, which since 2011 has directed federal agencies to leverage $4 billion of performance-based contracts for energy savings by 2016.

Ameresco proposed a holistic solution that exceeded GSA's established energy and water conservation goals for the National Deep Energy Retrofit (NDER) Program, meeting or exceeding the following NDER objectives:

- Savings greater than 50% when comparing pre-retrofit energy use to post-retrofit energy usage.
- Whole building retrofits that impact multiple end-uses and systems.
- Multiple energy conservation measures with interactive effects.
- Energy savings substantial enough cover the project investment.
- Increased overall facility value.

The project also aligns with the U.S. Department of Energy objectives to work towards overall energy reduction as well as the sustainability requirements associated with Executive Orders 13514 and 13423 and DOE's Guiding Principles for Federal Leadership in High Performance and Sustainable Buildings.

J.2.3.2. Project Energy Goals

- 60% energy reduction.
- 10% of remaining consumption from on-site renewable energy installations.
- 50% water reduction.

J.2.3.3. Short Project Description

Ameresco designed and is currently implementing a variety of traditional energy conservation measures (ECMs) and renewable energy technologies at both sites. ECMs include Lighting Upgrades and Advanced Lighting Controls (SSMC1 and NCFB); Complete Upgrade of Building System Controls (SSMC1 and NCFB); Premium Efficiency Motors (SSMC1 and NCFB); Water Conservation (SSMC1 and NCFB); Building Envelope Improvements (SSMC1 and NCFB); High Efficiency Transformers (SSMC1 and NCFB); Chilled Water System Improvements (SSMC1); Ventilation Air System Optimization (SSMC1); Heating and AC Upgrades to Chillers/Heater with Geothermal (NCFB); Solar PV System (NCFB); Solar Thermal System (NCFB); Domestic Water System Optimization (NCFB); Exhaust to Outside Air Energy Recovery (NCFB); Kitchen Exhaust Controls (NCFB); and Electric and Telephone Room Cooling System Upgrades (NCFB). Once completed, the project is expected to reduce greenhouse gas emissions by 20,000 metric tons annually.

J.2.3.4. Stage of Construction

Ameresco kicked off construction in March 2014, beginning the implementation of upgrades to the building control and HVAC systems and the installation of more than 11,000 individually addressable LED lighting fixtures and a network of 2000 new sensors. In the summer of 2014, Ameresco began construction on a geothermal well field at the NCFB in its north parking lot as part of the approximately one megawatt of on-site renewable energy to be installed under the ESPC, which also includes an 808 kW custom-designed solar canopy to cover all unshaded areas of the parking lot. Ameresco will also install a 67 kW carport structure in the south parking lot facing the New Carrollton Metro Stop and a new solar thermal heating system on the roof.

J.2.3.5. Point of Contact

Greg Caplan.
Director of Business Development.

Ameresco.
101 Constitution Avenue.
Suite 535 East.
Washington, DC 20,001.
202-738-8442.
gcaplan@ameresco.com.

J.2.3.6. Date of the Report

November 2014.

J.2.4. Site

Not available.

J.2.5. Building Description/Typology

J.2.5.1. Typology/Age

- Silver Spring Metro Center 1–27 years.
- New Carrollton Federal Bldg—17 years.

J.2.5.2. Type (Office, Barracks, Etc.)

Office.

J.2.5.3. Typology/Age

Post 1970—(Silver Spring Metro Center 1, 1987; New Carrollton FB, 1997).

J.3. Silver Spring Metro Center 1

Silver Spring Metro Center One (SSMC1) is located at 1335 East-West Highway in Silver Spring, MD. Construction of the nine story office building was completed in 1987. The exterior consists of double paned glazing and concrete between the floors. Below grade is a three-level parking garage. The building is generally occupied between the hours of 7 a.m. and 5 p.m. for 10 hours a day; however, there are some tenants who may work extended hours. Equipment typically starts at 5 a. m. to bring spaces to temperature before occupants arrive, and runs until 5 PM for a total of 12 h. 157,622 $feet^2$/14,644m^2.

J.4. New Carrollton Federal Building

This building is located at 5000 Ellin Road in Lanham, MD. The complex, completed in 1997, consists of three nine story office towers connected by the ground floor "public level." A service level is located one floor below the public level and connects to the underground parking garage. There is an additional tiered parking structure and ground level parking lot to the west of the building. The building facade is punched double and triple pane windows on the public levels through the 7th floor. The 8th and 9th floors are all glass. The building is generally occupied between the hours of 7 a.m. and 5 p.m. for 10 h a day; however, there are some tenants who may work extended hours. Equipment typically starts at 5 a.m. to bring spaces to temp before occupants arrive, and runs until 5 p.m. for a total of 12 h. Some equipment will run on a 24-7 basis. 1.097 million ft^2/101,910 m^2.

J.4.1. Architectural and Other Relevant Drawings

Not available.

J.4.2. National Energy Use Benchmarks and Goals for Building Type Described in the Case Study

J.4.2.1. National Energy Target for This Type of Building (If Any)

GSA's established energy and water conservation goals for the National Deep Energy Retrofit Program incorporate the following objectives:

- Savings greater than 50% when comparing pre-retrofit energy use to post-retrofit energy usage.

- Involves whole building retrofits that impact multiple end-uses and systems.
- Includes multiple energy conservation measures with interactive effects.
- Energy savings are significant to cover the project investment.
- Program increases overall value of the facility.

Additionally the project aligns with the DOE's objectives to work towards overall energy reduction as well as addresses the requirements of Executive Orders 13514 and 13423 and DOE Guiding Principles for Federal Leadership in High Performance and Sustainable Buildings (Table J.8).

Table J.8 National energy target

	Objective	Solution
Federal energy reduction goals[a,b]	30% reduction in annual energy use by 2015 (versus 2005 baseline)	NCFB ECMs for this project will result in energy use from 121 EUI to 50.5 EUI (guaranteed reduction) upon completion, which well exceeds this objective SSMC ECMs for this project will result in energy use from 120 to 64 EUI (47% reduction)
	Comprehensive energy and water evaluations in 25% of facilities each year	Investment grade audit completed at both facilities
	Ongoing program for verification of implemented efficiency measures annually	Customized long term plan included under ESPC Performance Period (M&V), supported with integration of advanced control system, and integration of dashboards in kiosk and energy management programs
	Deep retrofit energy reduction by 50% Incorporate smart building meters Smart building design and implementation for all new GSA construction, renovation and energy savings performance projects (draft 2 October 2012) Promote energy awareness with tenants and personnel	IGA results show a 60% reduction, which exceeds GSA program goal Meters and smart building design included with building controls ECMs Kiosk and visible energy measurements will promote overall energy awareness
GSA program specific goals	Renewable energy technology goals of 7.5% by 2013 and 25% by 2025	Photovoltaic (increases renewable energy contribution and provides visible benefit), Included under this IGA 10% renewable energy from solar sources and geothermal measures (percent of renewable energy post-implementation of this project) Onsite electrical energy counts double source for GSA

(continued)

Table J.8 (continued)

	Objective	Solution
Sustainability	Other sustainable features	Reduction of close to 22,000 metric tons of $\mathbf{CO_2}$ Utilization of Recyclable materials Installation of rain gardens as part of PV system to decrease storm water discharge
Facility improvements	Replace aging lighting control system and HVAC control systems. Other maintenance improvements	Included within this project: chiller replacement, HVAC controls, distribution system upgrades, comprehensive lighting replacement, building envelope improvements including new roof installation, repaved parking lot, and multiple plumbing fixtures

[a] Facility Energy Management Guidelines and Criteria for Energy and Water Evaluations in Covered Facilities (42 U.S.C. 8253 Subsection (f), Use of Energy and Water Efficiency Measures in Federal Buildings) 25 November 2008

[b] Note the total reduction of 60% is for both buildings. The renewable energy goal of 10% was only applied to NCFB since there is no renewable energy measure for SSMC1. Percentage of renewable energy is the percentage of renewable energy post-ECM installation that is provided by renewable energy sources. This would include the PV system, the solar thermal system and the geothermal field energy

J.4.3. Awards or Recognition

Not available.

J.4.4. Site Energy Cost Information (Electricity, \$/kWh; Gas, \$/m³, Etc.)

See Tables J.9 and J.10.

J.4.5. Pre-renovation Building Details

J.4.5.1. Heating, Ventilation, Cooling, and Lighting Systems

Silver Spring Metro Center

Table J.9 Energy cost, SSMC1—baseline utility rates

Utility	Units	Rate	Source
Water	$/kgall $/l	$6.31 $0.0001665	Current utility bill
Sewer	$/kgall $/l	$8.68 $0.000229	Current utility bill
Electrical demand (summer months)	$/kW	$2.9591832	Current utility bill/deregulated contract
Electrical demand (all months)	$/kW	$1.3109182	Current utility bill/deregulated contract
Electrical demand (PJM PLC ratchet)	$/kW	$5.8299625	Current utility bill/deregulated contract
Electrical energy	$/kWh	$0.095720	Current utility bill/deregulated contract
Natural gas	$/MMBTU $/MWh	$0.005719	Washington gas current utility bill

Table J.10 Energy cost, NCFB—baseline utility rates

Utility	Units	Rate	Source
Water	$/kgallon $/l	$6.66 $0.0001758	Current utility bill
Sewer	$/kgallon $/l	$9.24 $0.0002439	Current utility bill
Electrical demand (summer months)	$/kW	$2.06602008	Current utility bill/deregulated contract
Electrical demand (all months)	$/kW	$0.88020394	Current utility bill/deregulated contract
Electrical demand (PJM PLC ratchet)	$/kW	$5.8299625	Current utility bill/deregulated contract
Electrical energy	$/kWh	$0.0697325	Current utility bill/deregulated contract
Natural gas	$/cubic feet $/cubic meter	$0.1072 $3.7857	Washington gas current utility bill

HVAC

The building's HVAC system consists of two rooftop outside air units supplying conditioned outside air to the north and south plenums. The units are heated by a natural gas coil and cooled with a glycol loop chilled by the rooftop chillers through a plate heat exchanger. Each floor contains a north and west mechanical room housing an air handling unit serving that half of the floor. The units are equipped with VFDs allowing variable flow to heat and cool the space through a VAV system. There is a mix of induction and fan powered boxes. Heating in the building is supplied by the gas-fired heaters in the outside air unit, and by electric reheat coils in the VAV boxes. The cooling is supplied by chilled water and chilled glycol (OA units only) served by either a 400 (1400 kW) or 490 ton (1725 kW) chiller. The larger chiller is used during the summer and the smaller during the winter. Because there is no outside air economizer, and the manually-operated waterside economizer is rarely used, chillers are required year-round.

Lighting

Lighting in the building is generally provided by linear 4′ (1.22 m) T8 troffers. There is a mix of two, three, and four lamp fixtures. Lamps are also mixed between 32 and 28 W versions. The current protocol is to replace burned out units with 28 watt lamps.

Controls

HVAC controls are provided via a DDC system supplied by Pritchett Controls/TAC. The system uses both BACnet IP and proprietary TAC communications. When the VAV boxes were replaced in 2007, the new controllers supplied were the BACnet IP versions. The existing proprietary ASD bus was reused to control the AHUs on each floor.

New Carrolton Federal Building

HVAC

The buildings HVAC system generally consists of roof mounted dedicated outside air AHUs serving an OA duct that runs down the core of each tower. One variable flow AHU per floor (per tower) mixes the OA with return air from the plenum.

Cooling is supplied to each tower by a variable speed secondary chilled water loop. A constant speed primary chilled water loop circulates around the service level and is served by three 1250 nominal ton (4400 kW) Trane centrifugal chillers and one 230 ton (800 kW) machine. Two of the 1250 ton (4400 kW) chillers have been retrofit with VFDs on their compressor motors.

All building heating is done with electric resistance heat. Heating in the building is supplied by electric heating coils in the dedicated OA AHUs, by electric reheat coils in the VAV boxes, and by some electric resistance perimeter baseboard heaters.

Lighting

Interior office areas are lit using 2′ × 2′ (60 cm × 60 cm)parabolic fixtures with three 40 W biax lamps per fixture. Hallways use recessed cans with plug in style CFLs. Elevator lobbies use recessed incandescent, the main lobby has been converted to LED. Exterior and site lighting is done using a mix of metal halide and high pressure sodium technologies from 70 to 250 watts each. F40 T12 lamps light the pedestrian bridge from the metro stop over Ellin Rd to the main lobby.

Controls

The existing HVAC equipment is controlled using a Siemens/Powers system 600/Insight system. The system consists of a central workstation with distributed panels throughout the building. In each mechanical space the analog inputs and outputs from each controller are sent through an E/P transducer for pneumatic operation. The existing VAV boxes use standalone pneumatic controllers with pneumatic thermostats in each space. A lighting control system also exists that controls the lighting relay panels on the north and south end of each building (per floor).

J.4.6. Description of the Problem: Reason for Renovation (Non-energy and Energy Related Reasons)

To address federally mandated energy reduction goals and GSA program-specific energy reduction and renewable energy goals.

J.4.7. Renovation SOW (Non-energy and Energy Related Reasons)

Not available.

J.4.8. Energy Saving/Process Improvement Concept and Technologies Used

J.4.8.1. Building Envelope Improvement

- ECM SSMC1-08: Installation of the Thermolite Series supplemental interior energy window system with bronze color and ¼ in. (63.5 mm) LowE Tempered glass on 419 windows throughout floors three through nine of the Silver Spring Metro Center 1 facility. The Thermolite window system (see image above) is an additional window pane that will be added to the existing windows throughout the building. The two (2) windows (existing double pane windows and new Thermolite Window System) will be separated by a small air space that will serve to reduce the overall heat transfer coefficient. This will result in a reduction of heat losses to the outside environment and yield significant energy savings, as well greatly reducing the overheating on the south side of the building. In addition to the additional window pane, there will also be a set of blinds installed between the existing double pane windows and the new

Thermolite window system. The new Thermolite Window System will be installed on 419 windows throughout floors three through nine of the Silver Spring Metro Center 1 facility.

- ECM NCFB-09.
  - ***Roof Replacements***: Provides 11 new roof replacements (33,300 ft^2/ 3094m^2) in all three buildings.
- The new roof systems will include:
  - New hot-applied rubberized asphalt fiber reinforced membrane.
  - Re-using existing 2 in. (5 cm) board insulation, but possibly replacing roughly 10,000 SF (930m^2) of damaged insulation if determined to be unusable.
  - Add 3 in. (7.62 cm) board insulation on top of the existing/replaced 2 in. (5 cm) board insulation.
  - New concrete paver system in lieu of loose ballast. New pavers will be coated with a reflective coating meeting Energy Star 90% reflectivity rating or greater.
  - The resulting roof system will have an R-Value of almost 30.
  - New 0.018 Stainless Steel Reglet style Counter Flashing as indicated with new backer rod and sealant.
  - Isocyanurate insulation baselayer (approximately 5 in. (12.7 cm) thick on average) fully covered by ½ in. (1.27 cm) thick densdeck coverboard.
  - Fully adhered 80 mil TPO roof membrane with a reflective rating of 90% or greater as measured by Energy Star.
- ***Sliding Door Replacements***: Replace the two doors located between the corridor/courtyard with a revolving door system and install a double sliding door vestibule for the Building A South entrance location.
- ***Building C Exterior Windows***: Installation of the Thermolite Series supplemental interior energy window system with bronze color and ¼ in. (63.5 mm) LowE Tempered glass on all exterior windows in the Lobby A and Lobby C areas.

J.4.8.2. New HVAC System or Retrofits to Existing

- ECM SSMC1-03 Chilled Water System Improvements: Re-install a chilled water coil in an air-handling unit. Implement strategies to increase chilled water production and distribution throughout the building including the installation of a new 450 ton (1580 kW) high efficiency variable speed centrifugal chiller; new chiller with a fee cooling cycle to allow water side economizing using refrigerant migration; upgrade existing cooling towers with high efficiency fan motors and new variable speed drives; upgrade the chilled water distribution system to a variable primary system and install new motor with variable speed drive on the existing chilled water pump.

- ECM SSMC1-04 Ventilation Air System Optimization: Install an energy recovery ventilator on the roof and connect to exhaust air louver and outside air louver with duct work. The system will use an enthalpy wheel to transfer temperature and moisture properties from one airstream to another.
- ECM SSMC1-05 Building System Controls (HVAC related): Enhancements and optimization of existing Pritchett/TAC control system. Installation of variable frequency drives; demand control ventilation strategy including the installation of high quality carbon dioxide sensors and EMS programming; adjust indoor air temperatures to P100 standards and change night setback.
- ECM SSMC1-06 and ECM NCFB-06 Premium Efficiency Motors: Replacement of standard v-belts with notched v-belts on all belt-driven HVAC equipment.
- ECM NCFB-03 Heating and AC Upgrades to Chillers/Heater with Geothermal: Replace existing chillers with three new chillers (with one of the three set up to serve as a chiller and/or a heater). Replace three 1250 nominal ton (4400 kW) chillers with two 704 ton (2475 kW) water cooled centrifugal chillers and one 704 ton (2475 kW) water cooled centrifugal chiller set up as a chiller-heater. Re-purpose the large DOAS chilled water coils to function as heating hot water coils. Install 189 new geothermal wells to absorb a heat for the chiller-heater in the wintertime and to reject excess heat in the summer. Cascade energy between chiller-heaters to further improve heating and cooling efficiency.
- ECM NCFB-04 Building System Controls: Replace the existing control system with an open protocol 100% DDC system. New system will be GSAlink ready. Retrofit VAV boxes with new DDC controls. Carbon dioxide (CO_2) sensors in return air plenums for AHUs with outside air dampers. Space occupancy sensors. VFDs added to constant speed HVAC motors. Implement optimal start/stop sequence to reduce HVAC system run hours during temperate times. Reset AHU supply air temperature base on VAV box operation. Reset outside air AHU discharge temperature based on building demand.
- ECM NCFB-10 Exhaust to Outside Air Energy Recovery: Install coils in the exhaust and the Make-up Air Unit airstreams with a pump and glycol run-around loop circulating between them. This will recover a significant portion of the energy from the exhaust stream and transfer it to the make-up air stream.
- ECM NCFB-12 Electric and Telephone Room Cooling System Upgrades: Replace both the electrical room exhaust and telephone closet air-conditioning systems with a Variable Refrigerant Flow (VRF) system. Each system consists of a single outside condensing rooftop unit, circulating refrigerant to individual evaporator units installed in each of the respective tower's electric and telephone room closets.

J.4.8.3. New Lighting System

- ECM SSMC1-01 and ECM NCFB-01.
- ***Lighting Upgrades***: Replace or retrofit existing fixtures with new C LED fixtures or re-lamp and re-ballast existing linear fluorescent fixtures containing 32-W T8 lamps and normal-power ballasts with 25-W T8 lamps and low-power, high-efficiency electronic ballasts. Generally, most office areas will receive new LED fixtures where burn times are higher, while back of house areas will remain linear fluorescent. Based on evaluations, this is both the most cost effective and efficient strategy to reduce the energy consumption of the existing luminaires. The scope also includes a spare part inventory of materials.
- ***Advanced Lighting Controls***: The lighting control system will be integrated with the existing BACNET building automation system. This will allow the passing of occupancy sensor information from the lighting to the HVAC control system, enabling the existing VAV boxes to relax to unoccupied settings when occupants are not present. Advanced control of the existing VAV boxes results in maximum energy efficiency and increased savings.

J.4.8.4. New Generation/Distribution System

New Generation System—See item 12.5 for ECM NCFB-05 Renewable Energy Systems—Solar PV System.

ECM SSMC1-09 and ECM NCFB-07.

High Efficiency Transformers: Replacement of existing transformers with new Powersmiths ESAVER-C3L-A high-efficiency dry-type transformers with like capacity. This will improve the efficiency by approximately 10%.

J.4.8.5. Renewable Energy

ECM NCFB-05 Renewable Energy Systems—Solar PV and Thermal Systems.

This ECM includes sustainably designed Solar Renewable Energy systems including the installation of a PV system and a solar thermal domestic hot water system. There are no existing solar energy systems at New Carrollton Federal Building (NCFB) at this time. This ECM will provide 10% of onsite renewable energy for NCFB upon completion of this NDER project. This will allow GSA-NCFB to move towards its EISA 2007 goal of using at least 25% renewable power by 2025.

Solar PV System

North Parking Lot PV System—808 kW (DC) array to be installed in all unshaded areas of the north parking lot. It will include a custom designed structural system in an amphitheater themed layout. Although this new system will not impact storm water flow, as part of this ECM, rain gardens will be constructed along the existing storm water catch basins in this parking lot. This will help to reduce overall storm water runoff from the parking lot in support of other sustainability goals of GSA.

South Parking Lot PV System—Traditional single-cantilevered parking lot canopy with a total capacity of 67kW (DC).

Solar Thermal: This measure will collect heat from the sun and use it to heat domestic hot water for Building A cafeteria/kitchen hot water loads. Heating of the domestic water will be accomplished by circulating water through solar panels mounted on the southerly exposure of Building A rooftop and store the heated water in an insulated storage tank located on the Service Level of Bldg. A. Warmed water from the tank will then serve as the "cold water" make-up as well as re-heat for the hot water loop to the existing electric hot water heater, which is responsible for providing hot water to the cafeteria/kitchen within this building.

J.4.8.6. Daylighting Strategies

Not applicable.

J.4.9. Energy Consumption

J.4.9.1. Pre-renovation Energy Use (Total and per m^2/year)

For calendar years 2009 through 2011, GSA consumed, on average, 18,582 MMBtu (5,445,847kW) and spent $516,000 annually for electricity and natural gas at SSMC1 and consumed, on average, 132,786 MMBtu (38,915,740kW) and spent $3,334,000 annually for electricity and natural gas at NCFB. In addition to the expenditures for energy-related utilities, GSA also purchased an average of 2563 kgal (9700 L) annually for potable water and spent, on average, $38,000 for water and sewer services for the same time period at the Silver Spring Metro Center 1. At the New Carrolton Federal Building, GSA purchased an average of 29,126 kgal (110,250 L) annually for potable water and spent, on average, $460,000 for water and sewer services for the same time period. The following graphs show the relative relationship for the energy-related utilities for both consumption and cost per site.

Annual Average Energy Consumption

See Fig. J.18.

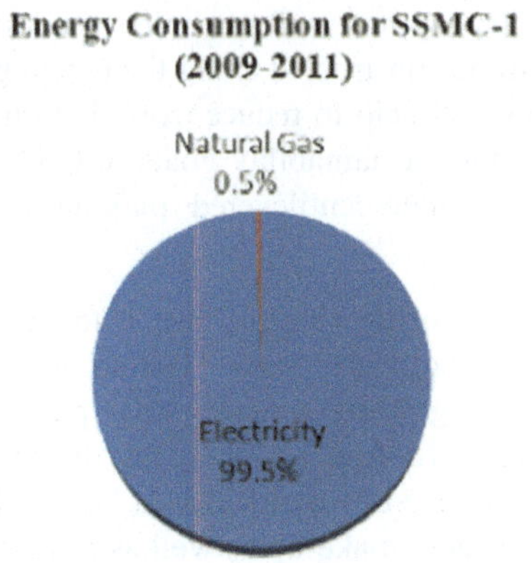

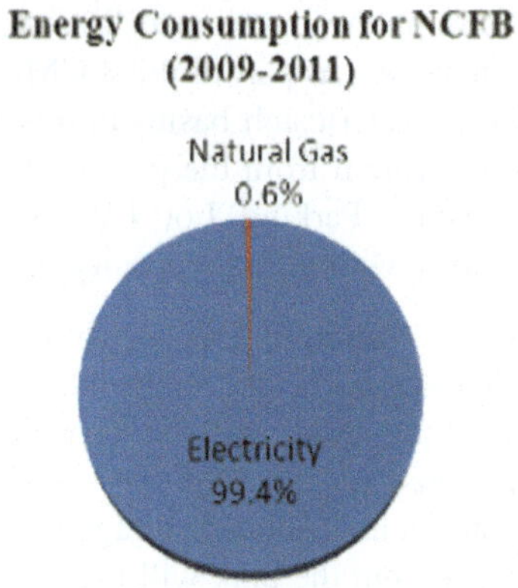

Fig. J.18 Yearly energy consumption

J.4.9.2. Predicted Energy Savings (Site, Source, GHG), Total and per m²/year

- Electricity: 27,714,088 kWh/year; $2,451,191.
- Water: 17,025 kgal/year (64,447L/year); $192,927.
- Estimated annual reduction of over 20,000 tons of GHG emissions.

J.4.9.3. Measured Energy Savings (Thermal, Electrical), Total and per m²/year

Project is currently in construction.

J.4.9.4. Annual Energy Use Reduction

- Silver Springs: 47%, 8,705 MMBtu (2,551,183 kW).
- New Carrollton: 61%; 81,919 MMBtu (24,008,092 kW).

J.4.10. Energy Cost Reduction

Not available.

J.4.11. Non-energy Related Benefits Realized by the Project, E.g., Improved Productivity, Increased Rent/Lease, Increased Useful Space, Reduced Maintenance, Etc.

- Reduces ongoing maintenance; O&M savings of over $68,000 per year.
- Promotes overall energy awareness.
- Reduces approximately 22,000 metric tons of CO_2.
- Creates/sustains approximately five hundred and fifty jobs.

J.4.12. Renovation Costs: Total and per m²

See Table J.11.

Table J.11 Renovation cost

IGA price	$586,172
Implementation price (including IGA)	$44,633,045
Financing procurement price	$4,387,077
Less: payment for IGA	($586,172)
Less: payment from savings during construction	($936,867)
Less: payment from utility incentives received during implementation	($4,335,414)
Total financed project price	$43,161,669

ECM number	ECM description	ECM implementation price
Silver spring metro center 1		
SSMC1-01	Lighting upgrades and advanced lighting controls	$1,004,284
SSMC1-03	Chilled water system improvements	$869,866
SSMC1-04	Ventilation air system optimization	$317,301
SSMC1-05	Building system controls	$604,513
SSMC1-06	Premium efficiency motors	$227,554
SSMC1-07	Water conservation	$70,116
SSMC1-08	Building envelope improvements	$651,777
SSMC1-09	High efficiency transformers	$135,395
New Carrollton federal building		
NCFB-01	Lighting upgrades and advanced lighting controls	$8,157,464
NCFB-02	Domestic water system optimization	$251,244
NCFB-03	Chiller system upgrade and geothermal field	$7,793,148
NCFB-04	Building system controls	$9,600,754
NCFB-05a	Renewable energy systems—parking lot solar PV system	$7,208,591

(continued)

Table J.11 (continued)

ECM number	ECM description	ECM implementation price
NCFB-05d	Renewable energy systems—solar thermal system	$164,036
NCFB-06	Premium efficiency motors	$227,806
NCFB-07	High efficiency transformers	$757,353
NCFB-08	Water conservation	$617,037
NCFB-09	Building envelope improvements with leaking roof scope	$2,424,461
NCFB-10	Exhaust air to outside air energy recovery	$1,123,842
NCFB-11	Kitchen exhaust controls	$66,384
NCFB-12	Electric and telephone room cooling system upgrades	$1,773,946

J.4.13. Business Models and Funding Sources

Not available.

J.4.14. Cost Effectiveness of Energy Part of the Project (NPV, SIR, Etc.)

See Table J.12.

J.4.15. User Evaluation

J.4.15.1. Description of User Training Programs Within the Refurbishment

ECMs SSMC1-01 & NCFB-01 Lighting Systems.

This will be a basic operations and troubleshooting-focused training session and safety course, using manufacturers' reference materials. The training session will be broken into two sessions of approximately 1 hour each, with one session pertaining

Table J.12 Cost effectiveness

Project simple payback (years)	14.2
Performance period term (years)	22

to indoor lighting applications and the other focusing on outdoor lighting. All O&M manuals and training session materials will be turned over to the appropriate Government personnel.

ECMs NCFB-03 & SCMC1-03 Chiller w/Heater & Geothermal Field.

Training for these ECMs will be an intense general functionality, operations and troubleshooting-focused training session and safety course, provided by the controls manufacturer, the chiller manufacturer, our integrator, and engineering team. The training session is estimated to require 4–6 h and will involve multiple sessions with sequences of operation and related controls drawings, field components (sensors, valves, HVAC and geothermal equipment, refrigerant monitoring and breathing apparatus, local controller hardware, etc.), and front-end system operations and troubleshooting. All O&M manuals and training session materials will be turned over to the appropriate Government personnel.

ECM SCMC1-03 Chiller Upgrade.

Training for this ECM will be chiller operations and troubleshooting-focused training session and safety course using manufacturers' reference materials. The training session is estimated to require 3–4 h, and will involve troubleshooting chilling system components, describing the product line system integration and all of the key hardware components. Control inputs (pressure sensors, motor feedback) and controller interfaces, as well as an in-depth overview of system operating parameters. All O&M manuals and training session materials will be turned over to the appropriate Government personnel.

ECM NCFB-04 & SMC1-05 Controls Upgrade.

Training for these ECMs will be an intense general functionality, kiosk programing, operations and troubleshooting-focused training session and safety course, provided by the controls manufacturer, our integrator and engineering team. The training session is estimated to require 3–4 h each and will involve multiple sessions with both field components (sensors, valves, HVAC equipment, local controller hardware, etc.), and front-end system operations and troubleshooting. All O&M manuals and training session materials will be turned over to the appropriate Government personnel. The training session will comply with the requirements of the TORFP. This includes 16 h of initial familiarization training and following construction acceptance, additional follow on training at the following sites will be 8 h bi-monthly for a period of 12 months. The six sessions are to incorporate both the heating and cooling seasons.

ECMs SSMC1-04 Ventilation Air System; NCFB-02 Domestic Water System; NCFB-04 Building System Controls; and NCFB-11 Kitchen Exhaust Controls.

Training for these ECMs will be a basic operations and troubleshooting-focused training session and safety course for the new equipment installed under these ECMs. This session will involve a basic review of the specific mechanical system components involved, function and design intent, and new system performance goals. Other topics include hands-on system troubleshooting, normal operational and safety indicators, and O&M best practices. The training session is estimated to require approximately 1–2 h at each facility. All O&M and training session materials will be turned over to the appropriate Government personnel.

ECM NCFB-05—Solar Thermal.

This session will involve a basic review of the installation and system performance goals. Proper O&M practices and procedures will be demonstrated for the solar panels, glycol heat loop components, heat exchangers, valve line-ups, and accompanying controls. The training session is estimated to require 1–2 h. All O&M and training session materials will be turned over to the appropriate Government personnel.

ECMs SSMC1-06 & NCFB-06—Motor Upgrades.

This will be a basic operations and troubleshooting-focused training session and safety course, using manufacturers' reference materials. The training session will be broken into two sessions of approximately ½ hour each, with one session pertaining to motor upgrades. All O&M manuals and training session materials will be turned over to the appropriate Government personnel.

ECMs SSMC1-07 & NCFB-08—Water Conservation.

This session will involve a basic review of the equipment installed and system performance goals. Proper O&M practices and procedures will be demonstrated for each new component type. The training session will require less than one hour. All O&M and training session materials will be turned over to the appropriate Government personnel.

ECMs SSMC1-09 & NCFB-07—Transformer Upgrades.

This will be a basic operations and troubleshooting-focused training session and safety course, using manufacturers' reference materials. The training session will be approximately 1–2 h, focusing primarily on proper and safe operation of the new transformers. All O&M manuals and training session materials will be turned over to the appropriate Government personnel.

ECM NCFB-12 Electric and Telephone Room Cooling Systems.

This session will involve a basic review of the installation and system performance goals. Proper O&M practices and procedures will be demonstrated for system rooftop condensing units, indoor ceiling cassettes, temperature sensors/controls, and operational and safety interlocks with existing electrical equipment. The training session is estimated to require 1–2 h. All O&M and training session materials will be turned over to the appropriate Government personnel.

J.4.16. Experiences/Lessons Learned

Not available.

J.5. Intelligence Community Campus, Bethesda, MD

J.5.1. Name of the Project, Location (City, Country)

Intelligence Community Campus—Bethesda (ICC-B), Bethesda, MD USA.

J.5.2. Pictures

Figures J.19, J.20, J.21, J.22, J.23, J.24, J.25, J.26 and J.27 show the building in its original and post-retrofit states, illustrating key features of the retrofit.

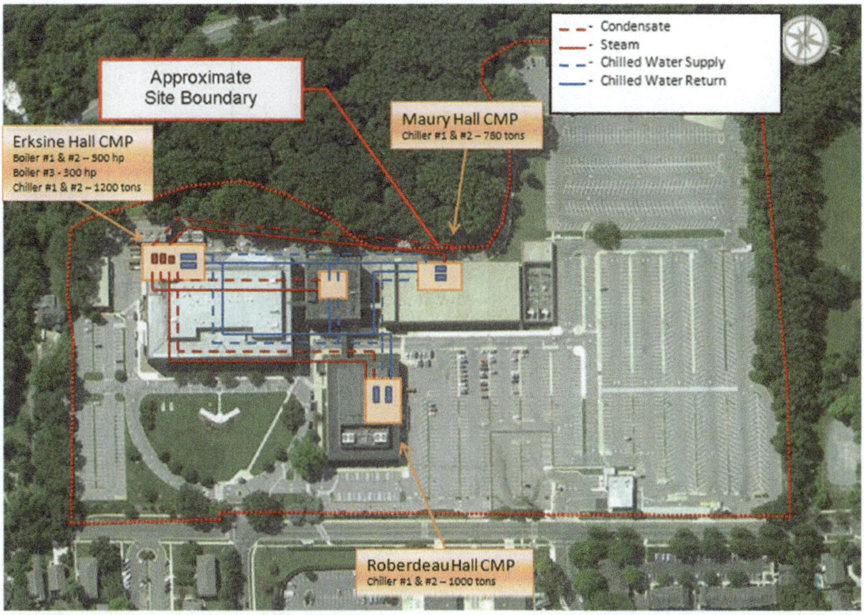

Fig. J.19 Former Sumner campus

Fig. J.20 Artist rendering of future ICC-B campus

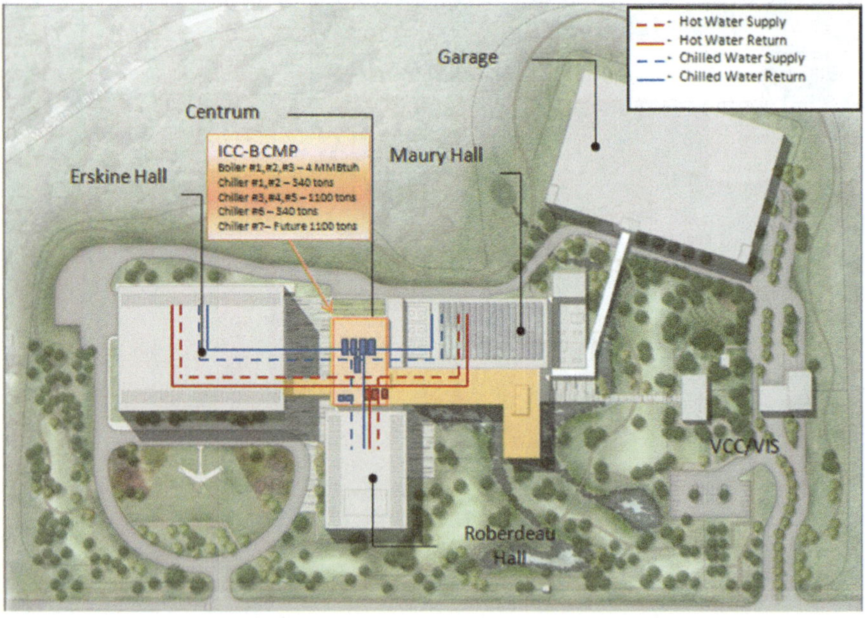

Fig. J.21 ICC-B campus vision

Fig. J.22 Centrum

Fig. J.23 Roberdeau hall

Fig. J.24 Removal of Erskine hall building envelope

Fig. J.25 Erskine hall

Fig. J.26 Utilities installation

Fig. J.27 CMP chillers

J.5.3. Project Summary

J.5.3.1. Project Objectives

Federal government agencies are faced with significant budgetary challenges when trying to meet requirements for infrastructure renewal and mandates for increased energy efficiency. Much of the building stock owned and operated by the Federal government is of the age where major renovation will be required to continue to meet the mission requirements. Additionally, there are numerous Federal mandates requiring that agencies meet prescribed goals for energy efficiency and sustainability such as net zero energy (NZE). Major facility renovation projects have traditionally been funded with sustainment, restoration, and modernization (SRM) funds requiring large capital appropriations. These funds are becoming increasingly difficult to obtain under current budget constraints. Alternative financing vehicles for energy projects have significant limitations in the ability to fund large scale renovations of buildings. These challenges require that the methods used to accomplish these requirements extend beyond what has traditionally been done.

One example of where this combination of requirements is being addressed in an innovative manner is the Intelligence Community Campus—Bethesda (ICC-B). The Defense Intelligence Agency (DIA) is currently embarked on an extensive campus redevelopment at the request of the Office of the Director of National intelligence. The 30-acre ($0.12 km^2$) campus will house several agencies of the Intelligence Community and had been previously occupied by the National Geospatial Intelligence Agency from 1946 until 2011. The campus was previously made up of six buildings originally constructed between 1946 and 1988 (small Visitor Center built in 2005) with a weighted average age of approximately 56 years. The redevelopment calls for the demolition of three buildings. The remaining buildings will have full scale renovations of both the interior and exterior shells. A new 220,000 ft^2 (20,400 m^2) building (Centrum) is to be added to serve as the center of the campus connecting the existing three buildings. The new construction and major renovation designs are required to meet aggressive energy and sustainability standards with a goal of achieving LEED Silver certification.

The overall campus redevelopment project began as a traditional design/build construction project for the North Campus consisting of a 1800 car garage, a Visitor Control Center, and Vehicle Inspection Station. All buildings on the North Campus were designed and built to be net zero energy buildings (produce as much renewable energy on site as they consume during the course of 1 year). The South Campus is being delivered as a traditional SRM construction project under a Single Award Task Order Construction Contract (SATOCC) administered by the Baltimore District, U.S. Army Corps of Engineers (USACE). The project was to be funded entirely relying on appropriated funds. It was quickly realized that budget constraints would significantly limit the ability to achieve the goals of the project. It was then decided to combine the traditional construction contract with an energy performance contract for the South Campus. This would allow a separate team to be brought on board to specifically focus on the core energy infrastructure. A separate Utility Energy Service Contract (UESC) was awarded by the Huntsville Engineer Center, U.S. Army Corps of Engineers to use this alternative financing vehicle leveraging energy and energy-related savings to fund significant portions of the project. This approach comes with numerous complications arising from the separate contract mechanisms and contractor teams.

The chosen approach has been successful in combining public and private funding streams to achieve the goals of the project within available funding constraints. The challenges associated with this unique approach are being identified and mitigated to ensure overall success. This paper will detail the business and engineering approaches employed to date as the project has evolved. The "lessons learned" will also be discussed in an effort to improve the implementation of this approach to other projects in the future. The results will show that the combined approach shown here provides a powerful mechanism for extending the reach of appropriated funds to achieve deeper energy savings and sustainability in Federal buildings.

J.5.3.2. Project Energy Goals

As mentioned previously, the redevelopment of the ICC-B South Campus was originally planned to be executed under a SATOCC contract using appropriated funds. The contract was awarded through a competitive solicitation for a design-build contract team (DBT). The DBT was to be made up of a general contractor along with design and engineering services. The successful DBT was selected based on a combination of qualifications and projected implementation budget for completing the campus redevelopment.

One of the principal motivations for considering the use of a private funding stream in conjunction with traditional appropriated funding was a shortfall in appropriated funds to accomplish the campus renovation. The original budget for the project was reduced and the reduction would have delayed many of the programmed elements of the redevelopment effort and hampered efforts to achieve significant reductions in future energy consumption. It was at this point that the use of alternative funding sources began to get serious consideration as a means to help keep the redevelopment effort on track.

The original concept for the UESC project was to replace the primary energy infrastructure for the campus. This was to include a central utility plant (CUP) and associated utility distribution system to provide chilled water and heating hot water to the campus. This approach had the following goals:

- Extend the reach of appropriated funds in terms of providing program floor space by leveraging alternative financing to construct the site energy infrastructure (this was, and remains, the primary goal).

- Help meet Federal energy conservation and sustainability mandates.
- Provide reliable/robust utilities service to the campus (enhance energy security at the site).
- Make energy-related improvements as unobtrusive as possible in response to community relations considerations.
- Evaluate the potential for achieving net zero energy consumption at all or part of the campus through a combination of renewable energy and a cogeneration component if economically viable.
- Have fixed accountability for energy systems performance.
- Provide phased implementation of energy infrastructure development to match the pace of construction/renovation of the campus buildings.

The projected energy usage reductions for this project have evolved through preliminary proposals and with each phase of development. There is the added complexity of determining a methodology for establishing a baseline from which to determine the energy and cost savings resulting from the UESC portion of the overall project.

This was accomplished through developing detailed hourly energy models of each building on the original campus and calibrating the campus model to the latest historical energy data representing full occupancy (2008). Once calibrated, the campus model was modified to account for the changes in building envelope composition and space use expected from the major renovation of the buildings to remain and the construction of the Centrum building. This revised campus model was treated as the Architecturally Adjusted Baseline with the buildings being served by the original campus primary HVAC equipment (boilers/chillers, etc.) and secondary HVAC equipment (AHUs, terminal devices) in the renovated buildings.

The energy and cost savings for each of the project phases were determined in subsequent model runs to account for the ECMs associated and the current level of design maturity. This analysis is now complete through Phase III. Figures J.28 and J.29 show the results of the energy modeling from the original campus model through Phase III.

The results project a 47% reduction in campus energy usage below the Architecturally Adjusted Baseline and approximately $1,100,000/year in energy cost savings.

J.5.3.3. Short Project Description

The campus redevelopment involves the renovation of the three largest buildings on the existing campus. The three smaller buildings are to be demolished and a new building constructed. The new Centrum building will be an approximately 230,000 ft^2 (21,367 m^2) structure designed to provide interconnection between the Erskine, Roberdeau, and Maury Hall buildings. The overall goal for the campus redevelopment is to achieve significant improvements in overall energy efficiency and sustainability for the buildings and campus overall. The renovation of the existing

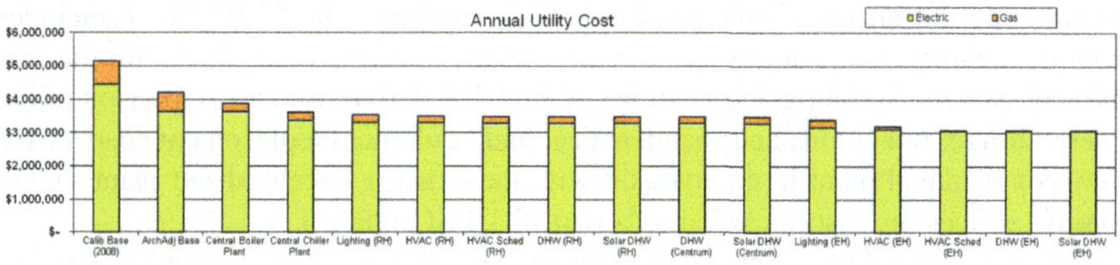

Fig. J.28 Modeled campus energy cost

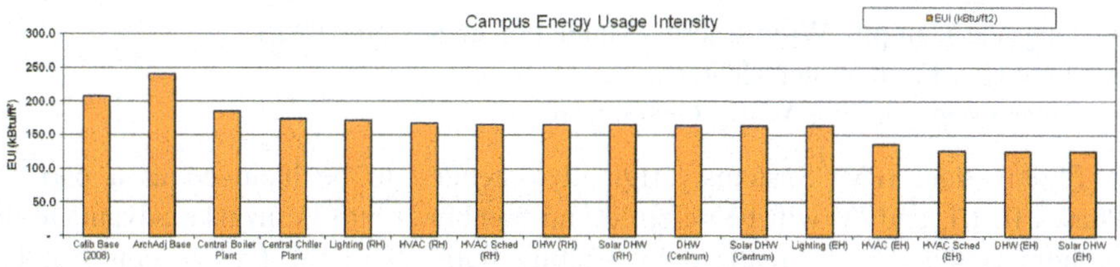

Fig J.29 Modeled campus energy intensity (kBtu/ft^2)

buildings seeks to minimize the internal and external load components and then meet them as efficiently as possible. The building facade of each structure is being replaced to increase the introduction of natural light while reducing the thermal loads from outdoor conditions. State-of-the-art LED lighting is being used predominantly throughout the buildings with an extensive intelligent lighting control system. These types of design features are designed to allow for significantly reduced mechanical equipment capacity and energy infrastructure. The campus and individual buildings will make use of extensive submetering to isolate energy usage into end use components. The submetering will allow continuous tracking of energy usage and direct billing of tenant organizations to raise occupant energy consumption awareness. All of these design elements combine to create the synergistic effect of substantial reductions in both implementation costs for the project and operational cost for the buildings going forward. The design standards for the construction and renovation projects called for achieving Silver certification under USGBC Leadership in Energy and Environmental Design. The long term goal is to achieve net zero energy consumption for as much of the campus as possible.

J.5.3.4. Stage of Construction

As mentioned in the preceding section, the development of the campus energy infrastructure was to be undertaken in phases. The centerpiece of the Phase I/Base portion of the proposed UESC project is the construction of a CUP to replace the existing decentralized plants on the ICC-B campus. The CUP ECM includes cooling, heating, and standby generation systems. Savings associated with the first two components (cooling and heating) of the CUP derive from a comparison of the energy usage, operations and maintenance, and equipment replacement costs of the new consolidated plant in comparison with the existing decentralized plants. CUP O&M services are also included as a part of UESC Phase I.

Three additional phases that will impact the buildings to be renovated as a part of the ICC-B redevelopment effort will be primarily comprised of the following ECMs:

- Lighting Upgrades and Lighting Controls.
- VAV Reheat AHU System with Energy Recovery DOAS.
- Upgrade Campus Wide Energy Management System (EMS).
- New Gas Fired Water Heaters.
- Solar Domestic Hot Water Generation.

Photovoltaic (PV) Systems, originally intended to be included as a part of Phases II, III, and IV will be deferred and combined into IV to take advantage of economies that can be realized by waiting until all of the PV systems can be installed simultaneously. The overall goal of the PV installation will be to produce sufficient renewable energy production to offset the energy requirements of Roberdeau Hall to render that building net zero energy.

Phase II will address measures to be implemented in Roberdeau Hall, Phase III will address measures in Erskine Hall, and Phase IV will address measures in Maury Hall. Operations and maintenance activities associated with the Centrum Building that are not covered as a part of Phase I (i.e., O&M not directly related to the CUP) will be included in Phase II. ECMs for Phase II will generally be implemented through providing the materials associated with the ECMs to the SATOCC contractor as Government Furnished Equipment (GFE) purchased as a part of the UESC project. One exception to the GFE concept is the EMS ECM, which will be a turnkey installation under the UESC project for all project phases.

While the GFE approach will be employed for elements of Phases III as well, some of the ECMs under these phases will also require the UET to take over both design and installation of the core mechanical/electrical ECMs. Under this approach, the UET will advance the design of the core mechanical/electrical systems for the building and be responsible for the turn-key installation of these systems. The tenant fit-out design in Erskine Hall will be advanced by the DBT and construction completed by them with the energy system equipment provided as GFE by the UET as in Phase II. In Phase IV, a different level of renovation is contemplated for Maury Hall. Much of the Maury Hall building infrastructure will be re-used and the UET will evaluate the existing energy systems for upgrade and replacement. While planning for this phase is still ongoing, this portion of the project will likely follow a more traditional energy performance contracting approach with the UET performing the majority of the development, design, and implementation separate from the DBT. As with Phase II, full building O&M services will be a part of Phases III and IV.

J.5.3.5. Point of Contact

Luis Ayala, DISL.
Senior Technical Expert (Facilities/Construction).
Defense Intelligence Agency.
Luis.Ayala@dodiis.mil.

J.5.3.6. Date of the Report

The project is still ongoing in phases. The Detailed Feasibility Study for Phase 3 Erskine Hall was submitted February 2015. The construction of the Phase 1 Central Mechanical Plant is nearly complete and expected to come on-line in August 2015.

J.5.4. Site

The ICC-B site is located in Bethesda, MD, USA (38.95, − 77.12) at an elevation of 249 ft. The site is located in Climate Zone 4. The design heating and cooling temperatures for the location are as follows (Fig. J.30).

Approximate calculation of degree-days to any base
for WASHINGTON DC REAGAN AP, VA, USA

Choice of base temperature

Requested base temperature °F 65

Approximate degree-days

	Mean temperature	Standard deviation of daily average temperature	Approximate Cooling Degree Days	Approximate Heating Degree Days
	°F	°F	°F-day	°F-day
January	36.3	9.7	0	890
February	39.4	8.5	0	717
March	46.7	9.1	2	570
April	56.6	8.1	19	271
May	65.9	7.3	104	76
June	74.8	6.0	298	4
July	79.7	4.8	456	0
August	78.0	5.0	403	0
September	70.8	6.7	196	22
October	59.4	7.5	31	204
November	49.7	8.3	3	462
December	40.1	9.0	0	772
Annual			**1512**	**3988**

Fig. J.30 Heating and cooling degree-days per month

Table J.13 shows the long-term average heating and cooling degree days for the area.

J.5.5. Building Description/Typology

The ICC-B is located in Bethesda, MD. The campus was originally occupied in 1946 by the Army Mapping Service and eventually the National Geospatial-Intelligence Agency (NGA), and has recently been turned over to DIA to transform the facility into a secure campus supporting U.S intelligence community activities. The facilities were last fully occupied and functioning in 2008. Table J.14 shows the list of buildings on campus as of 2008 and approximate size.

The information used to model the buildings was taken from a combination of facility drawings, interviews with facility personnel and detailed surveys of the buildings. The information was gathered based on how the buildings were occupied and operated during the 2008 baseline year. It was learned from facility personnel that the facilities were all operated 24 h per day, 7 days per week. These buildings were also reported to have high intensity plug loads. Several areas in the buildings were identified as computer rooms and modeled accordingly.

Table J.13 Design temperature

*Cooling design temperature—0.4% occurrence**	
Dry bulb temp C (F)	Mean coincident wet bulb temp C (F)
35(95)	24.4(76)
*Heating design temperature—99.6% occurrence***	
Dry bulb temp C (F)	
− 9.4(15)	

Table J.14 Original campus building list

Building name	Floor area (Gross)	Year built	Building status
Erskine Hall	< 400,000 (37,160 m^2)	1946	To be renovated
Abert Hall	< 95,000 (8826 m^2)	1962	To be demolished
Emory Building	< 15,000 (1394 m^2)	1963	To be demolished
Roberdeau Hall	< 140,000 (13,006 m^2)	1966	To be renovated
Maury Hall	< 155,000 (14,400 m^2)	1988	To be renovated
Visitor Center	< 1500 (139 m^2)	2005	To be demolished

J.5.6. Architectural and Other Relevant Drawings

See Figs. J.31, J.32 and J.33.

Fig. J.31 Sangamore road aerial view

Fig. J.32 Sangamore road master plan view

Fig. J.33 East Lawn plan view

J.5.7. National Energy Use Benchmarks and Goals for Building Type Described in the Case Study

The ICC-B campus redevelopment project has the overall goal of being a showcase for energy efficiency and sustainability. The design for the Centrum building and renovations of the legacy buildings to remain have been based on LEED requirements with the goal of Silver certification. The long term goal of the project is to provide sufficient renewal energy production to offset the energy requirements of Roberdeau Hall.

J.5.8. Awards or Recognition

Not available.

J.5.9. Site Energy Cost Information (Electricity, \$/kWh; Gas, \$/m^3, Etc.)

J.5.9.1. Electricity

Projected electrical energy costs were based on the following usage and demand rates:

- Electrical Energy: $0.0968/kWh.
- Electric Demand: $1.626/kW (based on annual peak demand).

These rates were derived from the tariff structure of the local electric utility provider and appropriate rate schedule.

J.5.9.2. Renewable Energy Feed-In Tariff

None.

J.5.9.3. Natural Gas

The projected natural gas energy costs were based on the following usage rate:

- $6.74/MBtu.

These rates were derived from a combination of commodity and local distribution charges.

J.5.10. Pre-renovation Building Details

J.5.10.1. Envelope Details: Walls, Roof, Windows, Insulation Levels

The components of the building envelope for each building were identified through existing drawings and visual inspection. The drawings indicated masonry facade construction (CMU, brick, terra cotta, and cast concrete) with various levels of batt or board insulation (including zero insulation) depending on the age of the building. Roofs are flat with layers of insulation and built-up roofing (BUR). The buildings were without windows with the exception of Erskine Hall with a limited window/wall ratio. Windows were generally fixed sashes with double glazing in aluminum frames. Table J.15 provides a comparison of the building envelope make-up of the original campus buildings in comparison to the planned renovation and new construction designs.

J.5.10.2. Heating, Ventilation, Cooling and Lighting Systems

The secondary HVAC systems in the buildings were generally variable air volume (VAV) air handling units with shut-off style reheat terminal boxes. The systems were generally in fair-condition. However, typical efficiency issues were persistent such as simultaneous heating and cooling in VAV reheat systems. The computer room spaces were served by chilled water computer room air-conditioner (CRAC)

Table J.15 Building envelope data

Building	Wall construction U-value (Btu/(ft2*oF))		Roof construction U-value (Btu/(ft2*oF))		Window system U-value (Btu/(ft2*oF))		Window coverage (gross window-wall ratio)	
	Pre	Post	Pre	Post	Pre	Post	Pre	Post
Roberdeau Hall	0.08	0.051	0.043	0.032	No windows	0.35	0	0.21
Erskine Hall	0.064	0.051	0.037	0.032	1.18	0.35	0.1	0.27
Maury Hall*	0.092	*	0.049	*	No windows	*	0	*
Centrum**	**	0.051	**	0.032	**	0.35	**	0.32

*Maury Hall currently now planned for envelope renovation
**Centrum Building is new construction

units. The primary HVAC systems consisted of chilled water and a steam plant. The chilled water system consisted of three separate chilled water plants located in the three campus buildings. The chilled water distribution system was connected to all the campus buildings such that all chillers could be used to provide cooling capacity to the loop. Each of the three buildings was equipped with three water cooled centrifugal chillers with a total nominal cooling capacity of approximately 6000 tons (21,101 kW). This level of connected capacity was due to each plant being made up of redundant chillers. Interviews with plant operators and chiller logs suggest that the peak cooling load for the campus was less than 2400 tons (8440 kW). The chilled water system equipment varied in age from 12–24 years and was reportedly of degraded efficiency. Several of the chillers were originally R-11 and had been converted to R-123.

Campus heating and domestic hot water loads were met using two 500-hp (4905 kW) and one 200-hp (1962 kW) steam boilers located in Erskine Hall. The steam was then distributed to the remaining buildings and generally converted to hot water for use in the buildings. The larger boilers were installed in 1975 and in generally poor condition. The smaller boiler was installed in 2001 to provide steam capacity during the summer months. The steam system overall was found to have significant energy losses in the ancillary devices and distribution systems.

The buildings were equipped with building automation systems made up of a blend of pneumatic and direct digital control (DDC) systems that managed the operation of the HVAC equipment, including the central plant, air handlers, and terminal devices. Control panels located within each mechanical room provide operational control of the major equipment such as air handlers, and pneumatic control valves to meet the space temperature requirements. The systems were integrated together for centralized control. No advanced control schemes for the purpose of energy optimization were reportedly in use.

J.5.11. Description of the Problem: Reason for Renovation

The Campus is in the process of being redeveloped into a modern, state-of-the art facility that will serve the United States Intelligence Community, a coalition of 17 agencies and organizations within the executive branch that work to gather the intelligence necessary to conduct foreign relations and national security activities. The purpose of the project is to develop a collaborative intelligence community campus for the relocation of roughly 3000 intelligence workers in the Washington National Capital area. The redevelopment is necessary because: (1) there is a shortage of secured administrative building space in the Washington National Capital area; (2) a shared intelligence community campus supports congressional desires for a collaborative community environment and the consolidation of an intelligence community facility strategy; and (3) it supports the reuse of existing government facilities.

The total budget for the campus redevelopment is approximately $359 million over a 5-year period (the original budget for the project was $559 million—there was a cut in the budget by $200 million, which is one of the reasons DIA is looking at Public Private Partnerships as a way for the Agency to be able to stretch the money they have for more of the needed renovations).

Goals of using an alternative financing vehicle in conjunction with the campus revitalization were discussed between DIA and the utility service provider, Washington Gas (the Washington Gas/Honeywell Team, or WG/HON). Goals include:

- Extend the reach of appropriated funds in terms of providing program square footage by leveraging alternative financing to construct the site energy infrastructure (this is the primary goal).
- Help meet Federal energy and sustainability mandates.
- Provide reliable/robust utilities service to the campus (enhance energy security at the site).
- Make energy-related improvements as unobtrusive as possible in response to community relations considerations.
- Have fixed accountability for energy systems performance.
- Provide phased implementation of energy infrastructure development to match the pace of construction/renovation of the campus buildings.

J.5.12. Renovation SOW (Non-energy and Energy Related Reasons)

Not available.

J.5.13. Energy Saving/Process Improvement Concept and Technologies Used—Mention Sub-Systems and Insert Boxes for Narrative Details as Appropriate.

J.5.13.1. New HVAC System or Retrofits to Existing

The UESC contract was to develop recommended energy conservation measures around the secondary HVAC systems, building automation system, lighting, and domestic hot water systems. The analysis also included the use of on-site energy production using solar thermal hot water and photovoltaic electrical production.

The UET developed an alternative HVAC design approach that offered significant advantages over the standard VAV design used in the existing building and the 35% design from the DBT. The proposed system uses dedicated outdoor air system (DOAS) air handling units to provide cooled and dehumidified primary air to the spaces. The DOAS unit would be equipped with a heat recovery wheel, chilled water cooling coil, and DX cooling coil (with hot gas reheat). The conditioned primary air would then be delivered via series fan powered terminal units in the zones throughout the building. The terminal boxes would draw return air from the plenum and blend with a small percentage of conditioned primary air (10–25%). The terminal unit would be equipped with a cooling coil to meet the sensible cooling loads in the space. The amount of primary air introduced will be modulated based on a combination of zone humidity and CO_2 levels. The total flow of air to the space will be modulated based on zone temperature requirements using variable speed ECM motors.

The proposed design has several significant advantages over the standard VAV design. The proposed design greatly reduces the overall fan power requirements by eliminating the need for local air handling units and ductwork to distribute the full flow of air to the spaces. The majority of the latent cooling would be handled in the DOAS units with combination cooling coils. The system provides outside air to each zone based on the specific requirements for occupancy and space humidity. The combination of these and other factors were shown in the energy model to provide an increase in energy savings of approximately 69% over the standard VAV design. In addition to increased energy efficiency, the proposed design is expected to provide benefits for implementation cost and building design. The design eliminates the need for separate air handling units and large duct runs throughout the building. This would reduce the cost to provide and install this equipment and frees up mechanical space for other space allocation. The UET proposed this alternative HVAC design along with the projected energy savings that would result from using this approach for the renovated buildings. The DBT had continued to progress the design of Roberdeau Hall to the 65% level based on the standard VAV system design. While the UET proposed design approach would result in a significant increase in energy savings, changing the basis of design fundamentally at 65% would significantly increase the time and expense required to complete the design. The decision was made to continue with the standard VAV HVAC design in an effort to maintain the projected occupancy date.

J.5.13.2. New Lighting System

The proposed lighting system uses light emitting diode (LED) extensively throughout the building including where we would more traditionally use linear fluorescent tubes. The UET provided the design requirements to the DBT for use in their development of the detailed design. The overall result shows the lighting connected load density of approximately 0.50 W/ft^2 (5.4 W/m^2). The design also includes an advanced integrated lighting control system to account for items such as scheduling, occupancy, daylight harvesting, and light level tuning.

J.5.13.3. New Generation/Distribution System

The projected cooling load for the campus was calculated by the DBT to be approximately 1890 tons (6645 kW). This projected cooling load excludes specialized mission critical cooling loads required to be met by systems with redundancy of both capacity and electrical power source. As a result, the chilled water system design for the CUP was based on three 1100-ton (3868 kW) variable speed chillers to meet the required N + 1 redundancy for the system. The design cooling efficiency for the chillers is 0.593 kW/ton (20.2 EER) at full load rated conditions. A fourth chiller was added to provide a combination of swing capacity and the ability to shift heat absorbed in the chilled water system to the heating hot water system. The heat recovery chiller is rated to provide 340 tons (1195 kW) with efficiency of 0.615 kW/ton (19.5 EER) at full cooling load rated conditions. Under full heat recovery operation, the rated capacity and cooling efficiency changes to 337 tons (1185 kW) and 0.853 kW/ton (14.1 EER) respectively while providing approximately 5000 MBH (1465 kW) of heating hot water. The chilled water distribution system is designed as variable primary flow where the CUP primary pumps modulate the flow of chilled water based on flow demands of the variable flow secondary distribution pumps at each building. The chiller units were selected with a 12 °F (6.7 °C) differential temperature between chilled water supply and return to decrease the pumping requirements for the system.

The projected heating load for the campus was calculated by the DBT to be approximately 10,320 MBH (3023 kW). As a result, the heating hot water system design for the CUP was based on three 4000 MBH (1172 kW) condensing hot water boilers. The N + 1 redundancy requirement is met through the use of the heat recovery chiller discussed previously. The heat recovery chiller will be used to meet the heating loads the majority of the year with the condensing boilers providing any required additional capacity. The thermal efficiency of the condensing boilers is in excess of 90% based on the design supply and return temperatures of the system.

J.5.13.4. Renewable Energy

As was mentioned previously, the overall campus redevelopment is planned to include the design and installation of solar photovoltaic panels as part of the Phase IV portion of the project. The goal of the PV project is to provide sufficient electrical production from the installed solar panels to offset the energy usage requirements of Roberdeau Hall. Under current projections, the installed capacity of PV panels will need to be approximately 1.2 MW to achieve this goal.

J.5.13.5. Daylighting Strategies

The advanced lighting control system described previously includes the ability to adjust the output of the lighting fixtures based on available sunlight in areas where daylight is present.

J.5.14. Energy Consumption

Table J.16 shows a comparison of the original campus energy usage to the Architecturally Adjusted Baseline and the renovated campus energy usage through Phase III. It is important to note that all savings were measured from the Architecturally Adjusted Baseline due to the changes in campus square footage (resulting from buildings either demolished or constructed), changes in space use and new building facade construction.

The resulting savings from the Architecturally Adjusted Baseline is estimated to be 97,717 MBtu/year representing a 47% reduction in overall energy usage for the campus. This is made up of 5,613,786 kWh/year electric and 78,557 MBtu/year natural gas savings.

J.5.15. Energy Cost Reduction

See Sect. J.4.14.

J.5.16. Non-energy Related Benefits Realized by the Project

Operations and Maintenance (O&M) services to encompass the entire campus are included as a part of the project. This approach establishes a single point of accountability for energy performance across the site and enhances energy security and reliability by reducing opportunities for conflict between parties that could negatively impact the delivery of utilities service to the supported buildings.

Table J.16 Phased energy usage profile(s)

Run	Electric Energy			Fuel		Electric Costs		Utility Costs			
	Lights (kWh)	HVAC (kWh)	Total (kWh)	Heating (Therm)	Total (Therm)	Energy ($)	Demand ($)	Electric ($)	Gas ($)	Total Cost ($)	EUI (kBtu/sf)
Calibrated Baseline - 2008 Calendar Year	5,439,076	10,279,926	38,048,968	466,803	480,984	$ 4,317,393	$ 112,657	$ 4,433,911	$ 704,614	$ 5,138,525	207
Architecturally Adjusted Baseline	3,250,741	10,873,144	36,172,908	802,670	826,957	$ 3,502,128	$ 132,741	$ 3,638,730	$ 557,067	$ 4,195,797	240
Through Phase III	1,651,878	6,708,325	30,559,122	32,099	41,383	$ 2,958,622	$ 111,633	$ 3,074,116	$ 31,243	$ 3,105,359	126

J.5.17. Renovation Costs: Total and per m²

The specific renovation cost breakdown has not been cleared for public disclosure by the DIA.

J.5.18. Business Models and Funding Sources

As mentioned previously, the redevelopment of the ICC-B South Campus was originally planned to be executed under a SATOCC contract using appropriated funds. The contract was awarded through a competitive solicitation for a design-build contract team (DBT). The DBT was to be made up of a general contractor along with design and engineering services. The successful DBT was selected based on a combination of qualifications and projected implementation budget for completing the campus redevelopment.

One of the principal motivations for considering the use of a private funding stream in conjunction with traditional appropriated funding was a shortfall in appropriated funds to accomplish the campus renovation. The original budget for the project was reduced and the reduction would have delayed many of the programmed elements of the redevelopment effort and hampered efforts to achieve significant reductions in future energy consumption. It was at this point that the use of alternative funding sources began to get serious consideration as a means to help keep the redevelopment effort on track.

DIA was selected to be the Executive Agent for the campus redevelopment effort. The challenges imposed on the redevelopment program by the budget constraint forced the DIA to identify alternative approaches to meeting the program goals under the revised conditions. Previous research conducted on the use of different private financing based performance contract vehicles seemed to provide a means of addressing many of the issues. As the budget constraints began to place the overall outcome of the redevelopment program in jeopardy, DIA's Senior Technical Expert for facilities and construction and the ICC-B project executive researched alternative strategies for energy infrastructure development that could be used to maintain the project goals and leverage appropriated funds.

One of the example projects researched involved the construction of a central utility plant (CUP) and associated utility distribution system at the Federal Research Center at White Oak (FRCWO) in Silver Spring, MD. FRCWO houses the headquarters and most of the centers of the U.S. Food and Drug Administration (FDA), and the campus was redeveloped and is managed by the General Services Administration (GSA) on property that formerly housed the Naval Surface Warfare Center. This Navy property became available pursuant to the Base Realignment and Closure (BRAC) legislation of 1993. The primary energy infrastructure at FRCWO was developed over the course of several phases using the Department of Energy's Energy Savings Performance Contract (ESPC). In addition to providing the primary energy infrastructure that serves the 3.7 million gross sq ft (374,000 m^2) campus, operations and maintenance (O&M) services for both the CUP and the supported campus buildings was also included by the energy service company (ESCO) as part of the ESPC contract.

The research also included a GSA redevelopment effort applying a similar approach in the development of the energy infrastructure at St. Elizabeths (St. E's), a property in the District of Columbia that was to be redeveloped as the headquarters for the Department of Homeland Security and several of its subordinate agencies. This project was also implemented using an energy performance based contract vehicle. In this case, the project was developed under a Utility Energy Service Contract (UESC). The project team was made up of the same ESCO used at the FRCWO teamed with the local utility providing natural gas to the campus. Both projects were used to compare the benefits of the different contract vehicles as applied to the ICC-B project. DIA ultimately decided to move forward with a UESC due to the greater flexibility and relatively quicker acquisition cycle time associated with that contracting vehicle in comparison with the DOE ESPC approach.

Because the redevelopment program was originally planned based on a single contract award, the transition to executing the project using separate contract vehicles and multiple contractors required significant coordination. The overall scope of the redevelopment was reexamined to determine what portions of the construction project would be "carved out" to be included in the UESC portion of the project. The UESC Energy Team (UET) was then required to identify energy conservation measures (ECMs) that would generate significant cost savings to be used to fund the implementation. The UET began examining the existing facilities and original design concepts while the DBT (the Designer of Record) continued to develop the design for the component pieces of the redevelopment. The challenge became coordinating with the DBT to integrate the ECMs developed by the UET into the already mature design concepts without significantly impacting the overall project implementation schedule.

J.5.19. Cost Effectiveness of the Energy Part of the Project (NPV, SIR)

The portion of the overall campus redevelopment including under the UESC is based on the portion of the project that can be funded through a combination of the energy and O&M cost savings with a simple payback of 20 years or less. The overall project value that will ultimately be included in the UESC in all four phases has not yet been determined.

J.5.20. User Evaluation

J.5.20.1. Description of User Training Programs Within the Refurbishment

As indicated elsewhere in this case study, Operations and Maintenance (O&M) is a key component of the overall UESC offering. UESC contractor personnel will be responsible for O&M activities both in the central plant and in the buildings themselves. Subcontractors and vendors are required to provide training to the O&M during start-up and commissioning. Additionally, the training will be videotaped to provide a foundation for training new members of the O&M staff and as refresher training for personnel who received the initial training.

J.5.20.2. Integration of Users Demands in the Planning Process

The UESC program has proven to be extremely responsive to the changing requirements of the end users of the affected facilities. Whenever new user requirements were identified or gaps in the coverage were uncovered, the UESC contractor would work with both the user and the other members of the campus development team to adjust procurement activities so that changes could be accommodated with minimal impact on cost and schedule. While the integration interaction was very challenging early on in the campus redevelopment program, both the timeliness and effectiveness of interaction has steadily improved as all of the parties became more comfortable working together.

J.5.21. Experiences/Lessons Learned

As the renovation of the ICC-B campus has progressed, lessons learned have surfaced over the course of the project. Unfortunately, the UESC Team experienced obstacles that could have been prevented if earlier calibration with all stakeholders would have taken place during the conceptual phase of the project. Coupling the traditional SRM contract with the UESC contract from the beginning would have required additional upfront planning, but resulted in a more cohesive product and potential for additional energy saving opportunities. One apparent hurdle was the contractual and scheduling constraints that limited the influence that the UESC Team could have on the DBT's designs. As mentioned above, the DBT had obligations and specific schedules to deliver completed phases of the campus. Therefore, it was necessary for them to continue progressing forward to meet their contractual obligations. In parallel, the UET was developing the Detailed Feasibility Study (DFS), which demonstrated viable energy conservation measures (ECM). The parallel engineering development along a condensed timeline created insurmountable issues when seeking to integrate solutions from both teams.

Throughout the development and completion of the UET Detailed Feasibility Study, the DBT progressed forward with their design obligations. When the two efforts reached maturity, stakeholders were in a position that left them few options. One option was to extend the schedule and pay for additional design fees to implement all of the UESC Team's ECMs and not sacrifice any potential energy savings demonstrated by the UET's DFS. Second, was to take a hybrid approach that was to pursue ECMs that had little impact on the overall schedule, but reduce the campus's carbon footprint and minimize additional cost to the project. As a result, all stakeholders agreed to move forward with the hybrid approach that fine-tuned the UET's ECMs to fit within the current DBT's design without impeding progress towards the ultimate goal of providing a state-of-the-art campus by contractual project completion date.

In future efforts such as this, the two contract teams should be identified and begin collaborating from the very beginning of the project development. Under this model, the energy project development team would work directly with the design team to integrate the energy conservation measures into the basis of design. This will allow the systems to be designed in a manner that maximizes the benefits achieved in both energetic performance and constructability. This integrated design team approach would also decrease the overall development costs over separately engineered solutions competing for schedule and budget resources. As the project moves from design into construction, the integrated team approach would greatly reduce the likelihood of coordination issues and scope gaps. The largest benefit may be that an integrated team approach would foster a more collaborative working environment working to achieve the best possible common result.

Projects such as these are highly complex and require instant communication between the team members. Regular and open communication will ensure scope gap conflicts are identified early. The complexity of the ICC-B project led the team to decide early on that weekly meetings would be necessary. Some weeks the teams met more than once and all meetings were held on site, away from home office distractions. The communication between all stakeholders is likely the single most important component to a successful project of this magnitude and complexity.

J.6. Johnson Braund Design Group, Seattle WA

J.6.1. Name of the Project, Location (City, Country)

Johnson Braund Design Group, Seattle WA United States.

J.6.2. Pictures

See Fig. J.34.

J.6.3. Project Summary

J.6.3.1. Project Objectives

The owners sought reduce the project energy use while maintaining a realistic budget.

J.6.3.2. Project Energy Goals

Reduce electrical grid consumption by 50%.

J.6.3.3. Short Project Description

In 2002, JBDG purchased a two-story office building built in 1984 to house its growing practice. The 8000 ft^2 office space consumed over 400 kWh of electricity per day, with a majority of that consumption coming from the building's original

Fig. J.34 JBDG two-story office building. *Source* Courtesy of Johnson Braund Design Group

HVAC system, which was nearing the end of its life cycle. The owners sought to reduce this energy use by half while maintaining a realistic budget.

J.6.3.4. Stage of Construction

Complete.

J.6.3.5. Point of Contact Information

Steve Allwine.
Johnson Braund Design Group, Inc.
Seattle, WA 98188.
253-709-2333.
http://www.jbdg.com/

J.6.3.6. Date of the Report

2012.

J.6.3.7. Acknowledgement

Northwest Energy Efficiency Alliance (NEEA).

J.6.4. Site

- Seattle, 47.65° N 122.30°W; Elevation: 450; WA Climate Zone 4C (Marine).
- HDD65: 4611.
- CDD65: 158 (Table J.17).

Table J.17 Design temperature

*Cooling design temperature—0.4% occurrence**	
Dry bulb temp C (F)	Mean coincident wet bulb temp C (F)
81 (F)	
*Heating design temperature—99.6% occurrence***	
Dry bulb temp C (F)	
23 (F)	

J.6.5. Building Description/Typology

J.6.5.1. Typology/Age

Office/1984.

J.6.5.2. Type (Office, Barracks, Etc.)

Office.

J.6.5.3. Typology/Age

1970.

J.6.5.4. General Information

- Year of construction: 1984.
- Year of previous major retrofit—if known: NA.
- Year of renovation (as described here): 2007.
- Total floor area (m^2): 743.
- Area of unconditioned space included above (m^2): NA.
- Other information as appropriate.

J.6.6. Architectural and Other Relevant Drawings

Not available.

J.6.7. National Energy Use Benchmarks and Goals for Building Type Described in the Case Study

J.6.7.1. Benchmark, According to the Standard National Average, min, and Max

Average for all U.S. office Buildings according to the Commercial Building Energy Consumption Survey (CBECS) and the average for comparable office buildings as determined by the Energy Star Portfolio Manager based on like type, size, occupancy, hours and climate.

J.6.8. Awards or Recognition

EPA Small Business Innovation Award.

J.6.9. Site Energy Cost Information (Electricity, $/kWh; Gas, $/m^3, Etc.)

J.6.9.1. Electricity

Total annual electric cost: $3860; average annual cost/kWh: $0.067.

J.6.9.2. Natural Gas

Total annual natural gas cost: $1871; annual cost/therm: $4.13.

J.6.10. Pre-Renovation Building Details

J.6.10.1. Heating, Ventilation, Cooling, and Lighting Systems

The building's prerenovation heating and cooling and ventilation was supplied by rooftop units (RTU's).

J.6.11. Description of the Problem: Reason for Renovation

The energy focus of this project was initiated in 2007 when the original HVAC rooftop units failed. In considering its replacement and upgrade, the firm saw an opportunity to increase efficiency by cutting energy consumption in half while meeting a return on investment (ROI) goal of between five and 6 years.

J.6.12. Renovation SOW (Non-energy and Energy Related Reasons)

At the time of the HVAC replacement, a larger list was drawn up and included other energy efficiency upgrades that could be implemented to further reduce the building's energy load. A majority of these items have been acted upon with a focus primarily on HVAC, lighting, and plug load measures. Envelope measures such as window replacement did not meet JBDG's 5-6-year ROI goal.

J.6.13. Energy Saving/Process Improvement Concept and Technologies Used

J.6.13.1. Building Envelope Improvement

A glazed entrance was added to the front of the building to create a buffered transition space between the interior and exterior.

J.6.13.2. New HVAC System or Retrofits to Existing

The project team worked with a major manufacturer to specify a residential multi-fuel, high efficiency heat pump in a side-by-side arrangement. The benefits of this selection included greater control flexibility, increased ventilation rates, and a wider selection of high-efficiency heat pumps. This system is responsible for a majority of the building's energy savings. JBDG has also implemented an energy recovery strategy that reduces the cooling load to the server and uses the recovered heat to warm areas within the office.

J.6.13.3. New Lighting System

Lighting was upgraded to T5-based fixtures with programmable start/stop ballasts that include occupancy sensors for the restrooms and kitchen area. In the open work area, individual work stations include built-in task lighting and an ambient uplighting element. The lighting control package includes daylighting and occupancy sensors. The connected lighting load is approximately 1.25 W/ft^2 (code at the time), but the daylighting and occupant controls significantly reduce the actual energy use.

J.6.13.4. Renewable Energy

A photovoltaic system on the roof provides an annual output of 7897 kWh, which offsets energy consumption by 14%.

J.6.13.5. Daylighting Strategies

Daylighting controls are incorporated in the open portion of the office.

J.6.14. Energy Consumption

J.6.14.1. Pre-renovation Energy Use (Total and per m^2/year)

- Pre-renovation EUI (kBtu/sf/year): 71.
- Pre-renovation total annual energy use (kBtu): 568,000.

J.6.14.2. Measured Energy Savings (Thermal, Electrical), Total and per m^2/year

- Total measured energy savings (kWh) = 1,146,432.
- Total annual measured energy savings (kBTU/sf/year) = 29.

J.6.14.3. Annual Energy Use Reduction

- Annual energy use reduction (kBtu/sf/year): 42.
- Annual energy use reduction (kBtu): 336,000.

J.6.15. Energy Cost Reduction

Not available.

J.6.16. Non-energy Related Benefits Realized by the Project, E.g., Improved Productivity, Increased Rent/Lease, Increased Useful Space, Reduced Maintenance, Etc.

As a design firm, these projects and improvements have provided JBDG with what Steve Allwine, owner representative for JBDG, calls "real world expertise" and a high level of legitimacy when it comes to encouraging clients to undertake such projects. Allwine pointed specifically to the heat recovery solution installed as part of the HVAC upgrade, which has since been included in two bank projects on which the firm has worked. He believed the strategy was especially innovative as it captured heat from the computer servers, reduced the need for cooling and recirculated the heat to reduce the heating load.

J.6.17. Renovation Costs: Total and per m^2

J.6.17.1. Total

- Total cost: $250,000 (after incentives).
- Cost per square foot: $31.

J.6.17.2. Business Models and Funding Sources

J.6.17.3. Decision Making Process Criteria for Funding and Business Models

Working with its in-house designers and engineers, JBDG had enough technically-qualified staff to identify, implement, and track key measures and successes. At the time of the HVAC replacement, the design team drew up a larger list of energy upgrades that could be implemented to further reduce the buildings energy load. Measures that did have a 5- to 6-year return on investment were not considered.

J.6.17.4. Description of the Funding Sources Chosen

Funding for the upgrades was provided through conventional bank financing typical to capital improvement projects. Incentives granted for the photovoltaic installation were:

- 30% Federal Tax Credit for photovoltaic installation.
- Washington Renewable Energy Production incentive: $5000/year.

J.6.18. Cost Effectiveness of Energy Part of the Project (NPV, SIR, Etc.)

The project team had a project goal of meeting a return on investment (ROI) goal of between 5 and 6 years. This basic principle was applied to all equipment replacements and upgrades.

J.6.19. User Evaluation

Not available.

J.6.20. Experiences/Lessons Learned

J.6.20.1. Energy Use

The firm encountered some bureaucratic barriers with their local utility coverage. Due to its location at the "end of the line," JBDG did not qualify for fuel switching, which would have enabled them to become an all-electric building and therefore would have offset more of their load through the use of on-site renewable power generation.

J.6.21. References

Johnson Braund Design Group: Steve Allwine, Facility and Project Manager.
JBDG Case Study—"Small Office Renovation."

J.7. Priest River, ID

J.7.1. Name of the Project, Location (City, Country)

Beardmore Building, Priest River, ID, USA.

J.7.2. Pictures

See Fig. J.35 and J.36.

Fig. J.35 After renovation. *Source* Brian Runberg, Runberg Architecture Group

Fig. J.36 Indoor after renovation. *Source* Brian Runberg, Runberg Architecture Group

J.7.3. Project Summary

J.7.3.1. Project Objectives

Restore the 1922 building using sustainable and energy efficient design principles.

J.7.3.2. Short Project Description

After decades of neglect under outside ownership, Brian Runberg, an architect and great-grandson of its original developer, repurchased the building in 2006 and began an extensive whole building historic restoration. The location currently functions primarily as office space, with tenant leased area comprising 85% of the building (when fully occupied) and a 4100 ft^2 theatre to be renovated at a later time.

J.7.3.3. Stage of Construction

Completed.

J.7.3.4. Point of Contact Information

Brian Runberg: One Yesler Way Suite 200 Seattle, WA 98104. Phone #: 206-956-1970.

J.7.3.5. Date of the Report

2012.

J.7.3.6. Acknowledgement

Northwest Energy Efficiency Alliance (NEEA).

J.7.4. Site

- Location: Priest River, ID, 48° N and 116° W, elevation: 2139 ft.
- Climate zone (e.g., ASHRAE 90.1-2004 Climate Zone): 6 B (dry).
- Cooling Degree Days (based on 65 F): 126, Heating Degree Days (based on 65 F): 7753 (Table J.18).

Table J.18 Design temperature

*Cooling design temperature—0.4% occurrence**	
Dry bulb temp C (F)	Mean coincident wet bulb temp C (F)
91.4 (F)	62.8
*Heating design temperature—99.6% occurrence***	
Dry bulb temp C (F)	
11.6 (F)	

J.7.5. Building Description/Typology

J.7.5.1. Typology/Age

Office/1922.

J.7.5.2. Type (Office, Barracks, Etc.)

Office.

J.7.5.3. Typology/Age

1910–1930.

J.7.5.4. General Information

- Year of construction: 1922.
- Year of renovation (as described here): 2008.
- Total floor area (m^2): 3019.3 (including theatre).
- Not including theater: 2675.6.
- Area of unconditioned space included above (m^2):
- Other information as appropriate: There is a 4100 ft^2 theatre in the building that was not included in the scope of the renovation.

J.7.6. Architectural and Other Relevant Drawings

Not available.

J.7.7. National Energy Use Benchmarks and Goals for Building Type Described in the Case Study

J.7.7.1. Benchmark, According to the Standard National Average, Min, and Max

Benchmark: 30–60% better than the average energy use for all U.S. Office buildings as determined by the Commercial Building Energy Consumption Survey (CBECs) and a comparable office average energy use as determined by the Energy Star Portfolio Manager Program based on like building type, size, occupancy, hours, and climate from statistical analysis of the CBECs dataset.

Modeled baseline building as described in the LEED NC project submittal.

J.7.8. Awards or Recognition

- LEED-NC Gold.
- 2010 American Association for State and Local History—National Award of Merit for Restoration.
- 2009 Pacific Coast Builders Conference—Grand Award for Best Adaptive Re-Use.
- 2008 The Idaho Historic Preservation Council—Orchid Award: Excellence in Historic Preservation.

J.7.9. Site Energy Cost Information (Electricity, $/kWh; Gas, $/m^3, Etc.)

J.7.9.1. Electricity

kBtu/sf/year = 32; Total site electricity (kWh) = 178,365; Total cost = $17,387.95.

J.7.10. Pre-renovation Building Details

J.7.10.1. Envelope Details: Walls, Roof, Windows, Insulation Levels

The existing above grade walls were uninsulated masonry (U-0.43) and the below grade walls were uninsulated masonry (U-0.193). The roof and floor was also uninsulated. The existing windows were single pane wood windows (U-1.2/SHGC 0.85) and there was some glass block.

J.7.10.2. Heating, Ventilation, Cooling, and Lighting Systems

- The existing main floor was heated with a propane fired furnace with the upper office spaces heated with electric resistance baseboard heaters. Both spaces are cooled with a split system air-conditioner.
- Vintage light shades were preserved and rebuilt with new fixtures using high-efficiency compact fluorescents.

J.7.11. Description of the Problem: Reason for Renovation

The building owner wanted to restore the building using sustainable and energy efficient design principles that he incorporates into his own architectural practice. Additionally, he wanted to play a part in revitalizing the Priest River community and economy by restoring the building to its former grandeur.

J.7.12. Renovation SOW (Non-energy and Energy Related Reasons)

The Beardmore building renovation is a mixed-use major renovation that is made up of two above grade levels and a below grade space. The scope of the project included adding insulation to the walls and ceiling and restoring and re-glazing all windows and the replacement of all heating, cooling and ventilation systems. Additionally the existing skylights were uncovered and refit with new glazing to enhance natural daylighting. Also, vintage light shades were restored and rebuilt with new fixtures. And upgrades were made to the structure to handle the increased snow loads that were a result of the increased insulation, without affecting the building's historic character.

J.7.13. Energy Saving/Process Improvement Concept and Technologies Used

J.7.13.1. Building Envelope Improvement

All the original wood windows were removed, restored and re-glazed with low-E, argon-filled insulated glass. Additional glazing was added to the interior of the leaded transom glass to preserve historic integrity and improve energy performance. Extensive insulation was added to the exterior walls, including R-50 for the roof cavities.

J.7.13.2. New HVAC System or Retrofits to Existing

High-efficiency, packaged rooftop heat pumps with economizers were installed. Demand control ventilation (DCV) with CO_2 sensors and modulating outside air dampers were also included, allowing ventilation to be based on actual occupancy rather than assuming full occupancy and thereby reducing energy for conditioning and moving the air. The mechanical engineer determined that the common area of the building did not require cooling and instead designed a barometric damper assisted by ceiling fans located at the curb of the skylights to exhaust air and create a convection-based air flow.

J.7.13.3. New Lighting System

Light fixtures exceed advanced lighting requirements of the local utility incentive program. Vintage light shades were preserved and rebuilt with new fixtures using high-efficiency compact fluorescents. The central lighting system has a night set-back to ensure low- to no-energy use during unoccupied times. Restrooms have occupancy sensors.

J.7.13.4. New Generation/Distribution System

The building is wired and ready for photovoltaic cell panels, but these have not been installed due to a payback period calculated to be 15–18 years and no available rebate.

J.7.13.5. Renewable Energy

The building is wired and ready for photovoltaic cell panels, but these have not been installed due to a payback period calculated to be 15–18 years and no available rebate.

J.7.13.6. Daylighting Strategies

After many years of being covered up, the original skylights were removed and refitted with new glazing to provide natural daylighting and ventilation.

J.7.14. Energy Consumption

J.7.14.1. Predicted Energy Savings (Site, Source, GHG), Total and per m^2/year

Predicted total site energy savings (kWh): 333,727.

J.7.14.2. Measured Energy Savings (Thermal, Electrical), Total and per m^2/year

- Measured total site energy savings when compared to the modeled baseline building as described in LEED submittal (kWh): 465,649; Measured EUI (kBtu/sf/year): 55.
- Measured EUI savings when compared to an average for a comparable office building as determined by Energy Star (kBtu/sf/year): 28.
- Measured EUI savings when compared to the average for all U.S. office buildings according to CBECs: 61.

J.7.15. Energy Cost Reduction

Not available.

J.7.16. Non-energy Related Benefits Realized by the Project, E.g., Improved Productivity, Increased Rent/Lease, Increased Useful Space, Reduced Maintenance, Etc.

According to Brian Runberg, tenants have been attracted to the Beardmore because of both the historic renovation and LEED certification. Brian believes his building has "sparked new economic life into the community, giving it a renewed sense of pride and entrepreneurial spirit. Tenants saw the potential of what could happen in the building and came with business ideas."

Due to the Beardmore's energy efficiencies and overall historic qualities, rents average about 35% higher than other local properties.

Runberg states in his article in the Daily Journal of Commerce: "This transformation of a decaying obsolete and lifeless shell into a high performance, healthy and vibrant environment carries many benefits. The initial investment has proven itself to be financially prudent, with substantially lower operation costs, greater lasting quality, and a healthy environment for its users. Yet equally important is the preservation of an important historic landmark, one in which my own history is tied."

The renovation of the Beardmore Building has helped Runberg in his own architectural practice when client discussions turn to the cost benefits of pursuing LEED and energy efficiency.

Renovation Costs: total and per m^2.

J.7.16.1. Total

$2,600,000; $105/sf (after incentives).

J.7.17. Business Models and Funding Sources

J.7.17.1. Decision Making Process Criteria for Funding and Business Models

A cost-benefit analysis was used to determine the economic impact of green building practices in terms of design, documentation, material salvage and construction. The owner/architect developed a methodology matrix to evaluate the sometimes contradictory requirements for Federal and local incentives, LEED certification and preservation standards, focusing on the most cost-effective strategies for energy, water and material use.

J.7.17.2. Description of the Funding Sources Chosen

Because the Beardmore Building was on the historic register, the owner received a tax credit for construction cost from the National Park Service. Amounting to $366,571, this was awarded for adhering to the Secretary of the Interior's Standards for Rehabilitation. The local utility provided a LEED certification and HVAC efficiency incentive of $1.25/sf, amounting to a $71,079 rebate. Due to recession market conditions at the time, construction loans were constrained and the project loan equaled only approximately 25% of total funding, with the rest supplemented by direct owner equity.

J.7.17.3. Funding Sources of the Business Model

Owner equity; construction loan; National Park Service tax credit; Utility incentives.

J.7.18. Cost Effectiveness of Energy Part of the Project (NPV, SIR, Etc.)

Not available.

J.7.19. User Evaluation

Not available.

J.7.20. Experiences/Lessons Learned

J.7.20.1. Energy Use

The ground-floor retail level had intricate leaded glass transom windows that provided almost no insulating properties. The architect first proposed to sandwich the leaded glass inside an insulated glazing unit, but this was rejected by the state historic preservation office. The approved solution allowed a separate insulated glazing unit to be applied in the interior, retaining the exterior character but providing the necessary U-value performance for the energy targets. The historic nature of the building did not allow for a vestibule to be added at the front doors, so ground-floor heat loss in the winter months is an issue.

J.7.20.2. Impact on Indoor Air Quality

Standard commissioning included testing of air infiltration, duct tightness, exhaust air flow rate and particulate and volatile organic compounds emissions prior to occupancy.

J.7.21. References

- Brian Runberg, Runberg Architecture Group.
- Ecotope: Jonathan Heller and Carmen Cejudo.
- "They Built the Beardmore," Cate Huisman, *Idaho Magazine*—January 2010.
- "Sustaining the Future while Restoring the Past," Brian Runberg, *Daily Journal of Commerce* April 2009.
- "Deep Energy Savings in Existing Buildings Case Study|The Beardmore Building," *New Buildings Institute*—April 2012.

- "The Beardmore LEED v.2.2 EAcr1 Documentation," *Ecotope Inc.* June 2007.
- www.beardmoreblock.com
- Photos: Maria Dominique Verdier.

J.8. Indio, Sunnyvale, CA,

J.8.1. Name of the Project, Location (City, Country)

435 Indio, Sunnyvale, CA, United States

J.8.2. Pictures

See Figs. J.37, J.38, J.39, J.40, J.41, J.42, J.43, J.44 and J.45.

J.8.3. Project Summary

J.8.3.1. Project Objectives

The project objectives were to create a health and positive work environment for the building occupants, which minimized its impact on the environment, all while accomplishing this at a price point that was operationally more profitable than the standard way of renovating to the city code.

Developers' goals:

- Create a healthy and positive work environment for the occupants:
  - Natural light
  - Thermal comfort
  - Natural ventilation
  - Exceptional acoustics
  - Constant connection to nature.
- Minimize the impact on the environment.
  - Net zero energy operationally
  - Carbon neutral
  - Minimize water use
  - Highly sustainable construction methods and materials.
- Accomplish this at a price point that will operationally be more profitable than the standard way of renovating to city code.

Fig. J.37 Existing conditions

Fig. J.38 Existing conditions

Fig. J.39 The vision

Fig. J.40 High performance envelope

Fig. J.41 High performance envelope

Fig. J.42 Integrated roof planning

Fig. J.43 Daylighting design

Fig. J.44 Lighting design

Fig. J.45 Passive thermal comfort. *Source* Kevin Bates, SHARP Development Company, Inc.

J.8.3.2. Project Energy Goals

The owner/developer set the energy goal as net zero energy usage on an annual basis. The project's EUI (kBtu/sf) goal of 21.5 was based on the amount of energy they could produce from the photovoltaic panels on the roof.

J.8.3.3. Short Project Description

Retrofit the uninsulated 30,000 ft^2 office warehouse to a net zero goal. This included upgrading the existing single pane windows, insulating the walls and roof and putting rooftop monitors in for better daylighting.

J.8.3.4. Stage of Construction

Complete.

J.8.3.5. POC

Kevin Bates, SHARP Development Company; 20 Prado Court, Portola Valley, CA 94028.

J.8.3.6. Date of the Report

7/24/15.

J.8.3.7. Acknowledgement

Kevin Bates and SHARPT Development, INTEGRAL Group.

J.8.4. Site

Sunnyvale, CA. 37 degrees N and 122 degrees W; Elevation: 125 ft; Climate Zone 3C; CDD: 220 and HDD: 2643 (Table J.19).

J.8.5. Building Description/Typology

J.8.5.1. Type

Tilt up Concrete/1970.

J.8.5.2. Type (Office, Barracks, Etc.)

Office/Warehouse.

Table J.19 Design temperature

*Cooling design temperature—0.4% occurrence**	
Dry bulb temp C (F)	Mean coincident wet bulb temp C (F)
33.6	19.5
*Heating design temperature—99.6% occurrence***	
Dry bulb temp C (F)	MCDB
2.0	9.8

J.8.5.3. Typology/Age

1970.

J.8.5.4. General Information

- Year of construction: 1970.
- Year of previous major retrofit—if known: NA.
- Year of renovation (as described here): 2014.
- Total floor area (sq ft): 31,000.
- Area of unconditioned space included above (sq ft): 31,000.

J.8.6. Architectural and Other Relevant Drawings

Not available.

J.8.7. National Energy Use Benchmarks and Goals for Building Type Described in the Case Study

J.8.7.1. Benchmark

Baseline model was a high performance building.

J.8.8. Awards or Recognition

Acterra Award.

J.8.9. Site Energy Cost Information (Electricity, $/kWh; Gas, $/m^3, Etc.)

J.8.9.1. Electricity (kWh)

200,066.

J.8.10. Pre-renovation Building Details

Not available.

J.8.11. Description of the Problem: Reason for Renovation

Take a dark, dingy, derelict and unrentable office building and transform it into a healthy and positive work environment.

J.8.12. Renovation SOW (Non-energy and Energy Related Reasons)

Insulate walls and roof, replace windows, install new light monitors, and upgrade mechanical, lighting and electrical systems. Add a roof mounted photovoltaic system.

J.8.13. Energy Saving/Process Improvement Concept and Technologies Used

J.8.13.1. Building Envelope Improvement

R-20 walls and R-40 roof, replace windows and add 43 rooftop monitors.

J.8.13.2. New HVAC System or Retrofits to Existing

Air source heat pump with Dedicated Outdoor Air System (DOAS) (Fig. J.46).

J.8.13.3. New Lighting System

New indirect/direct LED lighting system with wireless daylight dimming controls. 0.5 w/sf.

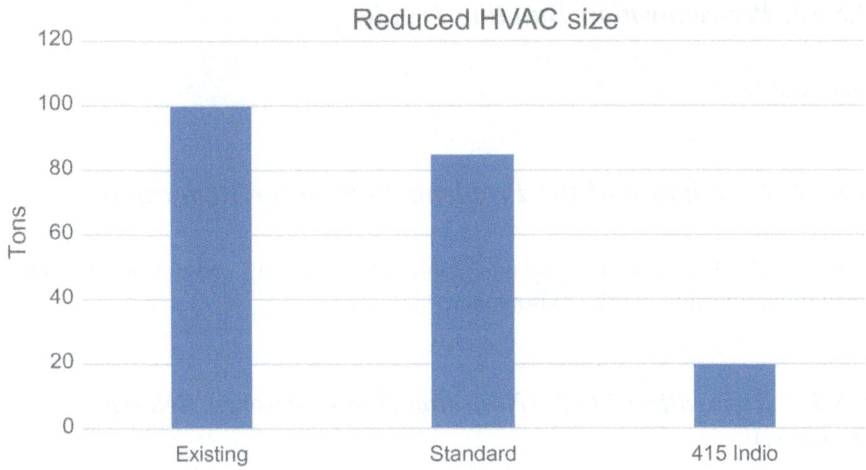

Fig. J.46 HVCA sizes before and after retrofit

J.8.13.4. New Generation/Distribution System

This project included a roof mounted photovoltaic system that in metered with the local utility.

J.8.13.5. Renewable Energy

148 kW Roof mounted photovoltaic system that is net metered with the local utility.

J.8.13.6. Daylighting Strategies

Install new rooftop monitors and include interior daylight sensors. Dynamic glass that can electronically reduce glare depending on sun position.

J.8.14. Energy Consumption

J.8.14.1. Predicted Energy Savings (Site, Source, GHG) Compared to Modeled Baseline

227,982 kWh and 24.5 kBtu/sf/year.

J.8.14.2. Measured Energy Savings (Thermal, Electrical), Total and per m²/year

Indio Net Zero Energy Budget (Figs. J.47 and J.48).

J.8.14.3. Annual Energy Use Reduction

24.5 kBth/sf/year.

J.8.15. Energy Cost Reduction

Not available.

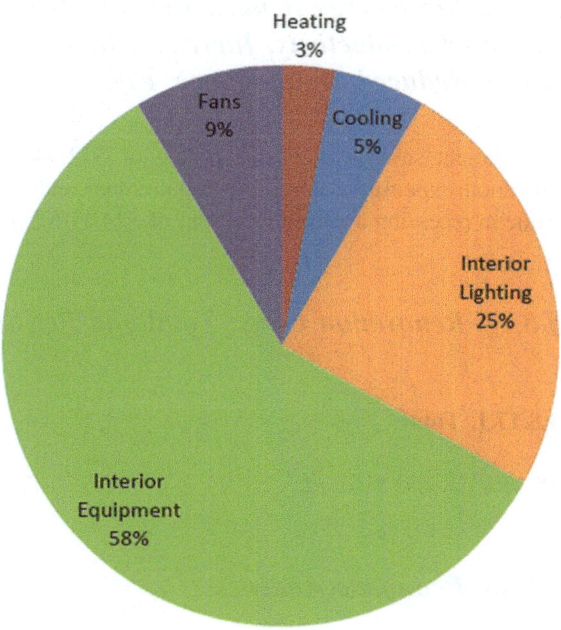

Fig. J.47 Percentage of measure energy savings per domestic service

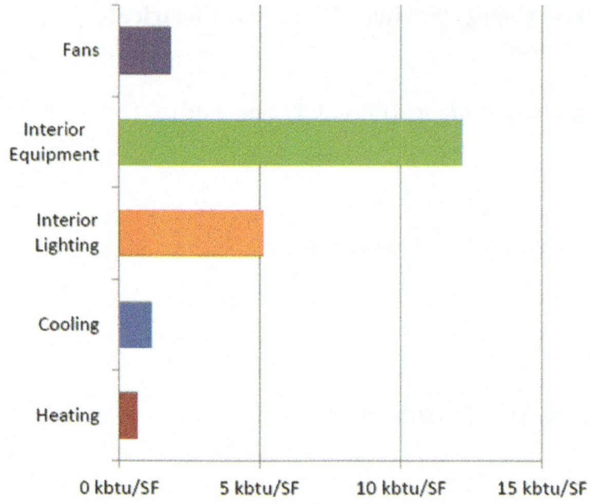

Fig. J.48 Measure energy savings per domestic service

J.8.16. Non-energy Related Benefits Realized by the Project, E.g., Improved Productivity, Increased Rent/Lease, Increased Useful Space, Reduced Maintenance, Etc.

The project saw an additional value of $52.94/SF as a result of reduction in operation expenses and reserve requirement and was able to generate additional value in receiving a premium in rent of $34.47/SF over the top of the market rents.

J.8.17. Renovation Costs: Total and Per Sq/ft

J.8.17.1. Total

$5,136,015 and $161.72.

J.8.18. Economic Analysis

See Fig. J.49.

Monthly Rent / SF	$2.75NNN	Actual		
435 Indio				
Cost Info	Standard	Sustainable		
	Renovation	Renovation	Difference	Per Sq. Ft.
Total A & E Cost	$260,800	$337,700	($76,900)	
Total Hard Cost	$2,865,944	$4,042,458	($1,176,514)	
Total Soft Cost	$426,550	$410,629	$15,921	
PV Cost	$0	$345,228	($345,228)	
Total Cost	$3,553,294	$5,136,015	($1,582,721)	($49.84)
Income Information / Valuation				
Annual rent net of operating expenses	$847,194	$1,086,848	$239,654	
Value of Rent Differential at 7.5% cap rate			$3,195,387	$100.61
Rent during first 18 mos. after completion	$0	$724,500	$724,500	$22.81
Net Additional Value if sold in 18 mos.			**$2,337,166**	**$73.59**
PG&E and Govt. Rebates	$0	$298,764	$298,764	$9.41
Additional Value if include rebates			$2,635,930	$83.00

Cash Flow				
Additional Cost			($1,582,721)	($49.84)
Less Early Lease-up			$724,500	$22.81
Net additional Cost				($27.02)
Net Cash Flow after Debt Service	**$585,350**	**$708,373**	**$123,023**	**$3.87**
Years to ammortize with cash flow				5.89
Rental Income Differential				NNN Rent
Operating Expenses	$280,220	$185,540	$94,680	$0.25
TI Replacement Reserves	$62,866	$47,638	$15,228	$0.04
Additional 326 SF of space	$0	$15,114	$15,114	$0.04
Premium over market rent & op. exp.	$0	$114,632	$114,632	$0.30
Total			$239,654	$0.63

Fig. J.49 Cost analysis of the renovation

J.8.19. Business Models and Funding Sources

J.8.19.1. Decision Making Process Criteria for Funding and Business Models

The developer compiled a pro forma comparing the cost of standard code base renovation and a sustainable, net zero renovation and realized when account for all the benefits of the sustainable, net zero renovation such as reduced operating costs and the ability to charge a premium, over market rent, it was only a 16 month simple payback to do the sustainable option.

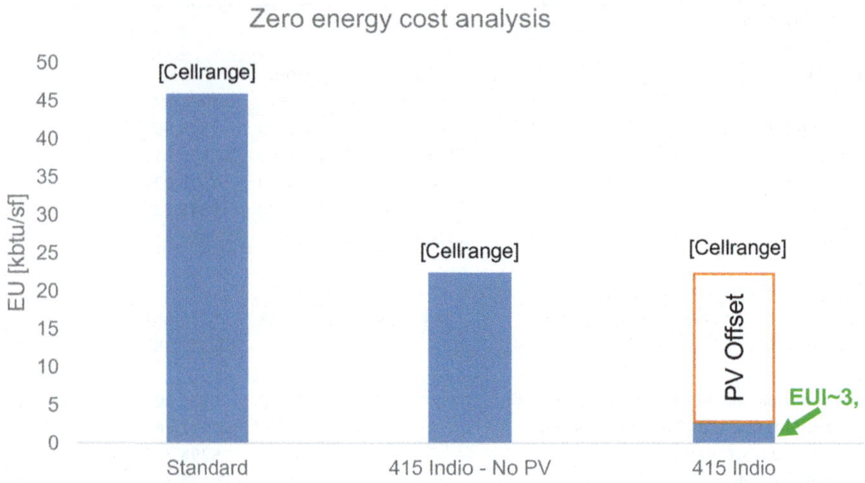

Fig. J.50 Energy cost analysis

J.8.20. Cost Effectiveness of Energy Part of the Project (NPV, SIR, Etc.)

See Fig. J.50.

Business case for zero energy cost:

- Additional cost to renovate sustainably versus Less expensive standard method of renovating = > ($49.84/sf).
- Additional value created due to a reduction in operating expenses and reserve requirements = > $52.94/sf.
- Additional value due to accelerated lease-up time vs. Average market down-time = > $22.81/sf.
- Additional value due to receiving a premium in rent over the top of the market rents = > $34.47/sf.

J.8.21. User Evaluation

Not available.

J.8.22. Experiences/Lessons Learned

Not available.

J.8.23. References

Kevin Bates presentation and information provided by Integral Group.

J.9 The Byron G. Rogers Federal Office Building, Denver, CO

J.9.1. Name of the Project, Location (City, Country)

Byron Rogers Federal Office Building, Denver, CO, USA

J.9.2. Pictures

These pictures show the building in its original and post-retrofit states, that illustrating key features of the retrofit (Figs. J.51, J.52, J.53, J.54, J.55, J.56, J.57, J.58, J.59, J.60, J.61, J.62, J.63, J.64, J.65 and J.66).

J.9.3. Project Summary

J.9.3.1. Project Objectives

The Rogers modernization exemplifies the 2009 American Recovery and Reinvestment Act goals of creating jobs, spurring the economy, and constructing energy efficient Federal facilities that deliver lasting progress toward building a more sustainable national infrastructure while reducing the Federal government's consumption of energy and water, and increasing the use of clean and renewable sources of energy.

1. Daylighting.
2. Economic catalyst: The project employed about 8300 workers and $59 million of the project's subcontracting work was performed by small business. This was a key project that put Americans back to work at a time our Country was experiencing an economic crisis.

Fig. J.51 Exterior Rogers building

J.9.3.2. Project Energy Goals

The project had an aggressive energy goal of less than 39.1 kBtu/SF/year, which achieves annual energy savings of approximately 55%, LEED GOLD certification, 50% better than ASHRAE.

J.9.3.3. Short Project Description

Funded by the American Recovery and Reinvestment Act, the $159M Byron G. Rogers Federal Office Building & Courthouse modernization project will deliver significant returns on sustainability investments and add 100 years of life to an important aging building. The18-story 494,156 ft^2 office building, home to 11 Federal agency tenants, was built in the 1960s and required a deep retrofit to reduce

Fig. J.52 OPM Entry way

energy use while preserving its historical significance. The facility is expected to attain a LEED® New Construction GOLD certification. The primary goal of the modernization was to upgrade all of the major building systems. This included replacement of mechanical, electrical, and plumbing systems.

J.9.3.4. Stage of Construction

Complete.

J.9.3.5. Point of Contact Information

- Cara Carmichael, Rocky Mountain Institute, ccarmichael@rmi.org.
- Jessica Higgins, General Services Administration, Jessica.higgins@gsa.gov.

Fig. J.53 OPM space

Fig. J.54 OPM conference

Fig. J.55 Café

Fig. J.56 2nd floor Cafeteria

Fig. J.57 LED lighting OPM space

Fig. J.58 Gallery

Fig. J.59 Lobby

Fig. J.60 Elevator lobby

Fig. J.61 Gym

Fig. J.62 Window detail

Fig. J.63 Recycling

Fig. J.64 Water storage tank

Fig. J.65 The lobby of the Byron G. Rogers Federal Building (courtesy of the GSA). *Source* General Services Administration

J.9.3.6. Date of the Report

September 2, 2015.

J.9.3.7. Acknowledgement

- Kim Bailey, General Services Administration.
- Bryan Zach, General Services Administration.

J.9.4. Site

- Location: 1961 Stout Street, Denver, CO.
- Latitude: 39.749763.
- Longitude: − 104.989046.

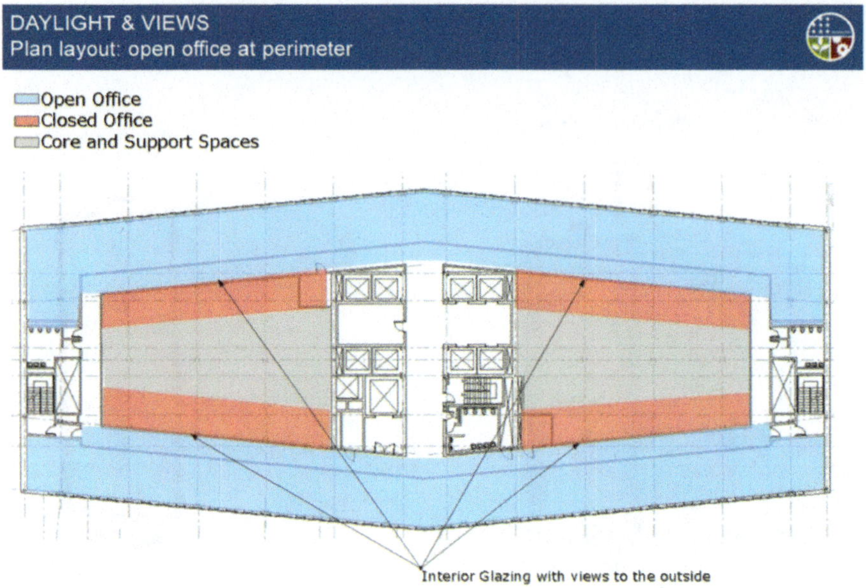

Fig. J.66 The first diagram is a typical floor plan illustrating outside view, which is a discussion point in the case study—open office along the perimeter and individual offices in the interior equipped with interior glazing to share natural daylight and provide views from one side to the other. *Source* General Services Administration

- Elevation: 5280 ft.
- Climate Zone: 5.
- CHD:719 (2014).
- HHD: 5836 (2014) (Table J.20).

J.9.5. Building Description/Typology

J.9.5.1. Typology/Age

51 Years.

J.9.5.2. Type (Office, Barracks, Etc.)

Office.

Table J.20 Design temperature

*Cooling design temperature—0.4% occurrence**	
Dry bulb temp C (F)	Mean coincident wet bulb temp C (F)
93.9 °F with 60.7° MCWB	Winter occupied mode (OAT < 50 °F)
*Heating design 48F MCWB temperature—99.6% occurrence***	
Dry bulb temp C (F)	Mean coincident wet bulb temp C (F)
− 1.4 °F	Summer occupied mode ((OAT > 60 °F)

J.9.5.3. General Information

- Year of construction: 1964.
- Year of previous major retrofit—if known: None.
- Year of renovation (as described here): 2013.
- Total floor area (m^2): 620,000 SF.

J.9.6. Architectural and Other Relevant Drawings

See Figs. J.66 and J.67.

J.9.7. National Energy Use Benchmarks and Goals for Building Type Described in the Case Study

J.9.7.1. Benchmark, According to the Standard National Average, Min, and Max

- The anticipated building energy use savings when compared to ASHRAE 90.1 2007 is expected to be 55%.
- Building will achieve LEED Gold, Energy Star 95.

J.9.7.2. National Energy Target for This Type of Building (If Any)

Target was 50% better than ASHRAE standards.

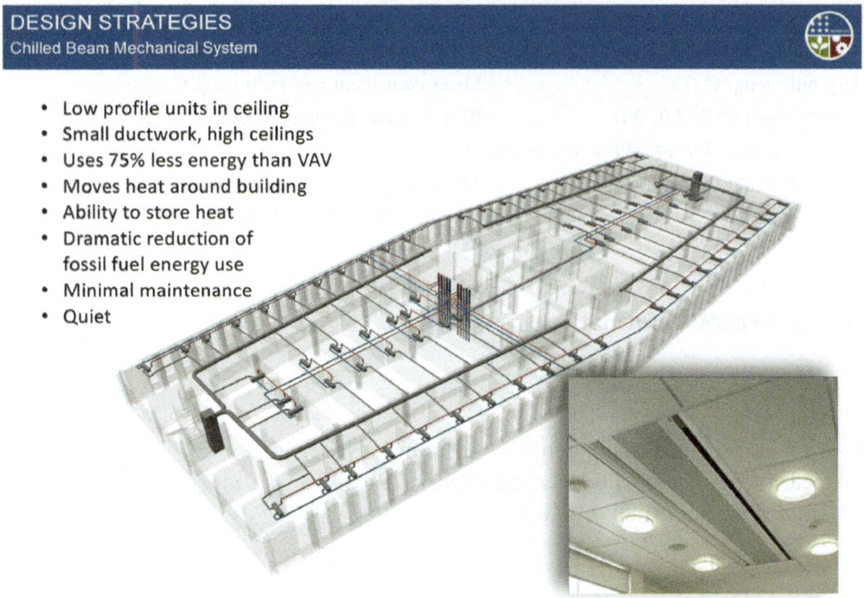

Fig. J.67 The second diagram includes some bullet points regarding the HVAC and illustrates the minimal amount of material to support the system allowing for higher ceilings. *Source* General Services Administration

J.9.8. Awards or Recognition

- LEED Gold.
- Energy Star 95.

J.9.9. Site Energy Cost Information (Electricity, \$/kWh; Gas, \$/m^3, Etc.)

- Electricity and Gas: \$1,020,284.80 per year.
- Electric (blended): \$0.094/kWh.
- Gas: \$.0066/ft^3.

J.9.10. Pre-renovation Building Details

J.9.10.1. Envelope Details: Walls, Roof, Windows, Insulation Levels

- Poor quality envelope, window coverings blocked natural light.
- R20 for the Roof, Envelope Was Evaluated at R5.

J.9.10.2. Heating, Ventilation, Cooling, and Lighting Systems

Standard conventional HVAC system.

J.9.11. Description of the Problem: Reason for Renovation (Non-energy and Energy Related Reasons)

- The Rogers modernization exemplifies the 2009 American Recovery and Reinvestment Act goals of creating jobs, spurring the economy, and constructing energy efficient Federal facilities that deliver lasting progress toward building a more sustainable national infrastructure while reducing the Federal government's consumption of energy and water, and increasing the use of clean and renewable sources of energy.
  - Daylighting.
  - Economic catalyst: The project employed about 8300 workers and $59 million of the project's subcontracting work was performed by small business. This was a key project that put Americans back to work at a time our Country was experiencing an economic crisis.
- A quote from GSA about the motivation for the project: "We are the largest commercial real estate agency in the nation, and we can leverage our buying power to call others to action. GSA is seizing opportunities to save taxpayer dollars by conserving energy, reducing water consumption, and implementing innovative technology in our buildings. GSA is a leader in sustainability. From energy and water efficient buildings, to a fuel efficient fleet, to smarter electronics disposal and green acquisitions, GSA is working to ensure that the agency is going green and saving green."

J.9.12. Renovation SOW (Non-energy and Energy Related Reasons)

See Table J.21.

J.9.13. Energy Saving/Process Improvement Concept and Technologies Used

J.9.13.1. Building Envelope Improvement

Super-insulated envelope achieves overall R-20 value.

J.9.13.2. New HVAC System or Retrofits to Existing

Active chilled beam system with thermal storage to take advantage of the northeast-southwest grid orientation in Denver.

Table J.21 Renovation SOW

1. Load reduction measures	
Daylighting and controls	The project team was able to maximize daylighting by removing the existing induction units and raising acoustical ceiling heights that were installed during previous renovations. The ceilings and induction units had obstructed 30% of the windows. Where open floor plans were not possible because tenants valued private offices, the team considered translucent or transparent walls to private offices
LED and task lighting	The building was fitted entirely with LED light fixtures. Even task lighting is LED
Insulation	The team super-insulated the exterior wall (without removal of the precast panels), which increased the overall R-value, including glass, from R5 to R20
High-efficiency glazing on windows	All windows were fitted with high-efficiency glazing with a thermal break. Window glazing was chosen to maximize the amount of visible light and isolative properties, and to minimize solar heat gain
Water savings	The project will achieve a water savings of 30% through the use of low-flow and infrared fixtures
Efficient plug loads	The team engaged tenants early on and published a sustainability guide that includes information about reducing plug loads through efficient appliances and shutting off devices when not in use
Efficient elevators	The team reduced vertical transportation energy by 15% using regenerative drive technology in the elevators

J.9.13.3. New Lighting System

LED lighting retrofits in all fixtures in building.

J.9.13.4. Renewable Energy

Solar thermal for water heating was installed.

J.9.13.5. Daylighting Strategies

Coverings that blocked windows were removed, and open office spaces were constructed, to maximize daylighting. Acoustical ceiling heights were raised that previously had obstructed windows. Where tenants did not want open offices, translucent or transparent walls were considered. Window glazing was chosen to maximize the amount of visible light and solar heating.

J.9.14. Energy Consumption

J.9.14.1. Pre-renovation Energy Use (Total and per m^2/year)

37,334 mmBTU and 75 kBTU/GSF. (It does not match with 24.14.4.)

J.9.14.2. Predicted Energy Savings (Site, Source, GHG), Total and per m^2/year

- 70%.
- Site: 22,954 mmBTU, source: 44,643.

J.9.14.3. Measured Energy Savings (Thermal, Electrical), Total and per m^2/year

- 68%.
- Elec.: 5600 mmBTU, 11,200 BTU/GSF.
- Thermal: 12,217 mmBTU, 24,434 BTU/GSF.

J.9.14.4. Annual Energy Use Reduction

EUI is currently 38.4 kBtu/GSF/Year, so 80.6 kBtu/GSF/Year are saved.

J.9.15. Energy Cost Reduction

J.9.15.1. Split in All Energy Forms—Electricity, Oil, District Heating

Electricity and gas = $422,701.35 (63.49% cost savings).

J.9.16. Non-energy Related Benefits Realized by the Project, E.g., Improved Productivity, Increased Rent/Lease, Increased Useful Space, Reduced Maintenance, Etc.

Quotes from Courtney Hoskins, Aging Services Program Specialist, Health And Human Services:

Moving back into my new office included getting an adjustable height desk. It has helped improve my mood, energy and productivity. This, and other workstation tools have been a wonderful addition to my work area.

I've worked in the Byron Rogers building for more than 20 years. I wasn't sure what to expect moving back into an old Federal building. But, I have been pleasantly surprised how I've been able to increase my overall well-being by taking advantage of the new gym, healthy food options, and more natural daylight. Also, overall agency productivity has increased greatly from having our agency partners co-located in the same building.

It's great to have most of our agency partners co-located in the same building. It allows us to be more productive and efficient being so close to one another. We're also improving our agency relationships with increased informal conversations happening as we share some of our office equipment like copiers and paper shredders.

J.9.17. Renovation Costs: Total and per m²

J.9.17.1. Total

$160,227.84.

J.9.17.2. Cost for Each Measure

See Table J.22.

Table J.22 Cost for each measure

Item number	Description	Cost [$] does not include mark-up, design, GCS and other soft costs
1	High efficiency HVAC ACB and thermal storage • active chilled beams • dedicated outside air system • high efficiency chillers, cooling towers and pumps • thermal storage system • solar thermal system (30% of service hot water) • condensing boilers, pumps • BACnet/LonWorks controls system • Heat recovery on outside air (sensible wheel) • DCV in all spaces tied in to occupancy sensors	15,500,000
2	High efficiency lighting System 1—Direct Lithonia RT LED • 0.6 W/sf "white box" configuration (1 × 4 recessed LED luminaire, 8 × 10 spacing) • 30 fc maintained illumination (assumed 80/50/20 reflections) • UGR = 19 (very uncomfortable) • Daylight sensors and dimming controls • Occupancy sensors in all spaces except back of house, which will have other automatic controls • Lighting controls extend to each luminaire via individual addressable control	4,200,000
3	Regenerative braking elevator and modernization	413,500
4	Office tower glazing Reglaze existing frames with 1 in. insulated glazing, interior dual pane blast window • 1/4 in. Crystal Gray w/ VE-2M #2 • Assembly U-0.14, SC-0.27, Tvis: 37%	4,650,000
5	Lav low flow fixtures	250,000
6	Urinals low flow	47,000
7	Toilets 1.0 gpf pressure assist	150,000
8	High efficiency roof and walls	815,000

J.9.18. Business Models and Funding Source

J.9.18.1. Description of the Funding Sources Chosen

The Rogers modernization exemplifies the 2009 American Recovery and Reinvestment Act goals of creating jobs, spurring the economy, and constructing energy efficient Federal facilities that deliver lasting progress toward building a more sustainable national infrastructure while reducing the Federal government's consumption of energy and water, and increasing the use of clean and renewable sources of energy. The total ARRA funding for the project was $146,961,349.96.

J.9.18.2. Description of the Business Model Chosen (Option)

Design-build was used. Quote from Bryan Zach, Project Manager at GSA: Some of our studies performed in central office concluded the design build process saved the government anywhere from 5 to 10% in overall costs and projects were more frequently completed on time and on budget. This can be said same for Rogers.

J.9.18.3. Funding Sources of the Business Model

ARRA provided most of the funding, and tenants provided the rest.

J.9.19. Cost Effectiveness of Energy Part of the Project (NPV, SIR, Etc.)

Not calculated yet because the building just became fully occupied this summer, so the energy savings data is not solidified yet.

J.9.20. User Evaluation

J.9.20.1. Description of User Training Programs Within the Refurbishment

Rocky Mountain Institute created a sustainability guide for tenants to help them learn how to minimize plug loads and optimize the performance of the building. Building operators were also engaged and taught how to control the building systems efficiently.

J.9.20.2. Integration of Users Demands in the Planning Process

Rocky Mountain Institute held design charrettes and workshops with tenants to include their input in the design process.

J.9.21. Experiences/Lessons Learned

J.9.21.1. Energy Use

68% energy savings, as listed above.

J.9.21.2. Impact on Indoor Air Quality

Improved, but the improvement has not been quantified.

J.9.21.3. Resulting Design Guidance

Quote from GSA: "The project provided government and taxpayer with an energy efficient building that promotes a healthy work environment. The project met or exceeded the expectations originally sought so there would be little to no change after all said and done, other than we may wish the project could have been implemented sooner."

J.9.21.4. Space Utilization Changes

More of the office space is now open offices, and some tenants practice hoteling in the space.

1.9.20.2 Integration of Users Demands in the Planning Process

1.9.21.2 Impact on Indoor Air Quality

The manufacturer's authorised representative in the EU is Springer Nature Customer Service Centre GmbH, Europaplatz 3, 69115 Heidelberg, Germany. If you have any concerns regarding our products, please contact ProductSafety@springernature.com

Printed and bound by CPI Group (UK) Ltd, Croydon, CR0 4YY
28/09/2026
02224057-0004